# Production, Properties, and Applications of Engineered Cementitious Composites

S. Praveenkumar
*PSG College of Technology, India*

J. Paulo Davim
*University of Aveiro, Portugal*

IGI Global
PUBLISHER of TIMELY KNOWLEDGE

A volume in the Advances in Chemical and Materials Engineering (ACME) Book Series

Published in the United States of America by
IGI Global
Engineering Science Reference (an imprint of IGI Global)
701 E. Chocolate Avenue
Hershey PA, USA 17033
Tel: 717-533-8845
Fax: 717-533-8661
E-mail: cust@igi-global.com
Web site: http://www.igi-global.com

Library of Congress Cataloging-in-Publication Data

Names: Shanmugam, Praveenkumar, 1989- editor. | Davim, J. Paulo, editor.
Title: Production, properties, and applications of engineered cementitious composites / edited by: Praveenkumar Shanmugam Shanmugam, and Paulo Davim J.
Description: Hershey, PA : Engineering Science Reference, [2024] | Includes bibliographical references and index. | Summary: "Production, Properties, and Applications of Engineered Cementitious Composites highlights the new and innovative ways of production, properties, and various applications of engineered cementitious composites. The main focus of the book is on the latest advancements, technical knowledge, tools, and solutions for engineered cementitious composites manufacturing, design, and technologies for construction from various perspectives. Covering key topics such as alternative materials, mineral admixtures, and testing of engineered cementitious composites, this premier reference source is ideal for engineers, industry professionals, researchers, academicians, scholars, practitioners, instructors, and students"-- Provided by publisher.
Identifiers: LCCN 2023022589 (print) | LCCN 2023022590 (ebook) | ISBN 9781668481820 (hardcover) | ISBN 9781668481837 (paperback) | ISBN 9781668481844 (ebook)
Subjects: LCSH: Cement composites.
Classification: LCC TA438 .P76 2023 (print) | LCC TA438 (ebook) | DDC 620.1/35--dc23/eng/20230627
LC record available at https://lccn.loc.gov/2023022589
LC ebook record available at https://lccn.loc.gov/2023022590

This book is published in the IGI Global book series Advances in Chemical and Materials Engineering (ACME) (ISSN: 2327-5448; eISSN: 2327-5456)

British Cataloguing in Publication Data
A Cataloguing in Publication record for this book is available from the British Library.
All work contributed to this book is new, previously-unpublished material.
The views expressed in this book are those of the authors, but not necessarily of the publisher.
For electronic access to this publication, please contact: eresources@igi-global.com.

# Advances in Chemical and Materials Engineering (ACME) Book Series

ISSN:2327-5448
EISSN:2327-5456

Editor-in-Chief: J. Paulo Davim, University of Aveiro, Portugal

**MISSION**

The cross disciplinary approach of chemical and materials engineering is rapidly growing as it applies to the study of educational, scientific and industrial research activities by solving complex chemical problems using computational techniques and statistical methods.

The **Advances in Chemical and Materials Engineering (ACME) Book Series** provides research on the recent advances throughout computational and statistical methods of analysis and modeling. This series brings together collaboration between chemists, engineers, statisticians, and computer scientists and offers a wealth of knowledge and useful tools to academics, practitioners, and professionals through high quality publications.

**COVERAGE**

- Wear of Materials
- Artificial Intelligence Methods
- Thermo-Chemical Treatments
- Polymers Engineering
- Chemical engineering
- Heat Treatments
- Biomaterials
- Materials to Renewable Energies
- Coatings and surface treatments
- Fracture Mechanics

IGI Global is currently accepting manuscripts for publication within this series. To submit a proposal for a volume in this series, please contact our Acquisition Editors at Acquisitions@igi-global.com or visit: http://www.igi-global.com/publish/.

The Advances in Chemical and Materials Engineering (ACME) Book Series (ISSN 2327-5448) is published by IGI Global, 701 E. Chocolate Avenue, Hershey, PA 17033-1240, USA, www.igi-global.com. This series is composed of titles available for purchase individually; each title is edited to be contextually exclusive from any other title within the series. For pricing and ordering information please visit http://www.igi-global.com/book-series/advances-chemical-materials-engineering/73687. Postmaster: Send all address changes to above address. 

## Titles in this Series

*For a list of additional titles in this series, please visit:*
*http://www.igi-global.com/book-series/advances-chemical-materials-engineering/73687*

***FlexPDE and Finite Element Method Applications in Thermal Energy Storage and Cavities***
Khashayar Hosseinzadeh (Sharif University of Technology, Iran) Mehdi Mahboobtosi (Babol Noshirvani University of Technology, Iran) and Davood Domairry Ganji (Babol Noshirvani University of Technoloy, Iran)
Engineering Science Reference • © 2024 • 380pp • H/C (ISBN: 9798369309322) • US $225.00

***Next Generation Materials for Sustainable Engineering***
Shrikaant Kulkarni (Padmashri Dr. V.B. Kolte College of Engineering, India) Boris Kharisov (Autonomous University of Nuevo Leon, Mexico) A.K. Haghi (University of Coimbra, Portugal) and Vipul Srivastava (Lovely Professional University, India)
Engineering Science Reference • © 2024 • 442pp • H/C (ISBN: 9798369313060) • US $295.00

***Advancements in Powder Metallurgy Processing, Applications, and Properties***
Shashanka Rajendrachari (Bartin University, Turkey) and Rayappa Shrinivas Mahale (Jain College of Engineering and Research, India)
Engineering Science Reference • © 2024 • 313pp • H/C (ISBN: 9781668493854) • US $275.00

***Sustainable Approach to Protective Nanocoatings***
Rajni Garg (Galgotias College of Engineering and Technology, India) and Ansar Anjum (Galgotias College of Engineering and Technology, India)
Engineering Science Reference • © 2024 • 434pp • H/C (ISBN: 9798369331361) • US $345.00

*For an entire list of titles in this series, please visit:*
*http://www.igi-global.com/book-series/advances-chemical-materials-engineering/73687*

IGI Global
PUBLISHER of TIMELY KNOWLEDGE

701 East Chocolate Avenue, Hershey, PA 17033, USA
Tel: 717-533-8845 x100 • Fax: 717-533-8661
E-Mail: cust@igi-global.com • www.igi-global.com

# Table of Contents

# Detailed Table of Contents



Concrete is a commonly utilized construction material in the engineering field. The low tensile strength and strain capacity of conventional concrete is a significant barrier to its actual integration in infrastructure. Cementitious composite is a combination of different materials that, when combined, form a composite that aids in the bonding of different components in a structure. It is a primitive combination that is still used in the construction industry. In contrast to brittle concrete, engineered cementitious composite exhibits high tensile strength with robust strain-hardening performance and multiple self-controlled crack. ECC could be used in any structural component in a building or structure because of its ability to be moulded into any shape. This emphasis the important role of engineered cementitious materials, and thus, in this chapter, the evolution of ECC materials and structures, including their characteristics, different kinds, and applications, will be thoroughly detailed, ranging from experimental investigations to enhanced numerical modeling methods and techniques.



Future structures based on cement-based composite materials must be more versatile, durable, and strong in addition to being environmentally friendly and sustainable. With the introduction of nanomaterials, it is possible to improve their compressive strength, flexural strength, tensile strength, elastic modulus, toughness, heat of hydration, thermoelectric conversion efficiency, thermoelectric figure of merit, power

Engineering cementitious composites (ECC) were discovered as a better alternative to conventional concrete due to their ability to achieve higher tensile strength. In subsequent years, additional studies were conducted to modify the materials in ECC to achieve a more sustainable and environmentally friendly composition. Incorporating nano-carbons such as fullerenes, carbon nanotubes, carbon nanoparticles, and graphene into ECC makes the advanced material highly advantageous. This chapter investigates the behavior of ECC reinforced with nano-carbons. The performance of ECC blended with nano-carbon is analyzed by comparing the fresh, mechanical and durability properties. In addition, the self-sensing and self-healing functionalities of ECC blended with nano-carbon are observed. The crack behaviour and healing performance of ECC are monitored due to the conductive ability of nano-carbon. Adding graphene in the range of 0.01 to 0.08% in ECC improves the young's modulus compared to conventional ECC.



Engineering cement composites (ECC) are a type of fiber cement composite material and possess high mechanical and durability properties. The main objective of this study is to develop a model for the relationship between compressive strength and split tensile strength. To develop these relationships, mechanical properties that are available in the literature for various types of ECC are collected in which 75% data is used for training and 25% for testing purpose. The study is further extended to check the relationships or models available in the standards, codes, and in literature for various mechanical properties are also evaluated. Sensitive assessment is performed to evaluate the new model developed for ECC based on mechanical properties. These relationships are required for designers, engineers, researchers, and academicians when one of the mechanical properties are not available for design purpose.

*Balpreet Singh Madan, Department of Art and Design, School of Design, Architecture, and Planning, Sharda University, Greater Noida, India*
*S. Logeswaran, Department of Civil Engineering, KPR Institute of Engineering and Technology, Coimbatore, India*
*M. Veerapathran, Department of Civil Engineering, Dr. N.G.P. Institute of Technology, Coimbatore, India*
*Pushpendra Rai, Department of Science and Humanities, K.J. Somaiya College of Engineering, Somaiya Vidyavihar University, Mumbai, India*
*Sanjay Kumar Singh, Department of Applied Sciences and Humanities, Institute of Engineering and Technology, Lucknow, India*
*Sampath B., Mechanical Engineering, Muthayammal Engineering College, Namakkal, India*

This chapter explores the integration of nano-composite materials into engineered cementitious composites (ECC), highlighting its significant impact on structural resilience, durability, and performance. It discusses innovative methodologies and advancements in ECC formulation, including nanoparticles like carbon nanotubes, graphene, and nano-sized additives. The advantages of ECC include enhanced strength, improved ductility, and increased resistance to environmental stressors. The chapter also examines the environmental sustainability quotient and potential challenges in large-scale implementation. It promises a sustainable future for construction practices.


*Chintada Chandrasekhar, Indian Institute of Technology Roorkee, Roorkee, India*
*G. D. R. N. Ransinchung, Indian Institute of Technology Roorkee, Roorkee, India*

Engineered cementitious composites (ECC) or strain-hardening cement-based composites (SHCC) were initially developed in the early 1990s, and later, many advancements were made in matrix composition. The development of ECC hinged on gaining a comprehensive understanding of how the materials integrated into the matrix performed in in-situ conditions. The field application provides an opportunity to capitalize on ECC's enhanced mechanical properties, particularly the structural performance caused by various loading conditions. This chapter briefly introduces

# Preface

Engineered Cementitious Composites (ECC) stand at the forefront of modern construction materials, offering a transformative alternative to conventional concrete and fiber-reinforced concrete. In this dynamic field, innovation thrives as we continually push the boundaries of what is possible in civil engineering applications. The collaborative efforts of researchers and industry professionals have propelled ECC into the spotlight, driven by its exceptional mechanical properties and durability.

In *Production, Properties, and Applications of Engineered Cementitious Composites*, curated by S. Praveenkumar and J. Paulo Davim, we delve into the intricacies of ECC, exploring its production techniques, inherent properties, and diverse applications. ECC represents a paradigm shift in construction materials, offering flexibility, sustainability, and resilience in equal measure.

Within the pages of this book, readers will find a comprehensive examination of the macro and micro mechanics underlying ECC, alongside insights into the incorporation of various materials to enhance its performance. By mitigating the inherent brittleness of traditional concrete, ECC opens avenues for environmentally sustainable construction practices with reduced embodied energy and carbon footprint.

Our aim with this volume is to provide a platform for the dissemination of cutting-edge research, technical expertise, and innovative solutions pertaining to ECC. From manufacturing advancements to design methodologies, this book offers a wealth of knowledge for researchers, engineers, and industry practitioners alike. It is our hope that this compilation serves as a beacon for the global academic and industrial communities, fostering collaboration and driving progress in the field of Engineered Cementitious Composites.

We extend our gratitude to the contributing authors for their invaluable insights and dedication to advancing ECC. Additionally, we would like to express our appreciation to the readers—graduate students, research scholars, scientists, and industry professionals—for their interest and engagement in this exciting and evolving domain.

Together, let us continue to explore, innovate, and build a sustainable future with Engineered Cementitious Composites.

## ORGANIZATION OF THE BOOK

In *Production, Properties, and Applications of Engineered Cementitious Composites*, we embark on a comprehensive journey through the multifaceted realm of Engineered Cementitious Composites (ECC). Authored by a diverse group of experts, this edited reference book serves as a compendium of knowledge, offering insights into the latest advancements, innovations, and applications shaping the field of ECC.

**Chapter 1:** "A Basic Outline on Engineered Cementitious Composites" sets the stage by providing a fundamental understanding of ECC. Authored by Muthuminal Ramu and Mohana Priya R from SNS College of Technology, India, this chapter delves into the mechanical properties of ECC, exploring the relationship between compressive strength and split tensile strength. The development of models presented here offers valuable insights for designers, engineers, researchers, and academicians seeking to optimize ECC compositions for various applications.

**Chapter 2:** "Multifunctional Contributions of Nanomaterials to Cementitious Composite Materials," authored by Rafael Vargas-Bernal from the University of Guanajuato, Mexico, explores the integration of nano-composite materials into ECC formulations. This chapter showcases the transformative potential of nanoparticles such as carbon nanotubes and graphene, elucidating their role in enhancing ECC's structural resilience, durability, and performance. Through innovative methodologies and advancements, ECC emerges as a sustainable solution poised to redefine construction practices.

**Chapter 3:** "Innovations and Applications of Engineered Cementitious Composites and Alternative Materials: Reimagining Construction," penned by a collaborative team from various institutions in India, offers a comprehensive overview of ECC and alternative materials. This chapter explores the composition, properties, and applications of ECC, shedding light on its potential to revolutionize construction practices. Real-world case studies underscore ECC's performance and underscore its viability as a sustainable construction material.

**Chapter 4:** "Harden and Self-Sensing Properties of Engineered Cementitious Composite Reinforced With Nano-Carbon," authored by Divya Sabapathi and colleagues from PSG College of Technology, India, delves into the enhanced properties of ECC through the incorporation of nano-carbon materials. This chapter investigates the fresh, mechanical, and durability properties of ECC blended with nano-carbon, highlighting its self-sensing and self-healing functionalities. The integration of graphene, in particular, offers promising improvements in ECC's young's modulus.

**Chapter 5:** "Development of Models for Mechanical Properties of Engineered Cementitious Composites," authored by a collaborative team from India, China, and Greece, presents a comprehensive study on developing models for various mechanical properties of ECC. By collecting and analyzing data from literature and standards, this chapter provides valuable insights for researchers and practitioners seeking accurate models for ECC design purposes.

**Chapter 6:** "Durability Properties of Engineered Cementitious Composites (ECC)," authored by Özlem Üstündag from Istanbul University-Cerrahpasa, Turkey, offers a detailed review of ECC's durability properties. This chapter examines ECC's performance under harsh environmental conditions, such as shrinkage, freeze-thaw, and chemical exposure, highlighting its superior durability and self-healing properties.

**Chapter 7:** "Corrosion Resistance and Permeability Properties of UHPC-ECC: Durability Studies of UHPC-ECC," authored by Athibaranan S and colleagues from various institutions in India, explores the corrosion resistance and permeability properties of Ultra High-Performance Engineered Cementitious Composites (UHP-ECCs). This chapter underscores the exceptional resistance of UHP-ECCs to corrosion and their low permeability, making them ideal for a wide range of applications in transportation infrastructure.

**Chapter 8:** "Engineered Cementitious Composites (ECC) Through Nano-Composite Integration: Enhancing Construction," authored by Balpreet Singh Madan and colleagues from various institutions in India, offers a deeper exploration of nano-composite integration into ECC formulations. By leveraging nanoparticles like carbon nanotubes and graphene, ECC emerges as a resilient, durable, and environmentally sustainable construction material.

**Chapter 9:** "Current Investigations and Field Performances of Engineered Cementitious Composites (ECC) in Transportation Infrastructure," authored by Chintada Chandrasekhar and GDRN Ransinchung from IIT ROORKEE, India, provides insights into ECC's applications in transportation infrastructure. Through field studies and real-world applications, this chapter demonstrates ECC's potential to enhance structural performance and durability in transportation-related projects.

**Chapter 10:** "Engineered Cementitious Composites for Diverse Applications," authored by Dhanya K R from Northamps, offers a comprehensive overview of ECC's wide-ranging applications in the construction industry. This chapter emphasizes ECC's unique properties, including shear capacity, crack control, and damage tolerance, positioning it as a superior alternative to conventional concrete for various structural applications.

Together, these chapters form a cohesive narrative, illuminating the vast potential of Engineered Cementitious Composites in modern construction practices. We extend our gratitude to the contributing authors for their invaluable contributions to this edited reference book, which we hope will serve as a definitive resource for researchers, engineers, and industry professionals alike.

## IN CONCLUSION

As we draw to a close on this journey through *Production, Properties, and Applications of Engineered Cementitious Composites*, we are filled with a sense of awe and appreciation for the depth of knowledge and innovation presented within its pages. Authored by esteemed researchers and industry professionals from around the globe, this edited reference book stands as a testament to the collective efforts driving the advancement of Engineered Cementitious Composites (ECC).

Throughout the chapters, we have explored the multifaceted nature of ECC, from its fundamental mechanical properties to its innovative applications in diverse construction contexts. We have delved into the integration of nanomaterials, the development of predictive models, and the exploration of ECC's durability and performance under various environmental conditions. Real-world case studies have underscored ECC's transformative potential in redefining construction practices and addressing the challenges of sustainability and resilience in the built environment.

The breadth and depth of research presented in this volume reflect the growing interest and investment in ECC as a sustainable, high-performance construction material. From enhancing structural resilience to mitigating environmental impact, ECC offers a promising path forward for the construction industry, providing solutions that marry technological innovation with environmental stewardship.

As editors, we extend our heartfelt gratitude to the contributing authors for their dedication, expertise, and passion in advancing the field of ECC. We also express our appreciation to the readers—researchers, engineers, industry professionals, and students—for their interest and engagement in this dynamic and evolving field.

As we conclude this preface, we are reminded that our journey with ECC is far from over. With each discovery, each innovation, and each application, we move closer to realizing the full potential of Engineered Cementitious Composites in shaping a sustainable and resilient future for construction.

May the insights and knowledge shared in this book inspire continued exploration, collaboration, and progress in the realm of Engineered Cementitious Composites.

*S. Praveenkumar*
*PSG College of Technology, India*

*J. Paulo Davim*
*University of Aveiro, Portugal*

# Chapter 1
# A Basic Outline on Engineered Cementitious Composites

**Muthuminal Ramu**
https://orcid.org/0000-0001-9931-2204
*SNS College of Technology, India*

**R. Mohana Priya**
*SNS College of Technology, India*

## ABSTRACT

*Concrete is a commonly utilized construction material in the engineering field. The low tensile strength and strain capacity of conventional concrete is a significant barrier to its actual integration in infrastructure. Cementitious composite is a combination of different materials that, when combined, form a composite that aids in the bonding of different components in a structure. It is a primitive combination that is still used in the construction industry. In contrast to brittle concrete, engineered cementitious composite exhibits high tensile strength with robust strain-hardening performance and multiple self-controlled crack. ECC could be used in any structural component in a building or structure because of its ability to be moulded into any shape. This emphasis the important role of engineered cementitious materials, and thus, in this chapter, the evolution of ECC materials and structures, including their characteristics, different kinds, and applications, will be thoroughly detailed, ranging from experimental investigations to enhanced numerical modeling methods and techniques.*

DOI: 10.4018/978-1-6684-8182-0.ch001

## INTRODUCTION

The term "engineered cementitious composite" (ECC), also known as "strain hardening cement-based composites" (SHCC) or frequently known as "bendable concrete," refers to a mortar-based composite that is effortlessly molded and strengthened with arbitrarily chosen short fibers, typically made of polymers. ECC differs from conventional concrete in that it has a tensile strain capacity of 3–7% as opposed to 0.01% for OPC paste, mortar, or concrete. Therefore, unlike OPC concrete, ECC behaves more like a ductile metal material than a fragile glass material, opening up a plethora of uses.

Engineered Cementitious Composite (ECC) is a special kind of cementitious matrix that has a special combination of low-volume fibers and various composites to give it great ductility, high tensile strength, and the capacity to repair itself. The endurance of structures is impacted by the delicate nature of traditional concrete and fiber-reinforced concrete, which rapidly cracks under both mechanical and environmental pressures.

Fiber-reinforced concrete has been through several phases during its development. Short steel fibers can effectively lessen the brittleness of concrete, according to Romauldi and colleagues' 1960s research. Various fibers, including glass, carbon, synthetic, and natural fibers, as well as hybrids that blend various fiber types or fiber lengths, have been added to this evolution over the years. RILEM TC 162-TDF has made structural design suggestions as a result of the continuously expanding understanding of fiber reinforcement efficacy. Interest in developing a fiber-reinforced concrete material with tensile ductility has been growing since the 1980s. The material's tenacity is improved in FRC, while the ductility remains unchanged. A criterion of tensile deformation (strain) capacities known as ductility is most often connected to materials like ductile steel but not concrete. The initial endeavors of Aveston and subsequently Krenchel and Stang, who showed that high tensile ductility may be reached with continuously aligned fibers, are examples of endeavors to achieve tensile ductility in concrete material. Concrete materials with textile reinforcement that can be prestressed serve as the current equivalent of continuous fiber reinforcement. Mobasher pioneered the study of continuous fiber-reinforced pultruded concrete. The usage of discontinuous fibers at high dosages (4–20%), such as in cement laminates and SIFCON (Slurry Infiltrated Fiber Concrete), was established concurrently and resulted in concrete composite materials that attain higher tensile strength than normal concrete and which are not as brittle, but with much less ductility than their continuous fiber and textile reinforced counterparts. Because all these materials reach varying levels of tensile ductility and frequently exhibit a strain-hardening reaction as opposed to the tension-softening response of FRC, they might be regarded as a sort of material apart from FRC. Such materials were categorized by Naaman and Reinhardt as High-Performance Fiber Reinforced Cementitious Composites (HPFRCC). In light of the fact that the majority of members of this class of material have matrices devoid of coarse aggregates, they ought to be categorized as fiber-reinforced cement pastes or mortars.

Contrary to conventional fiber reinforced concrete, ECC is a class of materials with nanoindentation design. A cementitious material can be referred to as an ECC as long as it was created using the principles of rupture mechanics and micromechanics. As a result, ECC does not comprise a static material design but instead a wide spectrum of subjects in various levels of innovation, investigation, and implementation. The family of ECC materials is growing. It takes extra work to produce an individual mix design for ECC by methodically developing the components at the nano, micro, macro, and composite sizes.

Except for its ability to distort (or bend) underneath pressure, ECC resembles regular Portland cement-based concrete in appearance. Research teams from the Universities of Michigan, California, Irvine, Delft University of Technology, University of Tokyo, Czech Technical University, University of British Columbia, and Stanford University are all involved in advancing ECC science. The development of ECC has been pushed forward by conventional concrete's limitations of endurance and breakdown under stress, some of which are caused by fragile behavior.

## Properties

ECC features a number of distinctive qualities, such as compact crack width, lack of lithographically weak planes, processing ease comparable to that of traditional cement, excellent tensile capabilities to other fiber-reinforced composites, and just a tiny volume fraction of fibers(2%). These characteristics are mostly attributable to the interactions between the reinforcing matrix and the fibers, which can be specifically adjusted via finite element designs. In essence, the fibers produce numerous microcracks with a very particular width as opposed to a small number of very massive cracks (as in conventional concrete.) As a result, ECC can flex without completely failing.

Due to the small diameter and density of the cracks, aggressive mediums have a tough time penetrating and attacking the structural bars. This micro-cracking behavior also promotes self-healing. Unreacted cement particles that have subsequently been revealed due to fracturing hydrate in the presence of water (after a rainfall, for example) and produce a multitude of products (calcium silicate hydrates (C-S-H), calcite, etc.) that extend and seal the crack. These items fill in the fissure like a white "scar" material. In addition to sealing the break to stop the flow of fluids, this self-healing action also restores mechanical qualities. Various types of conventional cement and concrete have shown this self-healing; however, over a certain crack width self-healing becomes less effective. All cracks are guaranteed to completely heal when exposed to the natural environment because of the strictly restricted crack widths exhibited in ECC.

All construction materials can enhance in conductivity and be utilized for fault detection when coupled with an additional conductive material. The basic idea behind this is that conductivity will alter when damage happens; adding conductive material is intended to increase conductivity to a point where such variations will be obvious. Semi-conductive ECC for damage detection is being researched, despite the fact that it is not an actual material property of ECC. The following are the categories that group these qualities together.

- **High-Temperature Resistance** – The structural qualities of regular concrete deteriorate with spalling damages in high temperatures. According to reports, when the temperature reaches 200 to 350 $^0$C, the concrete fragments will begin to strip off, and when the temperature climbs more, violent spalling may happen. The internal tensile stress that has built up inside the concrete is what caused it to spall when temperatures were high. According to this principle, it has recently been claimed that ECC performs better than other materials under extreme temperatures because of the flexibility that its inner, randomly oriented fibers maintain. It is possible to reduce ECC spalling, improving durability in high-temperature environments. It was discovered that the catastrophic spalling tendency of ECC was reduced by the introduction of PVA fibers. The high-temperature resistance of ECC has been further improved with the addition of fly ash. High temperatures cause ECC's strain-hardening capacity to diminish or disappear, which reduces its elasticity and increases the likelihood of premature failure or severe deformation. The variation in the number of microcracks after the high temperature directly contributed to the change in strain-hardening power of ECC. A moderate temperature (50-100 $^0$C) can significantly improve the characteristics of the fiber-matrix interface whereas a maximum temperature (200 $^0$C) weakens the yarn and fiber-matrix interface, resulting in the presence of the fibers as scattered composite flaws. While ECC lost ductility and began to exhibit a strain rate impact at 150 $^0$C, PVA-strain ECC's capacity increased with an increase in strain rate at 100 $^0$C. Steel fiber could enhance ECC integrity at high temperatures, as PVA-ECC produces four times as much debris at 600 $^0$C as steel fiber.ECC. Additionally, a high temperature has a considerable impact on the stiffness and compressive strength of ECC. Compressive strength and stiffness of ECC were practically consistent below 200 $^0$C. Furthermore, compressive strength and stiffness showed a linear decline and decreased by lot much than 75% when the temperature was elevated from 200 $^0$C to 800 $^0$C.

*Figure 1. Temperature's impact on the stiffness and strength of ECC (Engineered Cementitious Composites, n.d.-b)*

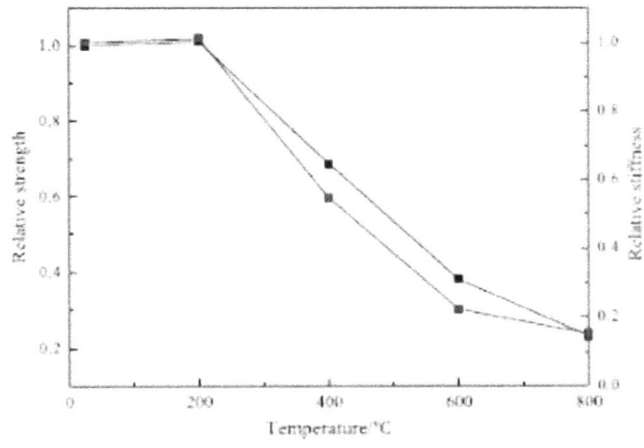


Additionally, the mechanical characteristics of PVA-ECC at sub-elevated temperatures were investigated (20 °C, 50 °C, 100 °C, and 200 °C). ECC tensile characteristics decreased as a result of PVA strength deteriorating under 200 °C exposure and ECC matrix toughness increasing on a micro-scale as shown in Figure 1. However, the negative impact of tensile characteristics was lessened because the fiber-matrix interface bonding was reduced.

*Figure 2. Microscopic images of tensile fracture surfaces of ECC under different temperatures (Engineered Cementitious Composites, n.d.-b)*

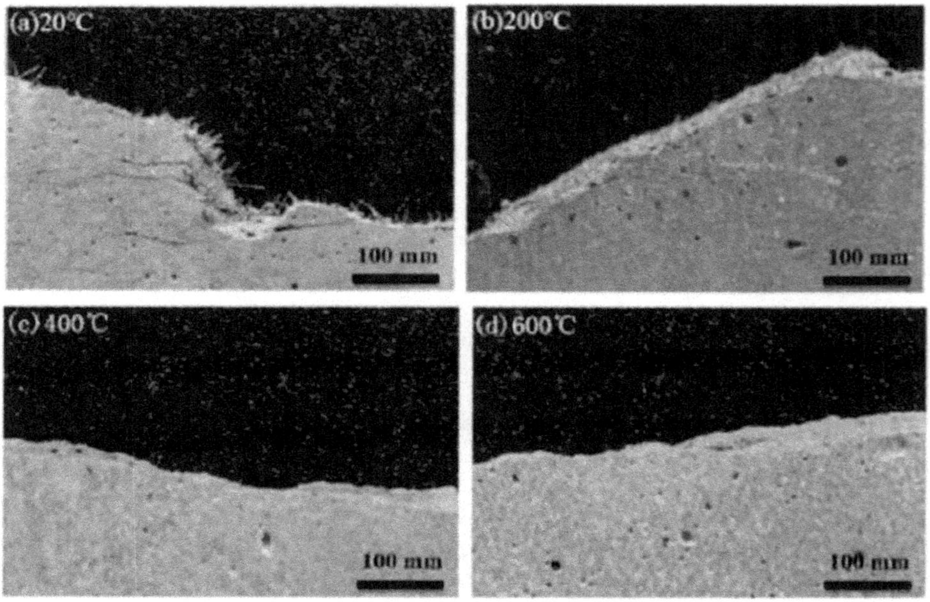


Figure 2 shows microscopic images of the tensile fracture surface of PVA-ECC specimens. The lengths and quantity of fibers at the worn surfaces were shown to steadily decrease with a rise in temperature, showing a considerable loss in the fiber bridging ability at a crack. The fibers were completely broken down in specimens under 400 0C and 600 0C, resulting in the fact that the strain-hardening activity of ECC vanished at high temperatures. There is currently a very small number of research on PE-high-temperature ECC's performance. It is clear that at 50 0C, PE-ECC obtains the highest tensile strain capacity, and at 200 0C, this almost seems to have no tensile strain capacity (equivalent to plain concrete). Up to 2000C, PVAECC's strain hardening and multi-cracking behaviour was sustained, but after that point, ECC started to become brittle because the PVA fibers started to melt and decompose at high temperatures.

- **Permeability Resistance** – Mesoporous structure connectivity, crack depth, and crack width are all specifically correlated to concrete's permeability resistance. The fissure has very little impact on concrete permeability when it is less than 100m wide. The optimum crack width of reinforced concrete was discovered to be inversely correlated with the square root of protective cover thickness (Kewalramani et al., 2017).

ECC's capacity for self-healing allows for further improvement of permeability resistance. The ratio of self-healing speed to ECC crack width is inverse. According to reports, a fissure with a width of less than 60m can stabilize the permeability of ECC in 3–4 days. Whenever the fracture width was greater than 100m, stabilizing ECC permeability required 7–10 days, perhaps even longer. ECCs with tight cracks can successfully fend off intrusions from moisture and corrosive ions by completing self-healing in a brief amount of time.

- **Shrinkage Resistance –** ECC's concerns with long-term endurance are brought primarily by excessive shrinking. ECC's drying shrinkage is influenced by a variety of variables, such as the water binder ratio, sand-binder ratio, cement percentage, fiber volume, and admixture. It was discovered that although fiber volume contents seemed to have no discernible impact on ECC shrinkage strain, ECC shrinkage enhanced with an upsurge in the water-binder ratio and decreased with an increase in sand-binder balance and fly ash concentration. The total shrinkage rate of ECC is decreased and the autogenous shrinkage and drying shrinkage tend to be equivalent with a rise in the sand-binder ratio and water-binder ratio (Nagaraj & Shanmugam, 2022).

The following table 1 shows the comparison between the other materials (Engineered Cementitious Composites, n.d.-a).

*Table 1. Comparison of properties*

| Properties | FRC | Common HPFRCC | ECC |
|---|---|---|---|
| Design Methodology | N.A. | Use high Vf | Micromechanics based, minimize Vf for cost and processibility |
| Fiber | Any type, Vf usually less than 2%; df for steel ~ 500 micrometre | Mostly steel, Vf usually > 5%; df ~ 150 micrometre | Tailored, polymer fibers, Vf usually less than 2%; df < 50 micrometre |
| Matrix | Coarse aggregates | Fine aggregates | Controlled for matrix toughness, flaw size; fine sand |
| Interface | Not controlled | Not controlled | Chemical and frictional bonds controlled for bridging properties |
| Mechanical Properties | Strain-softening | Strain-hardening | Strain-hardening |
| Tensile strain | 0.1% | <1.5% | >3% (typical); 8% max |
| Crack width | Unlimited | Typically several hundred micrometres, unlimited beyond 1.5% strain | Typically < 100 micrometres during strain-hardening |

## Types of ECC

- Lightweight Concrete
- Self Compacting Concrete
- Sprayable Concrete
- Extrudable ECC

### Lightweight Concrete

The use of lightweight concrete blocks for the load-bearing enclosure of cavity walls was approved in the UK at the beginning of the 1950s. Soon after, LWC of high strength, appropriate for structural work, was introduced thanks to the creation and production of new forms of artificial LWA (Lightweight aggregate).

These developments stimulated the use of LWA concrete in structural applications, especially when weight reduction in a structure was a crucial design or financial factor.

There are several types of lightweight aggregates suitable for structural reinforced concrete:-

1. **Pumice** – is employed for roof slabs made of reinforced concrete, primarily on industrial buildings in Germany.

2. **Foamed Slag** – was the first lightweight aggregate made in substantial quantities in the UK that was appropriate for reinforced concrete.
3. **Expanded Clays and Shales** – in a position to achieve a high enough strength for prestressed concrete. Aglite and Leca (UK), Haydite, Rocklite, Gravelite, and Aglite are well-known trade names (USA).
4. **Sintered Pulverised** – Fuel ash aggregate, marketed under the trade name Lytag, is used in the UK for a range of structural applications.

In recent scenarios, these lightweight concretes have been established through incorporating ECC technology. By including air gaps, glass bubbles, polymer spheres, and/or lightweight aggregate, low-density ECC has been created. Lightweight ECC has higher ductility than some other lightweight concretes. Examples include canoes, barges, and floating residences.

## Self-Compacting Concrete

Self-compacting concrete (SCC) is a concrete mixture with a low yield stress, high deformability, good segregation resistance (prevents separation of particles in the mix), and moderate viscosity (needed to guarantee homogeneous suspension of fine materials all through transportation, placement (without external compaction), and afterward until the concrete sets).

SCC is a very fluid mix that passes very conveniently inside as well as around formwork while pouring. It also has the capacity to flow all over corners as well as through impediments due to its "passing ability," is nearly self-leveling but not quite self-leveling, doesn't really necessitate vibration or tamping after pouring, and closely mimics the shape and surface texture of a mould (or form) once it has dried and set. As a consequence, laying SCC requires significantly less labor than mixing regular concrete. SCC is often comparable to regular concrete once it has been poured in terms of strength, setting time, and setting period (gaining strength). SCC doesn't require much water to become fluid; in fact, it could have less water than regular concretes. Instead, a disproportionately high amount of fine aggregates, such as sand (about 50%), mixed with superplasticizers (additives that make sure particles dispersion and do not settle in the fluid mix) and viscosity-enhancing admixtures, gives SCC its fluid qualities (VEA). Chemical admixtures were used to reduce viscosity, and mix proportioning was used to manage particle interactions to create self-compacting ECC.

### Sprayable Concrete

Instead of using the usual traditional pouring or inserting into formwork methods, sprayed concrete is concrete that is blown into position. The Sprayed Concrete Association lists Gunite, Shot Concrete, and Shotcrete as additional terminology that are frequently used.

Spray Applied Concrete, also known as Shotcrete, is concrete that is pneumatically projected at high speed onto a surface. After all of the materials have been combined to the project's requirements, it is often blown via a hand-held nozzle.

Sprayed concrete has many uses, notably both primary and secondary linings for geotechnical applications as well as for controlling layers. Additionally, it can be applied to pile facing, concrete repairs on walls, bridges, and other structures as well as freeform projects like skate parks and swimming pools. The development of sprayable ECC, which could be pneumatically dispensed from a hose, involved the use of several super plasticizing compounds and admixtures that reduce viscosity. In addition to its distinctive mechanical qualities, sprayable ECC provides improved pumpability as compared to certain other sprayable fiber-reinforced composites. Tunnel/sewer linings, retrofitting/repair work, and sprayable ECC have all been utilized.

### Extrudable Concrete

Concrete's extrudability refers to its capacity to continuously squeeze itself out of the nozzle. Knowing what flowability means, it goes without saying that increased flowability, up to a point, leads to better extrudability and conversely. This suggests that whichever elements have a significant negative impact on workability should also have a positive impact on extrudability. Extruded concrete often relates to thin, glass-fiber-reinforced concrete panels. The substance, also known as glass fiber-reinforced concrete, typically consists of 90 percent sand and 10 percent glass fibers, together with concrete additives and colorants (Singh & Munjal, 2020). In 1998, the first extrudable ECC for use in pipe extrusion was created. Compared to conventional extruded pipes, extruded ECC pipes have a better load capacity and a higher deformability than any other extruded fiber-reinforced composite pipes.

## ECC Durability Under Various Environments

The most recent research on long-term strain capacity and ECC subjected to several frequently encountered conditions is compiled in this section. These conditions include alkali-silicate reaction (ASR) resistance, hot-wet cycles, chloride immersion, deicing salt exposure, and freeze-thaw cycles (Li, 2008).

## Long-Term Tensile Strain Capacity

A building material's mechanical qualities must not deteriorate as time passes and fall short of minimal design requirements in order for it to be truly durable. A series of tensile tests were conducted to establish the long-term strain capacity in order to validate the long-term efficacy of ECC. Because cementitious materials continue to hydrate over time and because ECC has a delicate balance of cement matrix, fiber, and matrix/fiber interface characteristics, its strain capacity changes as it ages. Peak strain capacity is reached approximately 10 days after molding when a perfect equilibrium of matrix, fiber, and matrix/fiber interface characteristics is realized for oversaturated repeated cracking. As the matrix continues to hydrate, its hardness rises, decreasing the ductility of the composite. Eventually, as matrix and fiber/matrix characteristics mature, an ECC long-term steady strain capacity of 3% is produced, which is much less than the 5% capacity seen at early age but still significantly more than the distortion demand enforced by many structural applications. The long-term strain capacity is anticipated to stay at about 3% even though long-term tests have only been done for 180 days. The compressive strength growth curve in typical concrete can be compared to the strain capacity-age curve. However, as it is not increasing steadily over time, the long-term values ought to be utilized in design. For ease of use, the design strain capacity may be set at 90% of the 28-day strain-capacity value that asymptotically resembles the long-term behavior. The design process is simplified because the standard for compressive strength uses the same 28-day value.

## Freeze-Thaw Durability

According to ASTM C666, numerous freezing and thawing cycles were used to test the endurance of non-air-entrained ECC specimens. As calibration standards, concrete samples that weren't air-entrained underwent testing as well. Testing on concrete prism and ECC specimens was carried out concurrently during a 14-week period. After 5 weeks (110 cycles), the concrete samples had severely deteriorated, forcing removal from the freeze-thaw machine in compliance with the testing requirements. But during the 300-cycle test period, there was no decrease in the dynamic modulus of any of the ECC specimens. Calculations based on ASTM C666 indicate that this performance results in a concrete durability factor of 10 as compared to 100 for ECC. Wet-cured and freeze-thaw-exposed ECC tensile coupons were both tested under uniaxial tension at the same age. There was no obvious decline in strain capacity after 300 cycles. Either wet-cured and freeze-thaw specimens had a strain capacity of roughly 3%.

## Tropical Climate Exposure

Hot water immersion studies were carried out to model the long-term impacts of hot and humid surroundings, as opposed to the freeze-thaw testing mentioned above, which is intended to simulate temperature variations in winter conditions. Hot water immersion was used to assess the effects of environmental exposure on individual fibers, single fibers embedded in ECC matrix, and composite ECC material specimens. Both individual fiber pullout and composite ECC material samples underwent a 28-day room temperature curing period before spending up to 26 weeks submerged in 60°C hot water. Little change was seen in the fiber's strength, elastic modulus, or elongation following this exposure. The ECC's strain capacity did, however, decrease throughout time from 4.5% at a young age to 2.75%. Despite the fact that accelerated hot weather testing reduced the strain capacity of ECC, the 2.75% strain capacity—more than 250 times that of regular concrete—found after 26 weeks and equal to 70 years of weathering of rocks is still suitable for the majority of construction applications.

## Chloride Immersion

Chloride ions may change the fiber/matrix interface and consequently, the characteristics of the composite when ECC material is exposed to settings with high chloride concentrations, such as maritime structures or for pavements and bridge decks subjected to de-icing salt applications. ECC coupon samples were preloaded under uniaxial tension to various strain levels in order to study these effects. They were then subjected to a 3% NaCl solution at room temperature for 1, 2, and 3 months before being reloaded up to failure. Even though the mean crack width grew from 40 mm to 100 mm and the tensile strength decreased by around 10%, the reloaded specimens in every case maintained numerous micro-cracking behavior and tensile strain capacity of more than 3%. According to Kabele's single fiber pull-out test findings, a decrease in chemical bonding at the fiber/matrix interface may be the cause of the broader crack width and reduced tensile strength.

## Deicing Salt Exposure

Ahmaran and Li M looked at how well non-air-entrained ECC held up under mechanical stress and repeated freezing and thawing in the vicinity of de-icing salts. After 50 exposure cycles, the ECC's total mass of scaling residue and surface aspect visual rating were both maintained within ASTM C 672 permissible bounds. Even for specimens that were pre-loaded to fracture at high deformation levels, this level of durability held true. As reference samples, non-air-entrained mortar specimens were employed. Under the same testing conditions, these mortar prisms degraded significantly as was to be predicted. To ascertain their residual tensile behavior, pre-loaded and virgin (no pre-loading) ECC coupon specimens were also subjected to 25 and 50 cycles of freezing and thawing in the vicinity of de-icing salts. The preloaded specimens had barely any ductility loss, but they still exhibited numerous microcracking behavior and had a higher-than-3% tensile strain capacity. Numerous microcracks caused by mechanical loading were also discovered to be adequately repaired by freezing and thawing cycles in the presence of salt solutions, returning the samples to almost their previous stiffness. These findings demonstrate the durability of ECC, both virgin and micro-cracked, in the context of repeated freezing and thawing cycles and deicing salts.

### Alkali-Silicate Reaction Resistance

Li M and Ahmaran investigated the alkali-silicate reaction resistance of ECC (ASR). To measure the length change caused by the alkali-silica reaction, ECC bar specimens were submerged in an alkali solution at 80°C in accordance with ASTM C 1260. After a 30-day soaking period, the ECC bar specimens containing Class F or Class C fly ash did not exhibit any appreciable expansion. The crystalline structure of these sand particles reduces the responsiveness in an alkaline environment, despite the fact that very fine silica sand is utilized in ECC. Far less frequent is the alkali-silicate reaction when there is a lot of fly ash present since the pozzolanic reaction lowers the pH level. Furthermore, the PVA microfibers' presence helps to minimize any potential enlargement. These investigations demonstrate that ECC material won't degrade as a result of ASR.

## Cases of ECC Field Applications

FRC has a lengthy tradition of field uses. In the 1990s, Li M compiled a number of suitable processing instances, including two that involved PVA fibers. The incidences include requests for pavement base layer in Denmark, Canada, the United States, and the United Kingdom; repairs of pavement and bridge in Denmark, Germany, the United States, and Canada; floor/bridge deck slabs in France, the United States, Denmark, and Canada; wall panels in Japan, Australia, and Denmark; tunnel lining in Norway and France; and implementations for septic tanks and pipes in Australia, Belgium, and Denmark (France and Denmark). Thin strip components for cladding and stair treads were made with ECC. Practical uses of ECC in the US and Japan increased in the 2000s.

Nevertheless, because initiatives utilizing ECC were directed by experts in practice instead of experts in academics, there is less available documentation on the field uses of ECC. The implementation of ECC in various scenarios is demonstrated in the following.

## ECC Application to Pavement Overlay

According to Glavind, Ramboll, Balaguru, Shah, Van Mier, Ramakrishnan, and Johnston, FRC was applied for pavement overlay. ECC hasn't yet been applied to a complete pavement overlay, though. On the other side, LCA has been used to examine the viability and environmental and economic advantages of ECC overlay. In lab-size investigations, Zhang and Li M demonstrated that ECC may successfully stop the propagation of reflected cracking (Figure 3). Shamaran demonstrated how the use of ECC can considerably increase the binding strength between the concrete base and overlay.

Zhang did a life cycle analysis of pavement overlay systems and came to the conclusion that ECC can lower life cycle costs and GHG emissions. Lepech and Li, Zhang and Qian, and others who conducted LCAs reported similar findings.

*Figure 3. Comparison of high stress concentration effects on concrete and ECC repair materials, showing (a) brittle fracture in concrete layer and (b) ductile strain-hardening response in ECC layer (Park et al., 2020)*

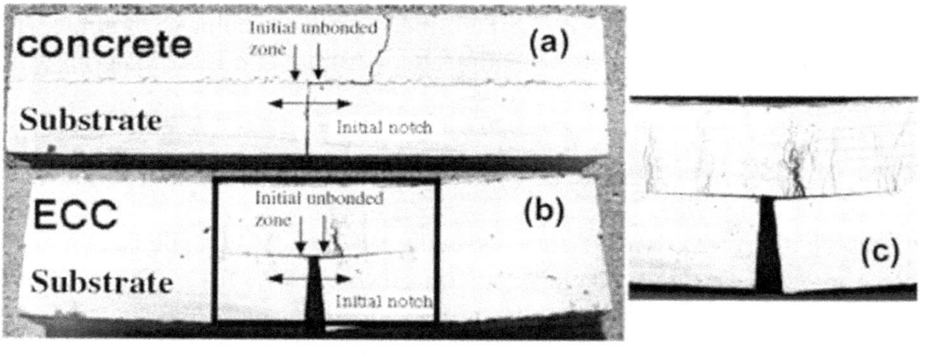


Li reported on the use of ECC for pavement restoration in the field. The ability to manage cracks in ECC and concrete patches on the pavement is shown in Figure 4. The concrete patch and ECC patch are laid on the bridge's surface, as seen in the left-hand image. As compared to a concrete fix, ECC exhibits better crack management.

*Figure 4. Comparison of ECC and concrete patches for pavement repair at Ann Arbor, Michigan (Park et al., 2020)*

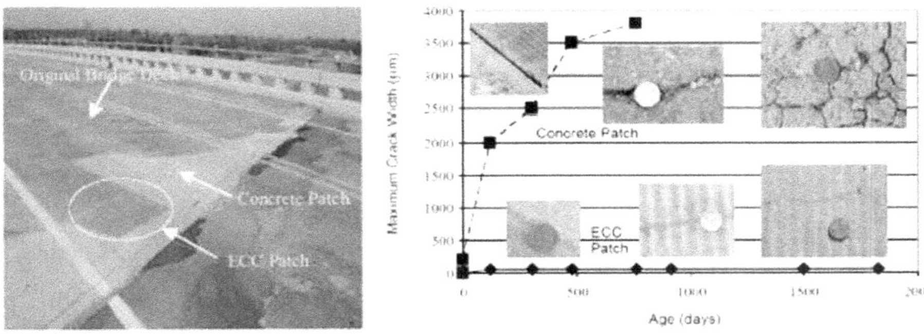


## ECC Application to Repair Concrete Structures

ECC is a useful repair substance because of its high ductility, capacity to disperse localized stresses, and self-healing properties. Bridge deck patchwork repair was one of the initial ECC uses, per the Li M. The cases of ECC repair applications are listed below (Li, 2008):

- The 60-year-old concrete Mitaka Dam in Hiroshima, Japan, suffered serious damage to the surface in 2003. Over 600 m$^2$ of the upstream dam surface, ECC with good workability was coated with just an approximate thickness of 20 mm.
- Early-age strength patch repair In Ann Arbor, Michigan, ECC was built on the Ellsworth Road Bridge over US-23.
- An earth retaining wall that had been cracked by ASR was repaired with cement mortar and ECC. On the damaged surface, a coating between 50 and 70 mm thick is applied. At 10 and 24 months following the repair, the condition of the repair layer was examined. The crack widths of ECC were smaller than those of cement mortar.

## ECC Steel Composite Bridge Deck

Using ECC, a slab of the Mihara bridge in Hokkaido, Japan, was built. Mihara Bridge is a recently built cable-supported bridge that is 1,000 meters long and has a maximal span length of 340 meters in the middle. It was erected in 2005. Steel girders and an ECC slab with composite behavior make up the bridge's superstructure.

The ECC slab had a 38 mm thickness and covered 20,000 $m^2$ in area. The ECC steel composite deck is predicted to have a much lower life-cycle cost and a maximum life span of 100 years.

*Figure 5. Construction of ECC Steel Composite Bridge Deck at Mihara Bridge in Hokkaido, Japan (Park et al., 2020)*

## ECC Application to Structural Members

In a high-rise structure in Osaka, Japan, ECC beams were used during construction, according to Kanda. The ECC coupling beams, which join the columns of a 60-story reinforced concrete structure, are crucial components for resisting seismic loads. Full-scale laboratory experiments have been carried out on various structural components built of ECC in the following ways, notwithstanding the fact that those are not field applications:

- Fischer, Li M., and Kim investigated bending beams under monotonic and cyclic loads. Tests on ECC-based shear beam (deep beam) components were conducted by Kanda, Fukuyama, and Shimizu.
- Fukuyama, Parra-Montesinos, Wight, and Yuan examined the behavior of ECC beam-column couplings under inverted dynamic loading.
- Kanda, Kesner, Billington, and Fukuyama studied the behavior of ECC wall elements under reversed cyclic stresses.

## ECC Application to Bridge-Link Slab

Bridge beams can deteriorate as a result of rainwater and other corrosives penetrating underneath the deck and seeping at expansion joints. This issue can be avoided by using continuous bridge decks without expansion joints, and link slabs are an appropriate place to use ECC. A connection slab was created by Lepech and Li M for a demonstration project in Michigan. Figure 6 depicts the ECC link slab, which is 5.5 m × 20.25 m in size. Figure 7 depicts the construction of the ECC link slab in 2005.

*Figure 6. ECC link slab design (Park et al., 2020)*

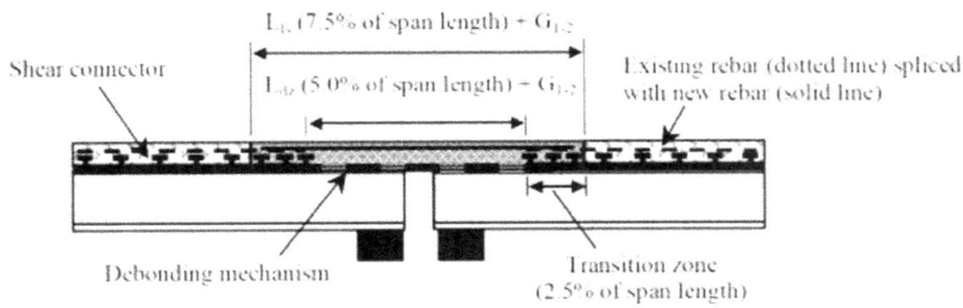


*Figure 7. ECC link slab ECC link slab constructed in Michigan (Park et al., 2020)*

With the assistance of the Michigan Department of Transportation, the operational demonstrations of the ECC link slab were carried out across three phases from 2003 to 2007. Using laboratory material studies, the first phase offered instructions for the design and construction of the ECC link slab. The Grove Street Bridge deck over Interstate 94 in Ypsilanti, Michigan, was built with the link slab. To guarantee the link safety, slab's full-scale mixing sequences, construction, and procedures were offered, and load testing was carried out. The third phase addressed mixture design and construction difficulties with an emphasis on contraction control to improve the ECC connection slab's long-term performance. Investigations done two years after the construction demonstrate that the ECC link slabs have superior wear and scaling tolerance to the typical concrete bridge deck. After ten years of use, Shamaran and Li M looked into the ECC connection slab's status and found that it had performed well without any maintenance. When compared with traditional expansion joints and conventional concrete link slabs, the ECC link slab saves cost and GHG emissions, according to Koeleian's LCA for bridge link slabs. Hou performed more lab-scale testing to confirm the effectiveness of ECC link slabs.

## CONCLUSION

In conclusion, Engineered Cementitious Composites offer a compelling solution for applications that demand high ductility, tensile strength, and durability. While challenges exist, ongoing research and advancements in construction practices may further broaden the scope of ECC applications in the field of civil engineering and infrastructure development. ECC's unique properties make it suitable for repairing and retrofitting existing structures, improving their performance, and extending their service life. ECC stands out for its high tensile strain capacity, exceeding that of both conventional and fiber-reinforced concrete. This property is particularly valuable in applications where the material needs to accommodate significant deformation, such as in seismic-resistant structures. It could also be used under seismic-prone areas due to its capability of undergoing large deformations. Despite its advantages, ECC may face challenges related to cost and construction practices. The specialized materials and mix designs involved in ECC production can sometimes result in higher costs compared to conventional concrete. Additionally, construction practices may need adjustments to accommodate the unique properties of ECC.

## REFERENCES

Engineered Cementitious Composites. (n.d.-a). https://en.wikipedia.org/wiki/Engineered_cementitious_composite

Engineered Cementitious Composites. (n.d.-b). https://www.engineeringcivil.com/engineered-cementitious-composite.html

Kewalramani, Mohamed, & Syed. (2017). Engineered Cementitious Composites for Modern Civil Engineering Structures in Hot Arid Coastal Climatic Conditions. *Procedia Engineering, 180*, 767-774. doi:10.1016/j.proeng.2017.04.237

Li, V. C. (2008). Engineered Cementitious Composites (ECC) – Material, Structural, and Durability Performance. In Concrete Construction Engineering Handbook. CRC Press.

Nagaraj, S., & Shanmugam, P. (2022). Mechanical properties of engineered cementitious composites (ECC) incorporating different mineral admixtures and fiber: A review. *Journal of Building Pathology and Rehabilitation.*, *7*(1), 40. doi:10.1007/s41024-022-00182-1

Park, P., Jones, R., Castillo, L., Vallangca, M., & Cantu, F. (2020). *Engineered Cementitious Composites (ECC) for Applications in Texas*. https://library.ctr.utexas.edu/hostedpdfs/utrgv/0-7030-1.pdf

Singh, S., & Munjal, P. (2020). Engineered cementitious composite and its applications. *Materials Today: Proceedings*, *32*, 797–802. Advance online publication. doi:10.1016/j.matpr.2020.03.743

# Chapter 2
# Multifunctional Contributions of Nanomaterials to Cementitious Composite Materials

**Rafael Vargas-Bernal**
https://orcid.org/0000-0003-4865-4575
*University of Guanajuato, Mexico*

## ABSTRACT

*Future structures based on cement-based composite materials must be more versatile, durable, and strong in addition to being environmentally friendly and sustainable. With the introduction of nanomaterials, it is possible to improve their compressive strength, flexural strength, tensile strength, elastic modulus, toughness, heat of hydration, thermoelectric conversion efficiency, thermoelectric figure of merit, power factor, Seebeck coefficient, electrical conductivity, piezoelectric coefficient, defect reduction and cracks, and reduction of liquid and gas permeation, among others. This chapter aims to review the multifunctional contributions that nanomaterials such as carbon nanotubes (CNTs), graphene, graphene oxide (GO), reduced graphene oxide (rGO), MXenes, titanium dioxide or titania, silicon dioxide or silica, lithium carbonate, calcium carbonate, and perovskite oxides are making in the improvement of cement-based composite materials. Without a doubt, this research topic will continue to drive technological and scientific innovation in the construction industry in this century.*

DOI: 10.4018/978-1-6684-8182-0.ch002

## INTRODUCTION

The social and economic growth of countries is associated with the development of civil infrastructure such as road networks, tunnels, bridges, dams, etc. that they have achieved (Irfan, 2021). To achieve this infrastructure, it is necessary to have materials based on concrete and cement because they are easy to build, economical, robust, can acquire complex geometries, and have excellent mechanical properties (Alvi, 2023). These materials have high compressive strength and quasi-brittle behavior, but to be used in infrastructure projects they require steel reinforcements to achieve tensile and flexural strength. Unfortunately, traditional concrete presents a set of limitations, design, and construction practices as well as severe exposure conditions that cause the deterioration of cementitious composite materials, which can result in aesthetic, functional, or structural problems that must be resolved but not on a volumetric scale. These structures are subject to degrading environments during operation and service that cannot be avoided such as sulfates, ice thawing, pressure or thermal differentials, $CO_2$ penetration, chlorine and/or acids, loading during service, and impact, as summarized in Figure 1. The effect of these environments produces microcracks that produce localized weakening and over time micro defects are produced that will reduce the integrity of the elements of the structures, compromising their original mechanical properties. To minimize maintenance and extend the service life of infrastructure built using cement-based composites, structural health monitoring (SHM) is necessary (Zhang, 2024). This monitoring must provide real-time information on the condition of structures using sensing schemes based on shape memory alloys, piezoelectric materials, optical fibers, or strain gauges, which are not necessarily compatible with cementitious materials. Traditional set cement presents porosity, unhydrated cement particles, crystals, and hydrated products (Metaxa, 2021). Through nanomaterials such as MXenes, graphene, carbon nanofibers (CNFs), graphene oxide (GO), hexagonal boron nitride (*h*-BN), and carbon nanotubes (CNTs), it is possible to develop self-sensing cement-based materials (Liu, 2020; Luo, 2023). The main challenge of using these fillers is to distribute them homogeneously throughout the volume of the material, giving them piezoresistive properties and doing so economically while retaining the mechanical properties. To achieve these properties using nanomaterials, it is necessary to embed them in polymeric materials to be exploited in cement-based composite materials. The use of polymeric materials in cementitious matrices lies in reusing waste polymers and changing properties such as moisture absorption, brittleness, toughness, fatigue, and density (Wang, 2022). Among the polymers most used to improve the performance of cementitious matrices are rubber, polyurethane resins, furan resins, epoxy resins, methyl methacrylate, unsaturated polyester resins, and urea-formaldehyde resins. The incorporation of polymers into cementitious matrices due to their viscoelasticity increases the toughness and reduces the fragility of the composite material obtained at a volumetric level (Irfan, 2021).

*Figure 1. Main environmental factors that degrade cementitious composite materials*

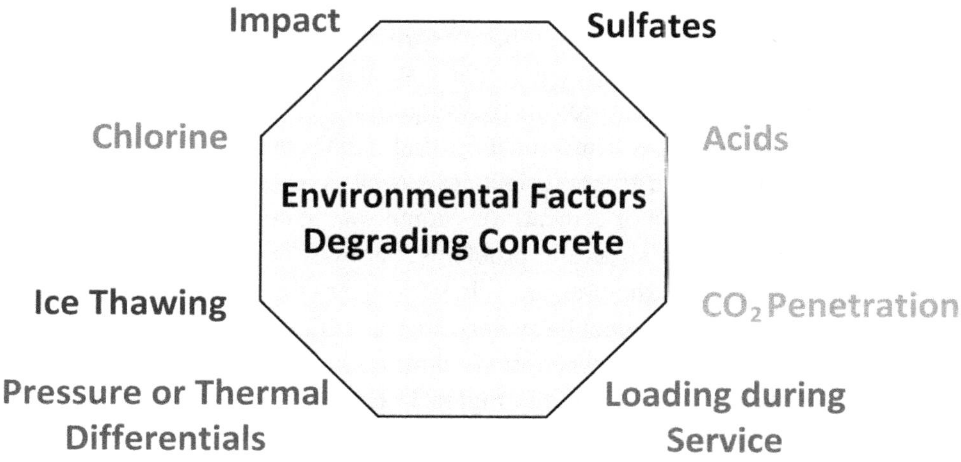


Concrete is the most consumed synthetic material worldwide, reaching a production of 30 gigatons per year (Krystek, 2019). The mechanical properties achieved by traditional cementitious materials until now must be improved to guarantee their reliability and optimize their performance (Metaxa, 2021). Traditional concrete has the following disadvantages: high tendency to fracture, reduced resistance to deformation, and low tensile strength, which do not guarantee its long-term durability in its applications. Additionally, the environmental conditions to which concrete is subjected modify its physicochemical properties, which leads to its degradation and a reduction in its service life (Zhang, 2023). The concrete surface, being porous, is susceptible to attack and thus to the gradual deterioration of its performance. The interfacial transition zone or ITZ is the region that surrounds the cement aggregates and where the concrete has high porosity, so this volume of material must be reinforced to increase its reliability (Alvi, 2023).

Because the amount of $CO_2$ produced per ton of cement ranges from 0.65 to 0.95 tons, researchers around the world are investigating new materials that reduce or replace the amount of cement used in the production of concrete and in particular the use of nanomaterials (Cosentino, 2020; Yoo, 2022). The use of nanomaterials in cement and concrete materials adds durability, and mechanical and structural properties to the design of conventional and advanced structures for sustainable applications in the construction industry (Bahari, 2022; Zhang 2024). The addition of nanomaterials to cementitious materials aims to improve their thermal, chemical, electrical, and mechanical properties (Metaxa, 2021). As the years go by, the cost of nanomaterials no longer represents a limitation to their use. Because the volumetric or weight percentages of nanomaterial should not exceed the percolation threshold required to improve the properties is small, its inclusion is not entirely costly either (Zhang, 2023). The inherent disadvantage of nanomaterials is that they must be uniformly dispersed in the matrix material of the cementitious composite, which is difficult to achieve without additional steps in the material synthesis process.

In this chapter, the main multifunctional contributions of nanomaterials to cementitious composite materials are discussed. Although the use of nanomaterials in cement-based composite materials is currently still low, however, the research carried out is making their incorporation increasingly promoted to achieve ultra-high-performance concretes (UHPCs) that are environmentally friendly, improve performance, and lowest cost, as will be described in this chapter. The chapter is divided as follows: in the section titled "*Technical Problems in Cementitious Materials*" the main areas of technical opportunity that cement-based composite materials must resolve to improve the performance of conventional materials are discussed. Below, in the section titled "*Properties that can be Improved using Nanomaterials*", the possible contributions that nanomaterials can make to improve the performance of cement-based composite materials are described. Next, the section titled "*Solutions achieved using Nanomaterials*" analyzes some of the proposals developed by researchers around the world to improve the performance of cement-based composite materials using nanomaterials. Subsequently, in the section titled "*Future Research Directions*" some of the possibilities of research topics that will be developed in the coming years are proposed to continue improving the performance of cement-based composite materials using nanomaterials. Finally, the section titled "*Conclusions*" provides the conclusions of the chapter regarding the research conducted on the multifunctional properties that nanomaterials have contributed to cement-based composite materials.

## TECHNICAL PROBLEMS IN TRADITIONAL CEMENTITIOUS MATERIALS

Cement is composed of four minerals that act as a binder in concrete: calcium oxide (CaO) or abbreviated as C, silica ($SiO_2$) or abbreviated as S, alumina ($Al_2O_3$) or abbreviated (A), and iron oxide (III) ($Fe_2O_3$) or abbreviated as F (Basha, 2023), as depicted in Figure 2. It is composed of four main compounds such as tetra calcium aluminoferrites ($C_4AF$), tricalcium aluminates ($C_3A$), dicalcium silicates ($C_2S$), and tricalcium silicates ($C_3S$) as illustrated in Figure 2 (Irfan, 2021). The main minerals found in the cement composition are summarized in Figure 2 (a), while the main compounds found in the cement are presented in Figure 2 (b). When the mixture of powdered minerals is hydrated, it undergoes a chemical reaction which, over time, transforms into solid fibrous crystals. Two of the products derived from the hydration process that fulfills transcendental functions are calcium silicate hydrate (CSH, $3CaO{\cdot}2SiO_2{\cdot}3H_2O$) and calcium hydroxide (CH, $Ca(OH)_2$) (Metaxa, 2021). The volume of hardened paste has a percentage in the range of 50 to 65% of CSH and 20 to 25% of CH. CSH provides durability and resistance to concrete, while CH maintains alkalinity with a pH > 12.0 which protects the steel with a layer of passivity around it. Minerals such as mono sulfate and ettringite that constitute 15 to 20% of the volume control the setting time and initial flow of the cement paste, and when these become hydrated, they give rise to swelling of the concrete which induces the formation of cracks. The rest of the hydration products include capillary voids with sizes from 10 microns to 10 nanometers as well as gel pores with sizes from 0.5 to 10 nanometers.

*Figure 2. Composition of cement: (a) main minerals and (b) main compounds*

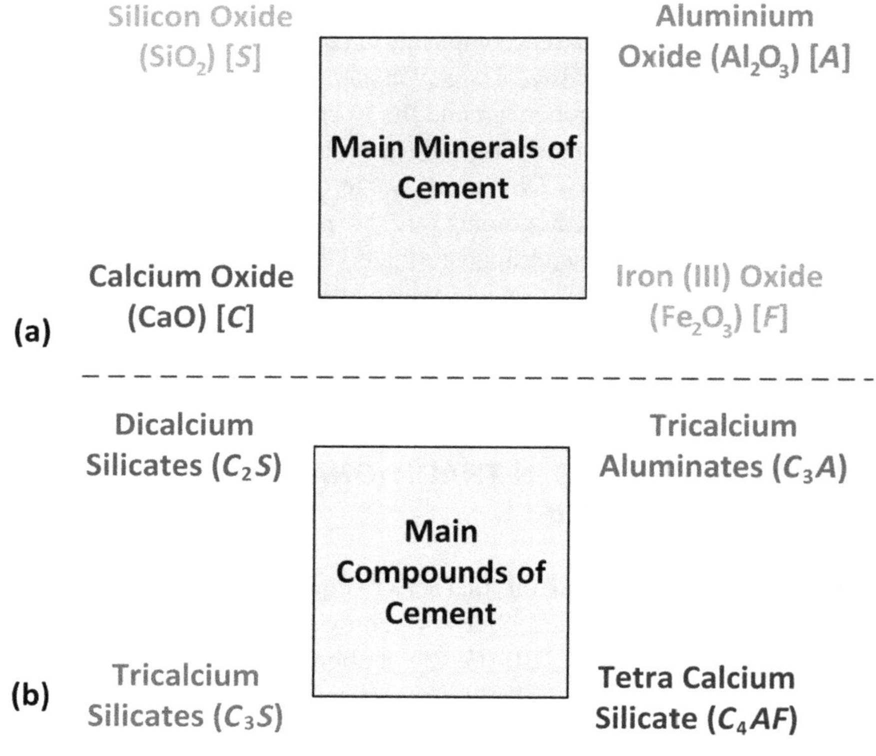

Traditional reinforcements are based on steel, carbon fibers, glass fibers, fiber-reinforced polymers, etc. (Krystev, 2019). They are used to achieve the shear strength that allows controlling the initiation and propagation of cracks at the macroscale. However, these reinforcements do not influence the brittleness and cracking that occurs at the nanoscale. Therefore, the introduction of nanomaterials as fillers to reinforce cement-based composite materials offers an interesting opportunity to reduce the effects found in conventional concretes. All components of a composite material must be dispersed to be uniformly distributed throughout the volume of the material to avoid their agglomeration which reduces mechanical performance leading to the generation of failure zones (Adesina, 2020). When the dispersing medium is water, hydrophilic materials can dissolve in it, while hydrophobic materials cannot dissolve, which reduces or makes their dispersion difficult. The dispersion of fillers requires adequate dispersants to separate the components and prevent them from agglomerating and settling while they are mixed.

The concrete has a plastic behavior or is green or fresh from the moment the cement is hydrated until it sets (Allujami, 2020). While the concrete is in this state, it can be molded into any shape required. Both the performance and final quality of concrete depend on its properties in its fresh state. Among the fresh properties of concrete are plastic shrinkage, heat of hydration, segregation, exudation, setting time, consistency, workability, and rheology, as shown in Figure 3. The possibility of tuning some of these properties opens the possibility of modifying the final quality of the concrete to achieve self-compaction, rapid hardening, three-dimensional printability, and other techniques required for structural design. In the hard state, concrete has achieved its long-term performance properties which include electrical conductivity, durability, mechanical strength, curing conditions, compaction, casting, and water-cement ratio, as depicted in Figure 3. From a mechanical point of view, properties such as toughness, stiffness, hardness, ductility, and strength achieved in the hard state of the concrete will determine its behavior to deformation and failure when it is subjected to external stresses and loads, as illustrated in Figure 4.

*Figure 3. Properties of cement: (a) properties in the fresh state and (b) properties in the hard state*

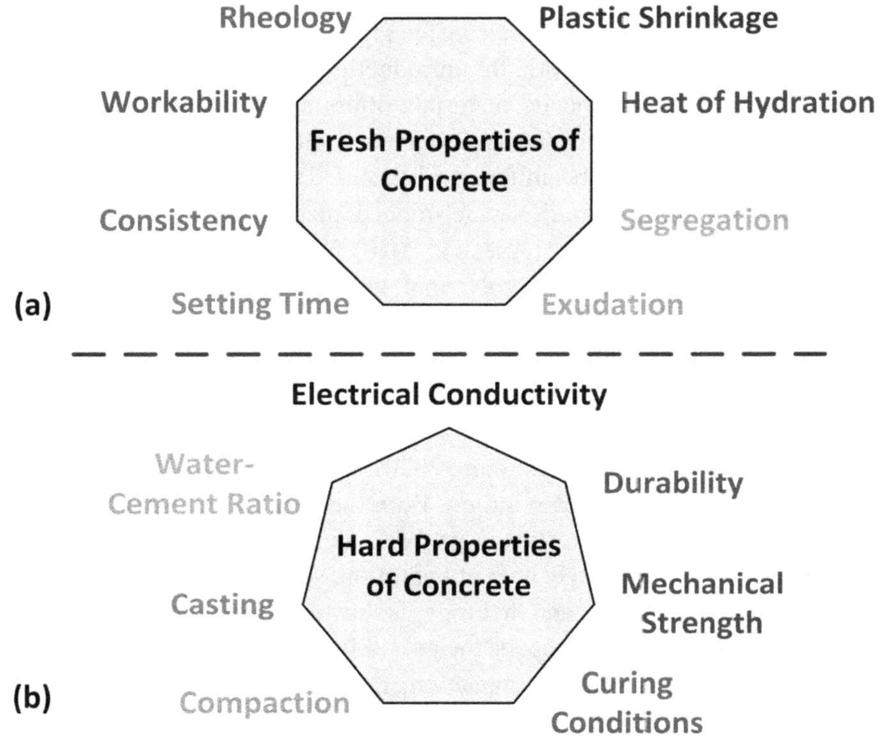

*Figure 4. Main mechanical properties of cementitious materials*

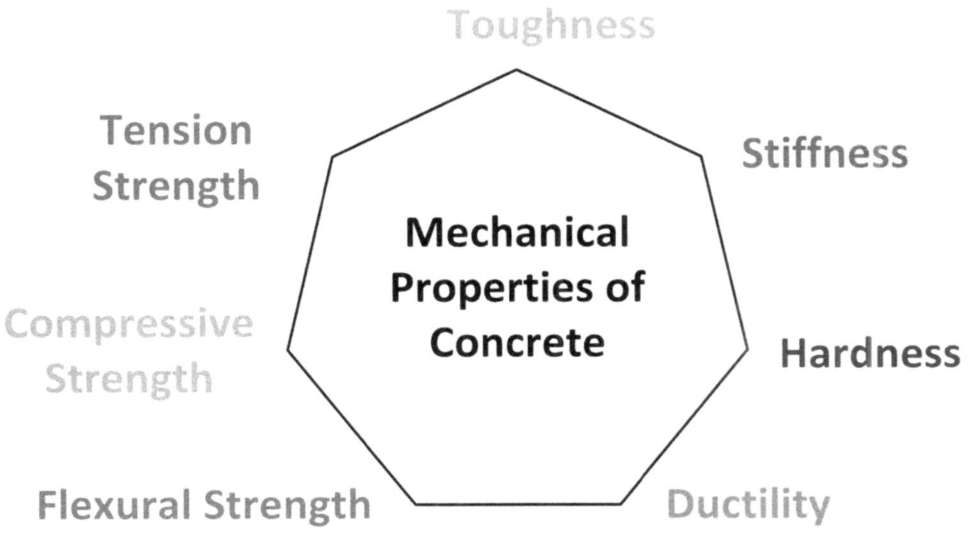


Concrete structures are exposed to climatic and environmental conditions such as dangerous species such as sulfate and acid attacks, which deteriorate them and over time produce irreparable failures (Alvi, 2023). Structural health monitoring (SHM) performs real-time, continuous measurements to determine the behavior of building structures for structural diagnosis (Safuta, 2023). Through monitoring and examining the technical condition of concrete structures, it is possible to guarantee long-term durability, safety, reliability, and sustainability. A visual inspection detects only visible changes, is time-consuming, and is extremely subjective to be used to monitor a concrete structure. Monitoring allows the identification of inadequate construction and design practices, inappropriate maintenance, as well as chemical and environmental degradation. Inadequate design can come from a lack of skill, improper curing, or calculation errors. Maintenance may arise from improper use, overloading of the structure, and unforeseen vibrations. Chemical and environmental degradation can occur due to the permeation of liquids and gases, as well as changes in temperature and humidity. Conventional implementation of structural health monitoring (SHM) makes use of piezoceramic transducers, thermometers, accelerometers, fiber optic-based sensors, and strain gauges. External sensing of large-scale cement structures has low compatibility with the structure, low durability, requires extensive equipment and connections, and has a high cost. Sensing embedded in the structure may experience interference in the concrete structure.

New trends in printable additive manufacturing technology for high-performance materials are being used to print three-dimensional concrete structures (Feng, 2023). Three-dimensional concrete offers unique properties such as no formwork, less labor, quick construction, lower material consumption, and flexible architectural design. The concrete must be deposited layer by layer until the required three-dimensional structure is obtained. However, to achieve resounding success this technology must be high quality and not be expensive. Three-dimensional concrete printing technology must offer environmentally friendly, sustainable, lightweight, and high-performance materials. When concrete is deposited, it must meet several attributes to be properly printed, including interlayer interval time, extrusion rate, nozzle speed and dimension, green resistance, yield strength, viscosity, and flow rate.

## PROPERTIES THAT CAN BE IMPROVED USING NANOMATERIALS

To be incorporated into cementitious materials, nanomaterials must first be dispersed in a suitable hydrophilic solvent depending on their nature, and then mixed with the cement in the preparation of concrete. Dispersion techniques involve the use of surfactants, sonication, ball milling, or mechanical or magnetic stirrers (Metaxa, 2021). To obtain a cementitious composite material, it is also necessary to mold the mixture to give the material the volumetric geometric shape required to be useful in a specific application. Then use curing in the mold by applying water or steam, always controlling the time and temperature in the process. The synthesis process of cementitious composite materials involving nanomaterials is illustrated in Figure 5.

*Figure 5. Synthesis process of cementitious composite materials using nanomaterials*

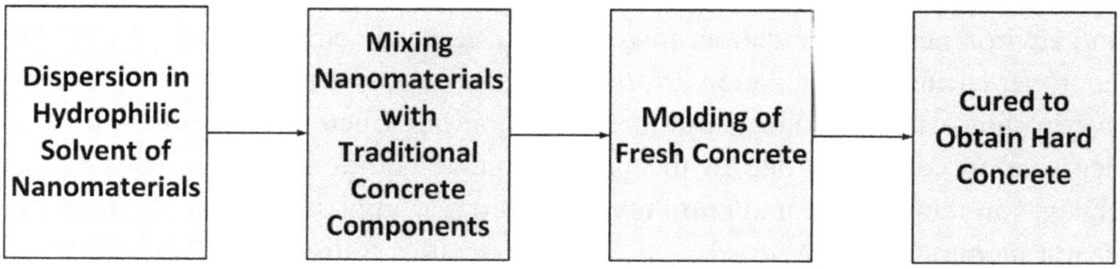


The inclusion of nanomaterials has two basic functions, the first being to facilitate the nucleation effect which increases the surface area due to the growth of CSH, and the second to act as a filler of gaps and pores at micro and nano levels which reduces porosity, and this improves the mechanical properties of durability and strength (Liu,

2020). Nanomaterials can provide strength, durability, sustainability, self-cleaning, antimicrobial, and anti-corrosion properties, with an air-purifying surface, or be able to develop sensing devices for physical variables associated with the state of structures or even structural members with self-sensing (Krystev, 2019). Nanomaterials exploit the following unique effects: dielectric confinement, quantum tunneling, surface and interface, and small size (Zhu, 2019). Nanomaterials are classified into three large groups: 1) 0-D, which has all their dimensions in the nanometer range, 2) 1-D, which has two of their dimensions in the nanometer range, and 3) 2 -D, which has one of their dimensions in the range of nanometers (Wang, 2022). Within the broad group of nanomaterials, graphene (type 2-D), carbon nanotubes functionalized with the hydroxyl ion (type 1-D), and nano titania (type 0-D) are those that present unique mechanical, physical, and chemical properties, especially because they have similar dimensions with the products obtained in concrete such as they are calcium silicate hydrate (CSH) as well as calcium hydroxide (CH). A summary of some of the references used in this chapter that are driving research into the impact of nanomaterials on improving the properties of cementitious materials has been compiled in Table 1.

*Table 1. Some references of nanomaterials that optimize cementitious composite materials*

| Nanomaterial | Purpose | Reference |
|---|---|---|
| Carbon-based nanomaterials | Reduce cracks, defects, and porosity. Improve electrical conductivity, corrosion resistance, workability, and mechanical resistance. | Krystev, 2019; Amini, 2021; Wei, 2021; Basha, 2023; Chen, 2023; Hosseini, 2023; Safuta, 2023; Salami, 2023; Zhang. 2024 |
| Carbonate-based nanomaterials | Improve compressive and flexural strength, toughness, elastic modulus, Seebeck coefficient, and thermoelectric conversion efficiency. | Wei, 2021 |
| MXenes | Improve microstructure, hydration process, compressive strength, and heat hydration. | Zhu, 2019 |
| Nanoclays | Improve tensile and flexural strength. | Xu, 2021 |
| Oxides | Improve wear resistance, microstructure, toughness, water absorption, and corrosion resistance. | Bahari, 2022; Yoo, 2022; Basha, 2023; Luo, 2023; Zhang, 2023 |
| All nanomaterials | A lot of properties are improved, and defects are reduced. | Adesina, 2020; Allujami, 2020; Liu, 2020; Metaxa, 2021; Alvi, 2023; Feng, 2023 |

The introduction of nanomaterials to cementitious composite materials aims to strengthen them at the nanoscale, which was not possible previously (Metaxa, 2021). The nanomaterials that have been used until now to reinforce cementitious materials can be grouped into six main groups: oxides, carbonates, graphene, and its derivatives, nanoclays, MXenes, and carbon nanotubes (CNTs). Nanomaterials improve the physical, thermal, mechanical, and electrical properties of cementitious composite materials, which lead to improved fire resistance, decreased porosity, accelerated hydration, and increased tensile and compressive strength. Nanomaterials improve fatigue performance, freeze-thaw resistance, corrosion resistance, abrasion resistance, resistance to sulfate attack, interfacial bonding, flexural, compression, and tensile strength, and packing density (Yoo, 2022). Nanomaterials prevent shrinkage, water absorption and permeability, and porosity. The multifunctional applications of nanomaterials in ultra-high-performance concrete (UHPC) allow taking advantage of properties such as self-sensing, electrical, and electromagnetic interference (EMI) shielding. Flowability is related to the hydrophilic (inversely) or hydrophobic (directly) nature of the components added to the mixture that makes up the cement-based composite material. Active nanomaterials tend to form hydrated products while inert nanomaterials tend not to react with water. The main nanomaterials to provide functional properties to cement-based composite materials are summarized in Figure 6. The following subsections describe the nanomaterials and analyze the properties that can be exploited in the construction industry.

*Figure 6. Main nanomaterials that functionally improve cementitious composite materials*

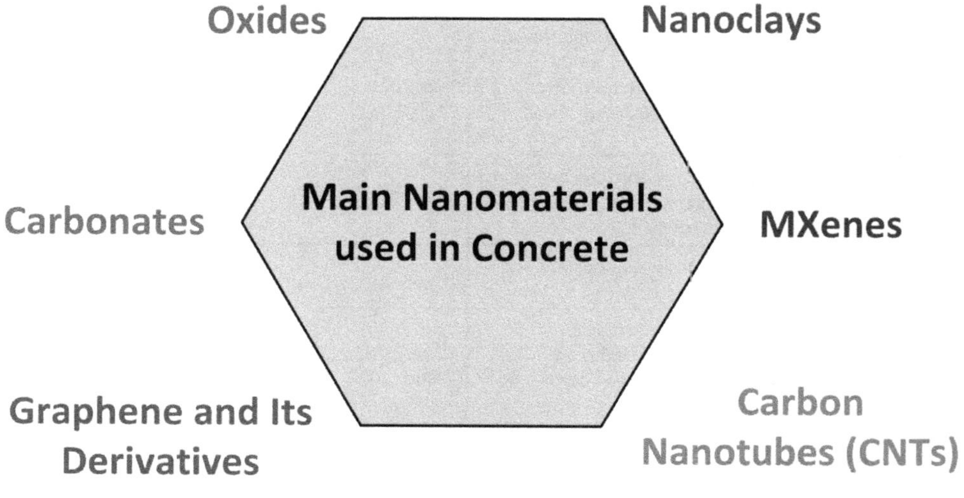

## Carbon-Based Nanomaterials

Synthetic allotropes of carbon such as graphene, graphene oxide, reduced graphene oxide, carbon nanotubes, and fullerenes exploit the $sp^2$ hybridization of carbon to achieve unique properties (Basha, 2023). These properties are achieved thanks to its dimensions, atomic structure, and strong chemical bonds. In addition, the applications of these materials are in the areas of construction, corrosion, manufacturing, medical, and electronics, among others. Carbon nanotubes discovered in 1991 can be classified as single-walled nanotubes (SWCNTs) and multi-walled nanotubes (MWCNTs) and have been the most studied nanomaterials for use in concrete technology (Krystev, 2019). Carbon nanotubes in cement composites refine porosity, reduce dry shrinkage, reduce cracks, and improve mechanical strength through the formation of CSH. In the case of carbon nanofibers (CNFs) and carbon nanotubes (CNTs) with a cylindrical shape with small diameters, they have large specific surface areas that can be exploited in composite materials with values of 45 to 230 $m^2/g$ and 60 to 400 $m^2/g$, respectively (Yoo, 2022). These nanomaterials perform functions in cement-based composite materials such as catalysts for cement hydration, fillers of macro and/or micropores, and bridges to hold micro cracks.

Graphene is a single layer of carbon atoms arranged in a hexagonal lattice structure with unique physical, chemical, thermal, mechanical, and electrical properties (Yoo, 2022). This has a mechanical strength of 200 times greater than steel, high stiffness, fracture strength, Young's modulus, high electrical conductivity, thermal conductivity, and specific area. Graphene, when functionalized and/or thanks to its derivatives, has a hydrophilic nature and improves its dispersion quality in solvents such as water and reduces electrical properties. These attributes can be exploited to provide interesting properties to materials used in construction and corrosion. The addition of graphene oxide (GO) densifies the microstructure of the composite material by reducing porosity and reducing dry shrinkage. Graphene oxide (GO) in cement paste increases plastic viscosity, and yield stress, improves fluidity, and reduces workability due to water absorption (Krystev, 2019). However, due to the possibility of flocculation and agglomeration of GO in the paste of cement, it is necessary to avoid this by adding functional groups, surfactants, and superplasticizers. Graphene exhibits a thermal conductivity of 5300 W/(mK), a large specific surface area of 2630 $m^2/g$, electron mobility of 15000 $cm^2/(V{\cdot}s)$, and an elastic modulus of 1000 GPa (Amini, 2021). Graphene oxide (GO) may contain functional groups such as carbonyl ($CO^{-2}$), carboxyl ($COOH^{-}$), epoxy ($C_2H_2O^{-2}$), and hydroxyl ($OH^{-}$). Graphene oxide has a maximum elastic modulus of 230 GPa and a minimum specific surface area of 1657 $m^2/g$. The presence of these functional groups on GO leads to a high dispersion of it in an aqueous solution to be used in composite materials. Conventional surfactants to facilitate the dispersion of graphene in water are polycarboxylate (PC), lignosulfonate (LS), sodium dodecyl sulfate (SDS), and naphthalene.

The rolled, cylinder-shaped structure of graphene, known as carbon nanotube, is a mechanically resistant material due to its aspect ratio between its diameter and length without adding much weight in addition to being flexible and having an electrical conductivity that can be tuned (Hosseini, 2023). Because these materials have a very small diameter and can be superficially modified using induced functional groups, they have a high chemical reactivity. The attributes of carbon-based nanomaterials have been exploited in the design of asphalt composites, geopolymers, and concrete in the construction industry. Among the properties that are improved in concrete composites thanks to the addition of carbon-based nanomaterials are energy storage, thermal and electrical properties, piezoresistivity, durability, mechanical strength, and freshness. In addition to concrete composite materials, coatings including carbon-based nanomaterials can be designed to improve the durability of concrete using the different versions of materials mentioned above. Adding carbon-based nanomaterials to cement beyond the percolation threshold instead of improving properties harms the concrete, so this natural and synthetic limit should not be exceeded and should be respected and verified before any implementation physical. Generally, carbon-based nanomaterials are dispersed in water before being incorporated into the dry mix of concrete components which will include superplasticizers. The solvent used to disperse the nanomaterials can be treated water extracted from wells, distilled water, and deionized water. Over the years different dispersing agents have been used including polyacrylic acid (PAA), CO809, PCE polycarboxylate superplasticizer, naphthalene sulfonate, polycarboxylate, polyvinylpyrrolidone (PVP), polyoxyethylene nonyl phenyl ether, darex super 20, pluronic F -127, gum arabic, carboxymethyl cellulose, high range water reducer (HRWRA), air entrainment agent (AEA), and sodium deoxycholate (DOC). Dispersion methods range from mechanical mixing, ultrasonication, and high-energy tip sonication, to high-speed shear. The dispersion time depending on the dispersing agent ranges from a few minutes to hours.

## Oxide-Based Nanomaterials

Nanomaterials based on perovskite oxides such as lanthanum strontium cobalt oxide ($LaSrCoO_3$) with a percentage of 1 to 4 weight percent can improve the microstructural, mechanical, and rheological properties of sustainable self-compacting concretes (SCC) (Bahari, 2022). This oxide improves the flexural and compression strength and densifies the microstructure of cement-based composites. Perovskite materials present chemical reduction to $NO_x$ and $SO_2$, high melting point, and high chemical activity. These properties are dependent on the morphology, shape, crystalline size, and structure of the nanomaterial. The cement mortar containing 2% by weight of perovskite oxide nanomaterial improved its flexural strength compared to a pure cement mortar. The results vary according to the concentration gradient, aggregation, and density of nanomaterial in the mortar samples.

Oxide-based nanomaterials such as ferric oxide ($Fe_2O_3$), zirconium oxide or zirconia ($ZrO_2$), copper oxide (CuO), titanium oxide or titania ($TiO_2$), aluminum oxide or alumina ($Al_2O_3$), clay, and silicon oxide or silica ($SiO_2$) are being used in self-compacting cement (SCC) to improve their mechanical properties (Bahari, 2022). These oxides have catalytic, dielectric, thermoelectric, magnetic properties, chemical and thermal stability, ion mobility, and electrical conductivity.

The pozzolanic reaction due to cement hydration can be promoted by active nanomaterials such as nano-alumina ($Al_2O_3$), nano-calcium carbonate ($CaCO_3$), and nano-silica ($SiO_2$) (Luo, 2023). Inert nanomaterials such as carbon nanotubes (CNTs), as well as titania nanoparticles ($TiO_2$), act as accelerators to produce nucleation centers, which increases the initial hydration of the cement and improves the initial compressive strength. Inert nanomaterials produce bridging, filling, and nucleation effects, while active nanomaterials give rise to the cement hydration reaction in addition to filling and nucleation effects (Yoo, 2022). A more exhaustive list of inert nanomaterials includes nano zirconia ($ZrO_2$, NZ), nano ferric oxide ($Fe_2O_3$, NF), nano titanium oxide ($TiO_2$, NT), graphene oxide (GO), carbon nanotubes (CNTs), carbon nanofibers (CNFs), and graphite nanofibers (GNPs). A more exhaustive list of active nanomaterials includes nano alumina ($Al_2O_3$, NA), nano calcium carbide ($CaCO_3$, NC), nano metakaolin (NMK), and nano-silica ($SiO_2$, NS). The silica nanoparticles, when hydrated in the cement, react with calcium hydroxide ($Ca(OH)_2$) forming calcium silicate hydrate (CSH). Metakaolin is made up mainly of silica ($SiO_2$) and alumina ($Al_2O_3$), both of which are important components to improve the hydration rate of cement. These nanoparticles react with free calcium hydroxide ($Ca(OH)_2$) forming secondary hydration products such as calcium silicate hydrate (CSH). Alumina nanoparticles ($Al_2O_3$) favor the formation of ettringite in the first stages of concrete formation, lead to the dissolution of gypsum, and modify the reaction with $C_3A$. This dissolution leads to the separation of aluminum and its incorporation into the CSH structure to produce calcium aluminum silicate hydrate (CASH). In the case of zirconia nanoparticles ($ZrO_2$), they do not react during the hydration of the cement, but they catalyze to accelerate hydration during the first day of setting because of nucleation, precipitation, and growth of CSH. The addition of ferric oxide ($\gamma$-$Fe_2O_3$) nanoparticles with a size of 40 nm provides electromagnetic interference (EMI) shielding to ultra-high-performance concrete (UHPC). The addition of 1% carbon nanotubes to concrete increases the electromagnetic interference (EMI) shielding effectiveness (SE) by factors of 8 and 25 for the frequencies of 10 KHz and 1 GHz, respectively. An absorption of more than 99% of microwaves in the 18

GHz band could be achieved when a plate with a thickness of 10 mm contained 0.3% by weight of graphene oxide (GO) and 1.3% by weight of $\gamma$-$Fe_2O_3$.

## Carbonate-Based Nanomaterials

Calcium carbonate ($CaCO_3$) chemically reacts with cement and accelerates its hydration process, increasing its initial resistance due to the formation of calcium silicate hydrate (CSH) (Cosentino, 2020). Calcium carbonate nanoparticles have porosity, are non-toxic and biocompatible, and their crystalline formation implies high water absorption rates. Calcium carbonate nanoparticles react with calcium hydroxide ($Ca(OH)_2$) and/or tricalcium aluminate ($C_3A$) generating hemi- and mono-carbo aluminate (Yoo, 2022).

Lithium carbonate ($Li_2CO_3$) does not form hydrates as it only exists in anhydrous form, it crystallizes using a saturated aqueous solution, is not easily decarboxylated, and decomposes at around 1300 degrees Celsius (Wei, 2021). This is used as an industrial chemical as a precursor to compounds for lithium-ion batteries. This ceramic material is used for the manufacture of refractory glass and as a low and high-fire ceramic glaze. Cement will be set more quickly when prepared with lithium carbonate.

## MXenes

An MXene is a two-dimensional crystalline material based on transition metal nitrides and/or carbides with a structure like graphite (Zhu, 2019). MXene has a multi-layer structure that in stacked form cannot provide improved mechanical properties. This must be delaminated to separate the MXene into multiple sheets to disperse them and achieve a larger surface area and better properties. The contribution of MXene in the cement hydration process is to promote a regular distribution of regularly ordered ettringite to form a crystalline structure which increases strength.

MXenes have been exploited as novel nanomaterials for a variety of applications since their discovery by Gogotsi et al. in 2011, including energy storage, electrodes, and environmental catalysts (Gogotsi, 2021). They have the chemical formula $M_{n+1}X_nT_z$, where $M$ is a transition metal element, $X$ is a chemical element of the third or fourth main group, and $T$ denotes a C or N atom. MXenes have been used as fillers to enhance the mechanical properties and thermal stability of polymers because of their strong mechanical strength and thermal conductivity. MXenes surfaces are typically terminated with O, OH, and/or F groups, resulting in MXenes with high hydrophilicity.

## Nanoclays

Nanoclays are multilayered mineral silicate nanoparticles, that have a unique chemical composition and morphology, so they can be classified as montmorillonite, bentonite, kaolinite, hectorite, and halloysite (Xu, 2021). These materials have 0.7 to 1 nm thick layers and very high mechanical properties. The layers are modified by surfactants to tune swelling capacity and plasticity. Furthermore, their surface properties depend on the exchangeable cations between layers, the nature of the surface atoms, and the electrical charge of the layers.

# SOLUTIONS ACHIEVED USING NANOMATERIALS

Nanofillers should be added to cementitious matrices to develop structural health monitoring (SHM) of materials at a functional level (Irfan, 2021). Conductive fillers should be added according to a specific percolation threshold that identifies structural problems in the materials using pressure/strain sensitivity that prevents mechanical failure. Reduced graphene oxide has been used to produce composite materials in the form of electronic textiles based on cementitious matrices modified with polymeric binders. These textiles are designed for fixed percolation thresholds that offer a piezoresistive response to cyclic and direct compression loads and are embedded in structural materials for monitoring. Changes in piezoresistance occur when the applied mechanical load is changed by the quantum tunneling mechanism and the percolation threshold occurs thanks to the inclusion of nanomaterials. When a structure is subjected to compressive stresses, the number of contacts between the nanomaterials increases and as the mechanical space is then reduced thanks to the tunneling effect generated by the nanomaterials, the electrical resistance is reduced, giving rise to a piezoresistive response. The piezoresistive behavior of the composite materials produced is that of a complex resistive network, in which the nanomaterials produce conductive electrical paths that modify the electric current that circulates through the material because of the mechanical stresses produced by the applied mechanical load. Concrete with self-sensing systems should include sensors designed based on materials used as functional fillers to convert physical changes in the cement structure produced by external stimuli (Safuta, 2023). Saturated cement composite materials can achieve high electrical conductivity due to ionic conduction phenomena that produce resistive electrical polarization sensitive to the aging of the composite material or environmental conditions. The addition of functional nanofillers into concrete produces contact conduction, field emission, and tunneling as electrical conduction mechanisms between the fillers once the percolation threshold has been reached.

Nanomaterials can play a leading role in extending the durability of cement-based composite materials by preventing permeability and volumetric changes as well as increasing resistance to physical and chemical changes (Adesina, 2020). The presence of nanomaterials in cement-based composite materials extends the length of the silicate chain which protects concrete against chemical attacks. A material is durable when it resists all the environmental conditions to which it is subjected. Cementitious materials are exposed to fluids or ions that deteriorate them, so the amount of penetration and the ease with which they do so must be taken care of to extend their operational life. The addition of nanomaterials leads to the refinement of the pores and thereby decreases the permeability of the composite materials. Furthermore, the inclusion of nanomaterials leads to better thermal and mechanical properties, and self-cleaning, among other benefits. Nanomaterials such as nano metakaolinite (NMK), nano silica (NS), nano alumina (NA), and carbon nanotubes (CNTs) decrease the permeability of concrete. Self-cleaning has been achieved thanks to nano titania (NT). Nano magnetite (NMT, $Fe_3O_4$) allows for self-sensing and reducing the penetration of chlorine ions from cementitious materials. Nanoclay improves the thermal properties of cementitious materials. Nanomaterials such as nano alumina, nano silicon, nano-titania, and maghemite ($\gamma$-$Fe_2O_3$) also help prevent the penetration of chlorine ions. Nanomaterials such as rice husk ash (RHA, $SiO_2$) and silica fume ($SiO_2$) provide resistance to attack by acids such as sulfuric ($H_2SO_4$) and nitric ($HNO_3$) acids. The protection of sulfates on concrete through nano-silica ($SiO_2$) reduces its weight loss, expansion, and strength. Sulfate attack tends to expand the volume of the material and dissolve hydrated products. The addition of silica ($SiO_2$), alumina ($Al_2O_3$), and clay nanoparticles improves resistance to freezing and thawing cycles thanks to the formation of hydrated materials that densify the concrete and make it less permeable to water. Cement composites require nanoparticles of silica ($SiO_2$) and nano titania ($TiO_2$) to increase their abrasion resistance when they are used for pavements, flooring, etc. which are subjected to the movement of load continuously. The stability of concrete at elevated temperatures can be achieved thanks to the use of alumina ($Al_2O_3$) and silica ($SiO_2$) nanoparticles. The main mechanical and thermal properties that nanomaterials improve in cement-based composite materials are summarized in Figure 7. The main electrical, magnetic, tribological, and chemical properties that nanomaterials optimize in cement-based composite materials are compiled in Figure 8. The following subsections describe some of the examples of solutions that have been proposed using nanomaterials to improve the properties of concrete-based composite materials by different research groups around the world.

*Figure 7. Properties of cementitious composite materials improved by nanomaterials: (a) mechanical and (b) thermal*

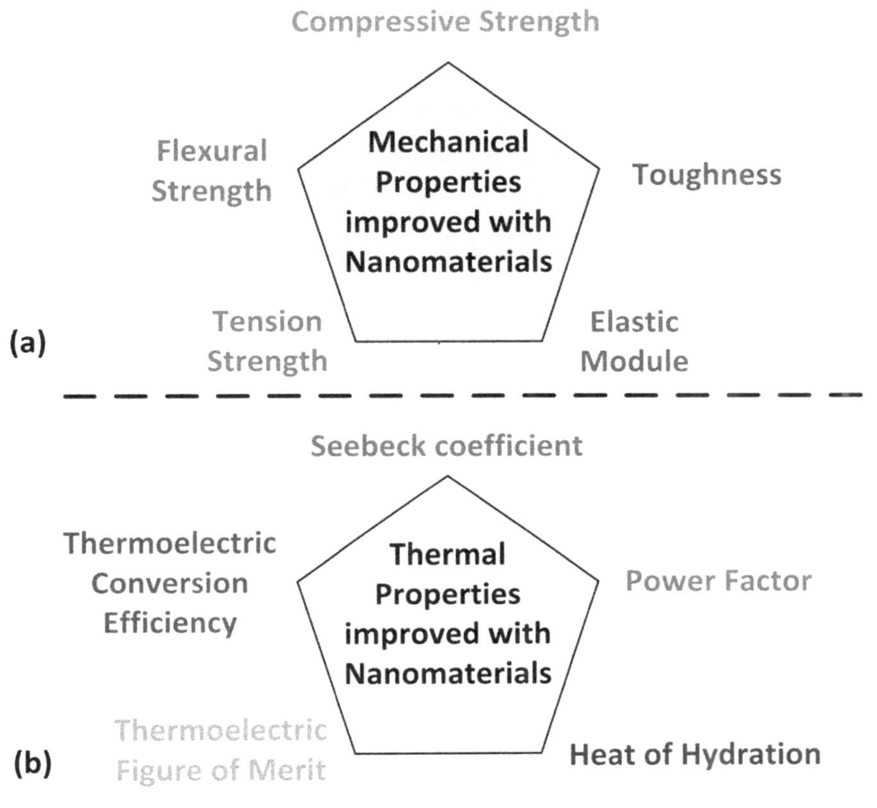

*Figure 8. Properties of cementitious composite materials improved by nanomaterials: (a) electrical and magnetic and (b) tribological and chemical*

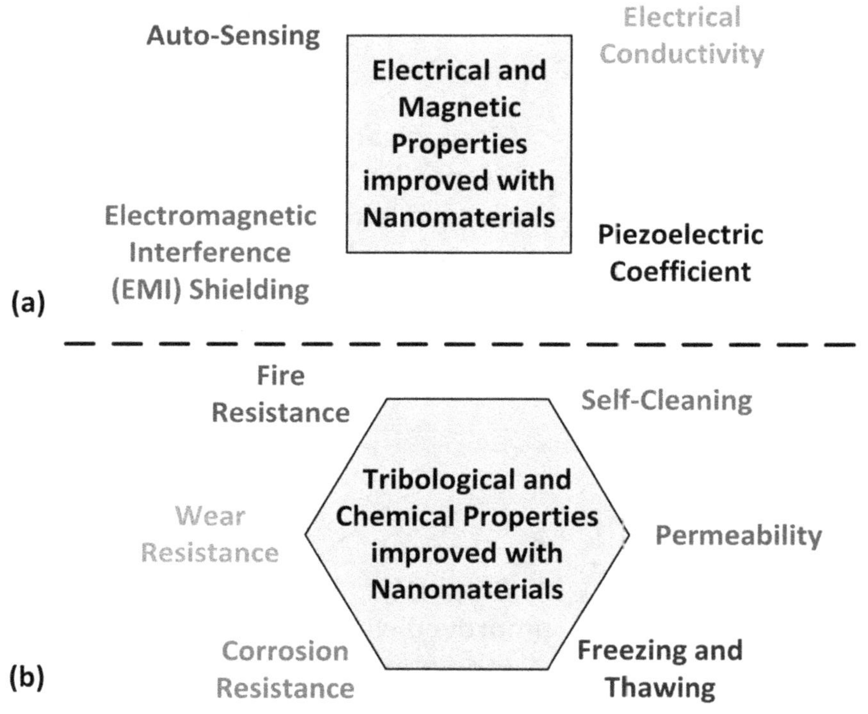


Physical and thermal properties such as fire rate, moisture content, material aggregates, type of aggregates, and type of structure directly influence the fire resistance of a material (Allujami, 2022). Exposure to high temperatures for a long time of concrete causes its strength and stiffness to be reduced. The loss of properties in concrete produces the phenomenon of spalling which implies that the steel reinforcement degrades, the cross-section of the structure is reduced, and the concrete surface is broken. The addition of silica nanoparticles (NS, $SiO_2$) as well as rice husk ash ($SiO_2$) have been used in percentages ranging from 1 to 3% by volume. The first option was not effective, and the second option was effective in protecting the concrete at high temperatures by increasing the residual compressive strength by at least 6%. The results achieved so far have been contradictory, so additional research must be developed to identify the best nanomaterial options to guarantee fire protection of cement-based composite materials.

The penetration of $CO_2$ into the concrete through the pores and its chemical reaction with calcium hydroxide to form calcium carbonate ($CaCO_3$) is known as carbonation (Allujami, 2022). Due to carbonation, concrete is affected in its chemical composition, pore distribution, porosity, strength, durability, and the protective layer of steel against corrosion. The addition of silica nanoparticles to protect concrete from exposure to $CO_2$ at levels of 20 ± 3% for periods of 3 to 28 days decreased carbonation depth by about 35% with further improvement for longer curing times. The improvement in carbonation resistance also increased the compressive strength of the concrete.

## Carbon-Based Nanomaterials

Three-dimensional printable concrete must achieve acceptable durability and mechanical strength through the proper selection of materials as well as the mix proportions necessary to control the print quality, buildability, as well as required rheological properties (Basha, 2023). Carbon-based nanomaterials such as carbon nanotubes, as well as graphene and graphene oxide nanoplatelets, are being incorporated into these composite materials to improve their electrical, thermal, chemical, and mechanical properties. Nanofillers are being used to improve the performance of concrete to improve its mechanical strength and durability through control of the packing ratio. The durability of concrete can be increased through the increase in mechanical resistance coming from the mechanical properties of the nanomaterials such as the mechanical modules and the aspect ratio, as well as the chemical bond which will prevent the presence of microcracks. The addition of graphene to concrete is aimed at solving the problems of conventional concrete such as permeability, low tensile strength, low resistance to harsh environments, and brittleness (Salami, 2023). Graphene provides remarkable strength, electrical conductivity, and heat of hydration, functional properties that can be provided to concrete. Graphene and its derivatives improve the tensile, flexural, and compressive strength of ordinary Portland cement (OPC). Thanks to these strengths, the structures are more resilient to stress and cracks, the service life is extended, and the required maintenance is simultaneously reduced. Graphene nanosheets facilitate the production of calcium silicate hydrate (CSH) by increasing the crystallinity of concrete due to the presence of a major number of nucleation centers. Using graphene, it is possible to reduce the orientation index of calcium hydroxide which produces a uniform and isotropic distribution of the crystals and reduces the stress concentration in cement-based composite materials. The addition of graphene increases the elastic modulus and toughness of the cement and reduces water permeability. From an environmental point of view, graphene as a concrete filler can contribute to reducing the environmental emission of $CO_2$, making the construction industry more sustainable.

The addition of graphene oxide (GO) at a maximum percentage of 0.08% reduces the setting time and fluidity and increases the hydration of the cement paste (Xu, 2021). The addition of certain doses and types of superplasticizers increases the fluidity of the cement paste. Also, the functionalization of GO with polyether amine adds fluidity to the cement paste, slump retention, and toughness; but the electrostatic repulsion and static obstacle produced by it act as dispersants of the cement particles. On the other hand, the addition of single-wall carbon nanotubes (SWCNTs) produces nucleation centers that favor the growth of CSH by increasing the size of the cement grains and reducing the fluidity of the cement paste to levels dependent on the type of nanotubes. By using AEA and DOC as dispersing agents for carbon nanotubes, the yield stress is increased, and the plastic viscosity of the designed concrete is decreased. The joint use of multi-walled carbon nanotubes (MWCNTs), carboxymethyl cellulose, and polycarboxylate ether plasticizer reduced the heat of hydration of concrete. The addition of reduced graphene oxide (rGO) to cement reduces the workability of the generated concrete due to the flocculation introduced by the functional groups and the upper surface area introduced by the nanomaterial.

The addition of graphene oxide (GO) to the cement up to percentages less than 0.03% and functional groups with activated oxygen such as hydroxyl, carboxyl, and sulfonic groups react chemically with the $C_2S$, $C_3S$, and $C_3A$ compounds forming hydration crystals that densify the microstructure, providing improved the flexural and tensile strength of concrete (Basha, 2023). However, additions greater than 0.03% of GO to the cement introduce compact crystals that increase the compressive strength but decrease the flexural and tensile strength. The addition of 0.03% GO improves the tensile, bending, and compression strengths by 79%, 61%, and 39%. Ingredients such as silica fume ($SiO_2$) in cement tend to separate the graphene oxide sheets and improve the compressive strength of concrete. Types of cement with fly ash including GO with a percentage of 0.125% improve the tensile and compressive strength in a proportion of 96% and 35%, respectively. The addition of PCE and GO to the cement improves the mechanical properties of the cement thanks to the stabilization effect introduced by the steric groups. The addition of polyether amine and GO to cement up to 0.03% improves toughness and flexural strength but produces the opposite when increasing the percentage of GO. Strain-hardening cementitious composites (SHCCs) exploit graphene oxide (GO) to improve microstructure, reduce porosity, and increase the bonding of polyvinyl alcohol (PVA) fibers with the composite. The addition of 0.08% GO in these composite materials further increased the compressive strength, tensile strength, flexural strength, and toughness by 24.8%, 37.7%, 80.6%, and 105%, respectively.

To improve the dispersion of GO in cement-based composite materials, it has been proposed to take advantage of the synergistic effect of metakaolin (MK) and polycarboxylate ether (PCE) in highly alkaline environments through the consumption of calcium hydroxide ($Ca(OH)_2$) (Amini, 2021). The composite material obtained after 28 days of curing showed an improvement in compressive strength, flexural strength, elastic modulus, and toughness of 27.6%, 36.3%, 49.8%, and 42.0%, respectively. Pozzolans, which include blast furnace slag, rice husk ash, fly ash, metakaolin, and silica fume, react with calcium hydroxide in an aqueous medium, allowing the production of calcium sulfate hydrate (CSH), which improves the mechanical properties and durability of cement-based composite materials.

Reduced graphene oxide (rGO) thanks to the functional groups and surface area accelerates the hydration ratio, with a percentage of 0.05%, improved the formation of CSH and the mechanical properties of cementitious composite materials, the compressive strength at three days by 113%, as well as the resistance to compression, flexural, tension at 28 days and the modulus of elasticity by 10%, 36%, 13%, and 18%, respectively (Safuta, 2023). The incorporation of 0.05% by weight of electrochemically exfoliated graphene (EEG) into ordinary Portland cement increased the tensile strength, the compressive strength, and the modulus of elasticity by 79%, 8%, and 9% (Krystev, 2019). These increases occur thanks to the hydration of the belite ($2CaO{\cdot}SiO_2$) and alite ($3CaO{\cdot}SiO_2$) of the cement to favor the formation of calcium silicate hydrate (CSH). Other researchers have reported that an rGO content of 2% can improve the compressive and flexural strength of concrete at three days by 44% and 49%, at 14 days by 43% and 52%, and at 28 days by 29% and 35%, respectively (Salami, 2023). A percentage of 0.009% of single-wall carbon nanotubes (SWCNTs) in DOC dispersing agent improved the compressive strength by 22% and the flexural strength by 33%, and in the case of using an AEA dispersing agent it increased the resistance to the compression by 17% and flexural strength by 12%, respectively. The addition of multi-wall carbon nanotubes (MWCNTs) to cement at a rate of 0.05% increased toughness by 54%, at a rate of 0.1% increased flexural strength by 22%, and at a rate of 0.2% increased compressive strength by 17%. The presence of calcium $Ca^{2+}$ ions in the cement tends to agglomerate the graphene oxide (GO) nanosheets. In the search to achieve a good dispersion of nanomaterials such as graphene oxide (GO) in cement-based composite materials, the use of a coating method can be used to achieve an interfacial transition zone (ITZ) when particles of cement and fly ash are covered with GO (Cheng, 2023). The previously mentioned GO coating increased workability, mechanical resistance, and impermeability by up to 21.6%, 38.4%, and 106.4%, respectively. In addition, this inclusion generated a hydration process to reduce the width of the ITZ and the crack ratio by providing the pore-filling effect. To achieve GO dispersion, it is necessary to use a polycarboxylate (PC) superplasticizer to coat the surfaces of the cement powder used in concrete. In addition, the application of 3-amino-propyltriethoxy silane is used as a steric-repellent.

The addition of GO to cementitious composites reduced dry shrinkage in a direct proportion (Chen, 2023). Graphene nanosheets can be added to cement to prevent the introduction of chlorine ions into the concrete through the reduction of porosity and elimination and/or reduction of microcracks. A percentage of 0.02% graphene decreased the penetration depth and diffusion coefficient of chlorine by 37% and 42%, respectively. The inclusion of 1.5% graphene nanosheets in the cement decreases water permeability by 80% and decreases the chlorine migration coefficient and the chlorine diffusion coefficient by 40% and 80%, respectively. The addition of GO to the cement reduced the carbonation depth by 20% by improving the degree of hydration by avoiding the carbonization of calcium hydroxide (CH) and calcium silicate hydrate (CSH) present in the concrete. Graphene oxide increases the durability and mechanical properties of cementitious mortar based on granulated blast furnace slag (GGBS) (Hosseini, 2023). The compressive strength was improved with 0.03% GO in a sample at 90 days. The inclusion of GO protects the concrete against chlorine and acidic environments as well as having less water absorption due to a denser structure and less susceptibility to cracks.

In real-time structural health monitoring of concrete structures, cement-based sensors placed externally to the structure are being extensively developed as they are more compatible than sensors embedded in the manufactured concrete structure (Zhang, 2024). The carbon nanofibers incorporated into the cement increase the electrical conductivity of the mixture with an increase in its content. The reinforcement of the cement with hybrid materials based on carbon fibers and carbon nanofibers in a percentage of 0.01% by weight allows the piezoelectric behavior to be improved thanks to the increase in repeatability, stability, and sensitivity achieved by the fractional changes in resistivity produced in the composite material obtained. The addition of carbon nanotubes modifies the electrical resistivity of cement composite materials because it depends on the type and conductivity of the nanotube, as well as on parameters of the manufacturing process such as flowability and dispersion of the nanotubes in the cement. By decreasing the flowability in the cement due to the nanotubes included in the cement (0.3%), the electrical resistivity is also reduced. The inclusion of 0.06% to 0.08% of GO reduces the electrical resistivity of the cement. Hybrid materials based on nickel nanofibers and nanocarbon black (7%), stainless steel gauze electrodes, and cement present electrical conductivity that can be tuned.

Composite materials based on graphene oxide (GO) and carbon nanotubes (CNTs) provide improved mechanical properties when incorporated into cement-based composite materials through three mechanisms (Salami, 2023). The first is that CNTs through their large surface area allow stronger interfacial bonding and greater charge transfer. The second implies that the composite material increases homogeneity, and this reduces the number of cracks and defects. Finally, the third is that the improved mechanical properties are derived from the spatial interlocking mechanism that prevents deformation due to load transfer.

## Oxide-Based Nanomaterials

The tensile, flexural, and compressive strengths of self-compacted cement mixtures containing these LSCO-based nanomaterials increased (Bahari, 2022). These properties reached a maximum value when the LSCO achieved 2%. The addition of lanthanum strontium cobalt oxide (LSCO) nanoparticles reduced the workability and increased the viscosity of the cement-based composite material. For 4% by weight of LSCO, an agglomeration of the nanoparticles is produced by the Van der Waals forces present in the cement composite material. Furthermore, the material obtained reaches an amorphous structure that gives it mechanical stability due to the chemical reaction between LSCO and calcium silicate hydrate (CSH). Additionally, these perovskite nanomaterials provide a reduction in weight loss after annealing producing a more thermally stable material.

The addition of alumina ($Al_2O_3$) makes the microstructure of the cement-based material more compact because it decreases hydration (Zhu, 2019). While the addition of titania ($TiO_2$) causes the compressive strength to be increased in the initial stage, an adverse effect is found in the long term. The use of hybrid material between carbon nanotubes (CNTs) and silica nanoparticles ($SiO_2$) improves the initial compression resistance, producing more CSH. The use of 3% silica nanoparticles increased the compressive strength, decreased the total crack area, and the interfacial transition zone (ITZ) between the aggregate and the paste was improved. The inclusion of 1% by weight of nano silica can improve fatigue performance, corrosion resistance, abrasion resistance, and water absorption (Yoo, 2022).

The high supplemental large cement materials (SCM) content, high polycarboxylate superplasticizer dosage, and low binder-water ratio slow the initial hydration of ultra-high-performance cement (UHPC) (Luo, 2023). Calcium silicate hydrate (CSH) seeds and silica ($SiO_2$) at the nanomaterial level can influence the initial hydration of ultra-high-performance cement (UHPC), affecting compressive strength. A percentage of 0.3% by weight of CSH seeds or silica nanoparticles increased the compressive strength at 1 day by 457% and 25%, respectively. The 3-day performance was better for the CSH seeds than for the silica ($SiO_2$) nanoparticles because the increase was 343% versus 12%.

When glass-fiber-reinforced-polymer (GFRP) or carbon-fiber-reinforced-polymer (CFRP)-based bars are used in concrete the objective is to prevent corrosion, but unfortunately, the bond between these and the cement, as well as the structure formed, are not completely satisfactory (Wang, 2022). The addition of titania in these composite materials can increase the ultimate bond strength between the components in the range of 16.2% to 37.8%, while the slip was reduced in the range of 28.7% to 35.4%, respectively. This effect is explained because the intrinsic composition was optimized, and the pore content was reduced by modifying the hydration.

Geopolymers are industrial wastes based on aluminosilicates that form long-range covalent and amorphous crystalline lattices, which can be exploited as additives for cement (Xu, 2021). However, their use is restricted. After all, they cannot be used because they crack easily, have low strength, and have low production efficiency. The joint use of geopolymers and nanomaterials is being investigated as an option to develop composite materials based on ultra-high-performance cement (UHPC) which enjoy unique properties thanks to nanomaterials. Nanomaterials improve the microstructure, fluidity, and mechanical properties of the concrete obtained.

The possibility of using recycled sand obtained from construction and demolition waste to reduce environmental effects is one of the possibilities for producing ultra-high-performance concrete (UHPC) and is a topic of interest in the construction industry (Feng, 2023). However, the inclusion of recycled sand (RS) in the concrete decreased its compressive strength, but options must be proposed to counteract this negative effect. The incorporation of nanomaterials in concrete has a positive effect on the fresh and hardened properties as well as on the microstructure of the hydration reaction products. The inclusion of 25% RS and 1% nano-silica (NS, $SiO_2$) improved the compressive strength of the concrete by 14%, reaching 152 MPa at 28 days. Ultra-high-performance concrete (UHPC) is characterized by excellent durability, high toughness, and compressive strength of more than 150 MPa. Recycled sand has sizes ranging from 0.15 to 5 mm, it is more angular, it has a gray color, it contains hard material that can present many cracks, it also contains material with superior water absorption, and low strength and density. Nanomaterials have micro-agglomeration and filling effects that make concrete more uniform and compact. Some nanomaterials can produce additional reactions during hydration generating hydrated materials to improve the interfacial transfer zone (ITZ) between the aggregates and the matrix. The inclusion of these nanomaterials must also be controlled, avoiding their accumulation so that the improvement events are not overshadowed.

Magnesium oxide as a nanomaterial has a high melting point, purity, and hardness. The addition of magnesium oxide (MgO) also favors the formation of hydrated products through nucleation and filling effects (Liu, 2020). A range of 3.0 to 8.0% by weight of MgO added to the cement decreased the initial compressive strength while the same strength at 28 days was increased in the concrete obtained. The hydration of MgO gives rise to magnesium hydroxide ($Mg(OH)_2$), which fills the voids in the concrete. The addition of MgO to cement during hydration can also give rise to magnesium aluminum hydrotalcite ($Mg_2Al\text{-}CO_3$) which tends to improve wear resistance, giving anti-abrasive properties to concrete.

## Carbonate-Based Nanomaterials

The addition of calcium carbonate nanomaterial ($CaCO_3$) allows the production of cementitious composite materials using almost zero $CO_2$ levels which have a high degree of purity and provide compressive and flexural strength as well as toughness in concrete samples at 7 and 28 days (Cosentino, 2020). The flexural strength, compressive strength, and elastic modulus were increased by 11%, 11%, and 29% in curing the concrete samples with 2% calcium carbonate at 7 days. Unfortunately, the previous mechanical advantages achieved by concrete were not maintained for the 28-day test.

Hybrid materials based on lithium carbonate ($Li_2CO_3$) and carbon nanotubes (CNTs) have been proposed to be added to cementitious composite materials with intelligent thermoelectric performance (Wei, 2021). These materials can convert the difference in temperature of a building between the interior and exterior environments into energy produced in summer and release it to remove ice and snow on the pavement in winter. Cement-based composite materials with this hybrid reinforcement exhibit high Seebeck coefficients and low electrical conductivity. Carbon nanotubes do not allow lithium carbonate to be dissolved by water. Lithium carbonate can be used as a filler with low water solubility as it is encapsulated in carbon nanotubes during cement curing. The difference in temperature in large cities between indoor and outdoor environments known as the heat island effect is in the range of 1 to 3 degrees Celsius. The maximum thermoelectric performance parameters such as thermoelectric conversion efficiency, thermoelectric figure of merit (ZT), and power factor for a mixing ratio of 0.1 at 70 degrees Celsius were $1.8x10^{-7}$, $1.75x10^{-5}$, and 0.039 $\mu W/(mK^2)$. Carbon nanotubes in this application must be used as carbonate containers thanks to their unique tubular structure, and lithium carbonate serves as a thermoelectric material that can be activated within the nanotubes. The Seebeck effect is the thermoelectric conversion principle used to obtain voltage from thermoelectric materials through the temperature difference and inversely these materials can be used to generate heat through the applied voltage.

## MXenes

The addition of delaminated MXene to cement composites improves their microstructure, hydration process, and mechanical properties (Zhu, 2019). A 1 wt% percentage of MXenes dispersed in a cement matrix improved the compressive strength, total heat of hydration, and the number of hydration products such as high-density (HD) calcium silicate hydrate (CSH) in the concrete. The addition of 0.04% MXene in cement improves hydration and strength in the initial stage of concrete formation. Thanks to its use in the form of nanosheets, MXene increased the heat of hydration for medium periods of 10 to 12 hours and long periods of 72 hours.

### Nanoclays

An addition of 8% by weight of nano clay in the cement increased the tensile strength of the concrete by 50% thanks to its ability to fill void spaces and the activity to react and form hydrated products (Xu, 2021). The joint use of 3% by weight of nano clay and 0.75% by weight of nano-silica in the cement increased the flexural strength of 9% of the concrete for a curing time of 90 days.

## FUTURE RESEARCH DIRECTIONS

Despite the advances achieved in cement-based composite materials that include functional nanofillers that optimize the properties or allow the monitoring of concrete structures, their performance will need to be optimized to achieve better results than those achieved until now. One of the strategies that undoubtedly must be developed is the reduction of the consumption of concrete components to mitigate the aspects of sustainability and environmental effects in current concrete production. Although nanomaterials in the construction industry have enormous application potential, aspects such as material durability, manufacturing scalability, and cost-effectiveness must be addressed to achieve comprehensive use for mass-level manufacturing. The construction industry requires that researchers continue to develop cement-based composite materials that are more versatile, more robust, and stronger. The advances achieved so far require that reinforcing mechanisms become increasingly more functional and better designed not only to overcome conventional problems found in cement-based composite materials but also to provide self-sensing capabilities and protection against other innovative aspects. Many nanomaterials must be incorporated so that multiple functionalities are incorporated into novel applications previously unimaginable. In the case of concrete reinforced with glass and carbon fibers, it is necessary to work on optimizing the type and content of nanomaterials, fiber embedment length, diameter and surface texture of the sample, and thickness of the cement cover. The effects of all types of nanomaterials on cement-based composite materials as well as different bond-slip models should be studied including nanomaterials should be developed and investigated. One of the challenges to be resolved continues to be the reduction in the production cost of nanomaterials to exploit them on a massive scale. Nanomaterials must also be exploited to provide safety to cement-based composite materials so properties such as dynamic modulus of elasticity, carbonization resistance, and fire resistance must also be studied. Furthermore, the effect of adding nanomaterials as seeds has not yet been fully analyzed to identify all the control parameters that must be established for their most effective use.

## CONCLUSION

This chapter reviews the main functional contributions that nanomaterials have made, are making, and will make to cement-based composite materials. Nanomaterials such as carbon nanotubes (CNTs), graphene, graphene oxide (GO), reduced graphene oxide (rGO), MXenes, titanium oxide ($TiO_2$), silicon oxide ($SiO_2$), lanthanum strontium cobalt oxide ($LaSrCoO_3$, LSCO), calcium carbonate ($CaCO_3$), lithium carbonate ($Li_2CO_3$), etc. are being used to functionally modify cement-based composite materials. Among the mechanical advantages that nanomaterials provide to cement-based composite materials are the increase in flexural, compression, tension strength, elastic modulus, and toughness. In the case of thermal properties, nanomaterials when used in cement-based composite materials increase the heat of hydration, thermoelectric conversion efficiency, thermoelectric figure of merit, power factor, and Seebeck coefficient. From an electrical point of view, nanomaterials in cement-based composites allow the possibility of tuning their electrical conductivity and piezoelectric coefficient. By integrating nanomaterials into concrete, researchers seek to reduce $CO_2$ production during cement production to achieve environmental improvement as well as achieve sustainability. At the level of catastrophic failures in cement-based composite materials, nanomaterials reduce or eliminate defects and cracks, in addition to avoiding the penetration of water, liquids, and gases, which degrade them to reduce the maintenance, repair, or replacement of structures, which in some applications are a guarantee of high quality for a high-performance infrastructure. Composite materials based on cement with nanomaterials seek to achieve safety and reproducibility long sought by users and engineers.

## ACKNOWLEDGMENT

The author is grateful for the support of the Universidad de Guanajuato for the development of this scientific research. He thanks the researchers who have shared their research publications to carry out this research work.

## REFERENCES


Adesina, A. (2020). Nanomaterials in Cementitious Composites: Review of Durability Performance. *Journal of Building Pathology and Rehabilitation*, *5*(1), 21. doi:10.1007/s41024-020-00089-9

Allujami, H. M., Abdulkareem, M., Jassam, T. M., Al-Mansob, R. A., Ng, J. L., & Ibrahim, A. (2022). Nanomaterials in Recycled Aggregates Concrete Applications: Mechanical Properties and Durability. A Review. *Cogent Engineering*, *9*(1), 2122885. doi:10.1080/23311916.2022.2122885

Alvi, I. H., Li, Q., Hou, Y., Onyekwena, C. C., Zhang, M., & Ghaffar, A. (2023). A Critical Review of Cement Composites containing Recycled Aggregates with Graphene Oxide Nanomaterials. *Journal of Building Engineering*, *69*, 105989. doi:10.1016/j.jobe.2023.105989

Amini, K., Ghasemi, A., Amiri, S. S., Mirvalad, S., & Korayem, A. A. (2021). The Synergic Effects of Metakaolin and Polycarboxylate-Ether on Dispersion of Graphene Oxide in Cementitious Environments and Macro-Level Properties of Graphene Oxide Modified Cement Composites. *Construction & Building Materials*, *270*, 121462. doi:10.1016/j.conbuildmat.2020.121462

Bahari, A., Sadeghi-Nik, A., Shaikh, F. U. A., Sadeghi-Nik, A., Cerro-Prada, E., Mirshafiei, E., & Roodbari, M. (2022). Experimental Studies on Rheological, Mechanical, and Microstructure Properties of Self-Compacting Concrete containing Perovskite Nanomaterial. *Structural Concrete*, *23*(1), 564–578. doi:10.1002/suco.202000548

Basha, S. I., Rehman, A. U., Aziz, M. A., & Kim, J.-H. (2023). Cement Composite with Carbon-based Nanomaterials for 3D Concrete Printing Applications – A Review. *Chemical Record (New York, N.Y.)*, *23*(4), e202200293. doi:10.1002/tcr.202200293 PMID:36782098

Chen, Z., Liu, Y., Wu, J., Guo, X., Chen, W., & Gao, Y. (2023). Graphene Oxide-Coated Fly Ash for High Performance and Low-Carbon Cementitious Composites. *Journal of Materials Research and Technology*, *25*, 6710–6724. doi:10.1016/j.jmrt.2023.07.141

Cosentino, I., Ferro, G. A., Restuccia, L., Bensaid, S., Deorsola, F., & Liendo, F. (2020). Nearly Zero $CO_2$ Cementitious Composites. *Material Design & Processing Communications*, *2*(6), e125. doi:10.1002/mdp2.125

Feng, Y., Zhang, B., Xie, J., Xue, Z., Huang, K., & Tan, J. (2023). Effects of Recycled Sand and Nanomaterials on Ultra-High Performance Concrete: Workability, Compressive Strength and Microstructure. *Construction & Building Materials*, *378*, 131180. doi:10.1016/j.conbuildmat.2023.131180

Gogotsi, Y., & Huang, Q. (2021). MXenes: Two-Dimensional Building Blocks for Future Materials and Devices. *ACS Nano*, *15*(4), 5775–5780. doi:10.1021/acsnano.1c03161 PMID:33906288

Hosseini, K., Atrian, M. A., Mirvalad, S., Korayem, A. H., & Ebrahimi, M. (2023). Influence of Ground Granulated Blast Furnace Slag on Mechanical Properties and Durability of Graphene Oxide-reinforced Cementitious Mortars. *Structural Concrete*, *24*(3), 6270–6282. doi:10.1002/suco.202200888

Irfan, M. S., Ali, M. A., Khan, K. A., Umer, R., Kanellopoulos, A., & Samad, Y. A. (2021). An Electronic Textile Embedded Smart Cementitious Composite. *Engineering Reports*, *4*(3), e12468. doi:10.1002/eng2.12468

Krystek, M., Pakulski, D., Patroniak, V., Górski, M., Szojda, L., Ciesielski, A., & Samorì, P. (2019). High-Performance Graphene-based Cementitious Composites. *Advancement of Science*, *6*(9), 1801195. 1002/advs.201801195 PMID:31065510

Liu, C., He, X., Deng, X., Wu, Y., Zheng, Z., Liu, J., & Hui, D. (2020). Application of Nanomaterials in Ultra-High Performance Concrete: A Review. *Nanotechnology Reviews*, *9*(1), 1427–1444. doi:10.1515/ntrev-2020-0107

Luo, Z., Zhi, T., Liu, X., Yin, K., Pan, H., Feng, H., Song, Y., & Su, Y. (2023). Effects of Different Nanomaterials on the Early Performance of Ultra-High Performance of Ultra-High Performance Concrete (UHPC): C-S-H Seeds and Nano-Silica. *Cement and Concrete Composites*, *142*, 105211. doi:10.1016/j.cemconcomp.2023.105211

Metaxa, Z. S., Tolkou, A. K., Efstathiou, S., Rahdar, A., Favvas, E. P., Mitropoulos, A. C., & Kyzas, G. Z. (2021). Nanomaterials in Cementitious Composites: An Update. *Molecules (Basel, Switzerland)*, *26*(5), 1430. doi:10.3390/molecules26051430 PMID:33800797

Safuta, M., Ciesielski, A., & Samori, P. (2023). Controlling the Formation of Electroactive Graphene-based Cementitious Composites: Towards Structural Health Monitoring of Civil Structures. *Chemistry (Weinheim an der Bergstrasse, Germany)*, *29*(71), e202301816. doi:10.1002/chem.202301816 PMID:37725392

Salami, B. A., Mukhtar, F., Ganiyu, S. A., Adekundle, S., & Saleh, T. A. (2023). Graphene-based Concrete: Synthesis Strategies and Reinforcement Mechanisms in Graphene-based Cementitious Composites (Part 1). *Construction & Building Materials*, *396*, 132296. doi:10.1016/j.conbuildmat.2023.132296

Wang, X., Ding, S., Qiu, L., Ashour, A., Wang, Y., Han, B., & Ou, J. (2022). Improving Bond of Fiber-Reinforced Polymer Bars with Concrete through incorporating Nanomaterials. *Composites. Part B, Engineering*, *239*, 109960. doi:10.1016/j.compositesb.2022.109960

Wei, J., Zhang, M., Wang, Y., Qiao, S., Zhang, H., & Li, X. (2021). Synergistic Optimization of Thermoelectric Performance in Cementitious Composites by Lithium Carbonate and Carbon Nanotubes. *International Journal of Energy Research*, *45*(2), 2460–2473. doi:10.1002/er.5940

Xu, Z., Huang, Z., Liu, C., Deng, H., Deng, X., Hui, D., Zhang, X., & Bai, Z. (2021). Research Progress on Key Problems of Nanomaterials-modified Geopolymer Concrete. *Nanotechnology Reviews*, *10*(1), 779–792. doi:10.1515/ntrev-2021-0056

Yoo, D.-Y., Taekgeun, O., & Banthia, N. (2022). Nanomaterials in Ultra-High-Performance Concrete (UPHC) – A Review. *Cement and Concrete Composites*, *134*, 104730. doi:10.1016/j.cemconcomp.2022.104730

Zhang, L., Bian, M., Xiao, Z., Wang, X., & Han, B. (2023). A Comprehensive Review of Cementitious Composites modified with Nano Silica: Fabrication, Microstructures, Properties and Applications. *Construction & Building Materials*, *409*, 133922. doi:10.1016/j.conbuildmat.2023.133922

Zhang, L., Li, L., Ashour, A., Ding, S., & Han, B. (2024). Monitoring Damage of Concrete Beams via Self-Sensing Cement Mortar Coating with Carbon Nanotube-Nano Carbon Black Composite Fillers. *Journal of Intelligent Material Systems and Structures*, *5*(6), 633-648. doi:10.1177/1045389X23122112

Zhu, J., Li, G., Feng, C., Wang, L., & Zhang, W. (2019). Effect of Delaminated MXene ($Ti_3C_2$) on the Performance of Cement Paste. *Journal of Nanomaterials*, 2019, 3074206. doi:10.1155/2019/3074206

## ADDITIONAL READING

Adesina, A. (2019). Durability Enhancement of Concrete using Nanomaterials: An Overview. *Materials Science Forum*, *967*, 221–227. doi:10.4028/www.scientific.net/MSF.967.221

Kewalramani, M. A., & Syed, Z. I. (2018). Application of Nanomaterials to Enhance Microstructure and Mechanical Properties of Concrete. *International Journal of Integrated Engineering*, *10*(2), 98–104. doi:10.30880/ijie.2018.10.02.019

Nair, S. R., & Rahim, A. (2017). Effects of Nanomaterial on Concrete. *International Journal of Earth Sciences and Engineering*, *10*(1), 5–10. doi:10.21276/ijee.2017.10.0102

Rattan, A., Sachdeva, P., & Chaudhary, A. (2016). Use of Nanomaterials in Concrete. *International Journal of Latest Research in Engineering and Technology*, *2*(5), 81–84.

Srinivas, K. (2014). Nanomaterials for Concrete Technology. *International Journal of Civil, Structural, Environmental and Infrastructure Engineering Research and Development*, *4*(3), 79–90.

Yin, X., Zhang, M., Wang, L., & Liu, Y. (2020). Interface Debonding Performance of Precast Segmental Nano-Materials based Concrete (PSNBC) Beams. *Materials Express*, *10*(8), 1317–1327. doi:10.1166/mex.2020.1734

Zhao, Z., Sun, R., Xin, G., Wei, S., & Huang, D. (2013). A Review: Application of Nanomaterials in Concrete. *Applied Mechanics and Materials*, *405-408*, 2881–2884. doi:10.4028/www.scientific.net/AMM.405-408.2881

## KEY TERMS AND DEFINITIONS

**Carbonate:** Salt used to produce ceramic materials obtained by the reaction of a metal with carbon dioxide containing the anion $CO_3^{2-}$.

**Cement-Based Composite Materials:** Hardened material based on a cement paste obtained by hydrating cement, water, and ceramic, metallic, and polymeric materials used as reinforcements.

**Graphene:** Two-dimensional monolayer material formed by carbon atoms packed in a honeycomb structure with excellent mechanical, electronic, and thermal properties.

**Graphene Oxide (GO):** Material based on a simple monolayer of carbon obtained by oxidized graphite with oxygen-containing functionalities.

**MXenes:** Two-dimensional inorganic material with layers of thickness of a few atoms containing transition metal carbides, nitrides, or carbonitrides.

**Nanomaterial:** Material with one or more of its dimensions in nanometers with unique physical and chemical properties.

**Oxides:** Materials obtained by oxidizing elements of the periodic table with unique magnetic, mechanical, thermal, electrical, optical, and electrochemical properties.

**Plasticizer:** Substance or material based on mineral or chemical mixtures that are added to concrete to reduce its water requirements, improve its strength and workability, as well as to soften it.

**Reduced Graphene Oxide (rGO):** Material based on a simple monolayer of carbon obtained by chemically treating graphene that contains residual oxygen, heteroatoms, and structural defects.

**Surfactants:** Chemical compounds that decrease the interfacial tension or surface tension between a liquid and a solid, a liquid and a gas, or two liquids, and these operate as dispersants, foaming agents, wetting agents, detergents, or emulsifiers.

# Chapter 3
# Innovations and Applications of Engineered Cementitious Composites and Alternative Materials:
## Reimagining Construction

**Balpreet Singh Madan**
*Department of Art and Design, School of Design, Architecture, and Planning, Sharda University, Greater Noida, India*

**Sampada Viraj Dravid**
*Department of Mechanical Engineering, Vishwakarma Institute of Information Technology, Pune, India*

**G. Deenadayalan**
*Department of Mechanical Engineering, Meenakshi College of Engineering, Chennai, India*

**Praveen Rathod**
*Department of Mechanical Engineering, Vishwakarma Institute of Information Technology, Pune, India*

**Geetha Arumugam**
*Department of Chemistry, Kongu Engineering College, Erode, India*

**S. Boopathi**
https://orcid.org/0000-0002-2065-6539
*Mechanical Engineering, Muthayammal Engineering College, Namakkal, India*

## ABSTRACT

*The construction industry is set for a transformative era with advancements in engineered cementitious composites (ECC) and alternative materials. This chapter explores their composition, properties, and applications, as well as alternative materials like geopolymers and recycled aggregates. It highlights the potential of these innovations to redefine construction practices, their environmental benefits, and*

DOI: 10.4018/978-1-6684-8182-0.ch003

*economic viability. The chapter also discusses challenges in widespread adoption and offers potential solutions. Real-world case studies demonstrate successful applications of ECC and alternative materials, highlighting their performance and potential for further research.*

## INTRODUCTION

Construction materials have evolved significantly, moving from traditional substances to advanced solutions like Engineered Cementitious Composites (ECC) and alternative materials. This shift reflects a desire for improved performance, sustainability, and innovation in the industry. Traditional materials like concrete and steel have limitations in durability, tensile strength, and environmental impact, prompting the exploration of novel materials like ECC and alternative options (Shoji et al., 2022).

Engineered Cementitious Composites (ECC) mark a paradigm shift in material engineering. Unlike conventional concrete, ECC comprises a unique blend of fine-scale fibers (often polymer-based) dispersed within the cementitious matrix. This incorporation significantly enhances its tensile properties, ductility, and durability compared to traditional concrete. ECC exhibits strain-hardening behavior, enabling it to flex and deform without catastrophic failure, thus contributing to improved resilience in structural applications. Furthermore, ECC's ability to self-heal micro-cracks due to its unique composition ensures long-term durability, reducing maintenance costs and increasing the lifespan of structures (Ma et al., 2021).

The evolution toward alternative materials involves a broader spectrum beyond ECC, encompassing geopolymers, recycled aggregates, and other innovative compositions. Geopolymers, for instance, are synthesized from industrial waste materials or naturally occurring substances, offering reduced carbon footprints and enhanced chemical and thermal resistance compared to traditional binders. Recycled aggregates derived from demolished concrete or industrial by-products provide sustainable alternatives to virgin aggregates, contributing to waste reduction and promoting circularity in construction (Ding et al., 2022a).

The shift toward Engineered Cementitious Composites and alternative materials is driven by several compelling factors. Sustainability stands as a pivotal driver, compelling the industry to reduce its environmental impact. ECC and alternative materials often employ recycled components, reduce carbon emissions during production, and promote resource efficiency. Moreover, their improved mechanical properties enable the construction of lighter and more resilient structures, fostering safer and more adaptable built environments. However, challenges persist in the widespread adoption of these materials. Factors like higher initial costs, limited awareness, and the need for specialized expertise in handling and design hinder their mainstream implementation (Al-Fakih et al., 2021a). Additionally, standardization and codes governing these materials require further development to ensure their seamless integration into existing construction practices.

In conclusion, the evolution from traditional construction materials to Engineered Cementitious Composites and alternative options represents a progressive shift towards sustainable, resilient, and innovative building solutions. While challenges exist, the potential benefits in terms of performance, durability, and environmental sustainability make these materials integral to the future of construction, driving a transformative change in the industry's landscape.

Innovation in construction materials plays a pivotal role in shaping the industry towards sustainability and enhancing overall performance (D. Zhang et al., 2020). This proactive approach not only addresses existing challenges but also sets the stage for creating structures that are more durable, environmentally friendly, and economically viable.

At the heart of this importance lies the need to mitigate the environmental impact of the construction sector. Traditional materials, like concrete and steel, have significant carbon footprints due to their energy-intensive production processes and high use of raw materials. Innovation aims to counter this by developing materials that reduce reliance on non-renewable resources, minimize energy consumption during production, and promote recycling and reuse (Cho et al., 2020).

Sustainability, a core tenet of innovative construction materials, is achieved through various means. Engineered Cementitious Composites (ECC) and alternative materials, such as recycled aggregates and geopolymers, exemplify this shift. These materials often incorporate recycled components or use industrial by-products, reducing the demand for virgin resources and diverting waste from landfills. Moreover, they exhibit improved durability and longevity, resulting in reduced maintenance and replacement needs, thereby enhancing the overall life cycle sustainability of structures (Lu & Zhong, 2022). Innovative materials also address performance-related concerns prevalent in traditional construction. Engineered Cementitious Composites, for instance, offer enhanced mechanical properties, such as higher tensile strength and ductility, compared to conventional concrete. This enables the construction of more resilient structures capable of withstanding extreme conditions, such as earthquakes or harsh weather, reducing the risk of damage and improving safety (Li et al., 2020).

Furthermore, innovative materials often pave the way for more efficient construction processes. Their unique properties, such as self-healing capabilities in ECC or the ease of use of certain alternative materials, streamline construction operations, reduce construction time, and optimize resource utilization. This not only enhances project efficiency but also contributes to cost-effectiveness in the long run. Innovation in construction materials is closely intertwined with technological advancements. Integration of nanotechnology, 3D printing, and smart materials into material development facilitates the creation of materials with enhanced functionalities (Li et al., 2020; Zhu et al., 2022). Self-monitoring and adaptive materials can detect and respond to structural stresses or environmental changes, ensuring optimal performance and longevity of constructions. Challenges persist in the widespread adoption of innovative materials, including high initial costs, limited awareness, and the need for updated building codes and standards. However, concerted efforts from researchers, manufacturers, policymakers, and construction professionals can mitigate these challenges (Malathi et al., 2024; Subha et al., 2023).

Innovation in construction materials is crucial for sustainability and performance. The development and adoption of these materials are essential for transitioning towards an environmentally conscious and resilient built environment. These innovations address current challenges, shape a future with durable, eco-friendly structures, and ensure a sustainable legacy for future generations.

## ENGINEERED CEMENTITIOUS COMPOSITES (ECC)

Engineered Cementitious Composites (ECC) represent a groundbreaking evolution in construction materials, offering a paradigm shift from traditional concrete. ECC is a class of advanced cement-based materials designed with a unique combination of cement, fine aggregates, additives, and high volumes of short, discrete fibers, typically polymer-based. What distinguishes ECC from conventional concrete is its exceptional tensile ductility and strain-hardening behavior. Unlike traditional concrete, which is brittle and prone to sudden failure under tension, ECC exhibits remarkable tensile strength and ductility. This enables it to deform significantly before failure, making it highly resistant to cracking and enhancing its structural resilience (Al-Fakih et al., 2021a; Zhu et al., 2022).

The incorporation of fine-scale fibers, often ranging from a few micrometers to a few millimeters in length, within the cementitious matrix forms a three-dimensional network. This network enables load transfer and effectively distributes stresses, enhancing the material's ability to withstand tensile forces. The fibers arrest crack propagation, promoting multiple micro-cracking instead of catastrophic failure, thus improving the material's toughness and durability. Moreover, ECC possesses self-healing properties at the micro-scale (Karthik et al., 2023; Kavitha, Malini, et al., 2023; Murali et al., 2023). When micro-cracks occur due to applied loads or environmental factors, the material's inherent properties facilitate the autogenous healing of these cracks through hydration processes or the movement of fine particles within the matrix. This self-healing ability contributes to the material's longevity and reduces maintenance requirements over time (Ma et al., 2021).

The manufacturing process of ECC involves careful control of material proportions, fiber content, and mixing techniques to ensure uniform dispersion of fibers within the matrix. This meticulous production methodology is crucial for optimizing ECC's mechanical properties and performance characteristics. The applications of ECC span across various domains within the construction industry. Its enhanced properties make it suitable for structural elements, such as bridge decks, earthquake-resistant buildings, overlays for deteriorated structures, and retrofitting projects. ECC's ability to accommodate significant deformations without compromising structural integrity makes it particularly valuable in seismic-prone regions (Shoji et al., 2022).

While ECC offers remarkable advantages in terms of durability, resilience, and sustainability, challenges remain. Issues related to material cost, construction practices, and standardization pose hurdles to its widespread adoption. However, ongoing research and development efforts continue to refine ECC, addressing these challenges and expanding its applicability in diverse construction contexts (Boopathi, Venkatesan, et al., 2023; Gowri et al., 2023; Karthik et al., 2023). Engineered Cementitious Composites (ECC) are a revolutionary construction material that enhances structural performance, durability, and sustainability in the built environment, positioning them as a key player in creating resilient and long-lasting infrastructure.

## Composition and Properties of ECC

Engineered Cementitious Composites (ECC) are unique materials with exceptional properties, distinguishing them from traditional concrete, making their composition and properties crucial for understanding their importance in the construction industry (Ding et al., 2022b).

**Composition:** ECC is primarily composed of four main constituents:

- **Cementitious Materials:** ECC relies on cementitious binders like Portland cement, fly ash, or other supplementary cementitious materials as the primary binding agent. These materials form the matrix that binds the composite together.
- **Fine Aggregates:** Similar to conventional concrete, ECC contains fine aggregates such as sand, which contribute to the matrix formation and overall structure (Boopathi, 2022b).
- **Admixtures:** Various chemical admixtures are incorporated to modify the properties of ECC. These additives enhance workability, control setting time, improve flow characteristics, and enhance bonding between components.
- **High-Volume Discrete Fibers:** This is the defining characteristic of ECC. Short, fine, and evenly dispersed fibers, often made of materials like

polyethylene, polyvinyl alcohol, or metallic fibers, are added in high volumes. These fibers serve as reinforcement, providing the material with remarkable tensile ductility and crack control capabilities (Boopathi, 2024a).

**Properties:** The unique composition of ECC contributes to several exceptional properties (Al-Fakih et al., 2021b; Ding et al., 2022b):

- **Tensile Ductility:** ECC displays remarkable tensile strength and ductility, allowing it to undergo significant deformation without fracturing. Unlike traditional concrete, which is brittle under tension, ECC can bend and flex without failure, making it highly resistant to cracking.
- **Strain-Hardening Behavior:** When subjected to loads, ECC exhibits strain-hardening behavior. This means that instead of sudden failure upon reaching its tensile capacity, it continues to carry load and deform, gradually increasing its resistance.
- **Self-Healing:** ECC possesses self-healing capabilities at the micro-scale. When small cracks occur due to external factors, such as loading or environmental stresses, the material can autonomously heal these cracks through various mechanisms, maintaining its integrity and prolonging its lifespan.
- **Enhanced Durability:** The incorporation of fibers and optimized matrix composition contributes to increased durability. ECC structures are less susceptible to cracking, have improved resistance to fatigue and environmental degradation, and require minimal maintenance over their lifecycle.
- **Reduced Weight:** Despite its enhanced properties, ECC is often lighter than traditional concrete due to the reduced need for thick sections, offering potential benefits in seismic applications and ease of handling.

## TYPES OF MANUFACTURING PROCESSES IN ECC PRODUCTION

ECC's unique properties, including durability, flexibility, and self-healing, make it a promising material for creating resilient and long-lasting infrastructure in construction practices.

- **Discrete Manufacturing**: This involves the production of distinct items or units. Each product is separately identifiable, and production usually occurs through assembly or machining processes.

- **Process Manufacturing**: This method is used for producing goods that result from a combination of ingredients or components. It's common in industries like chemicals, pharmaceuticals, and food and beverage (Boopathi, Khare, et al., 2023; Ingle et al., 2023).
- **Repetitive Manufacturing**: This process involves the continuous production of the same or similar products, often in a fixed sequence. It's suitable for high-volume production with minimal variation.
- **Make-to-Order (MTO)**: Goods are produced based on specific customer orders. Production starts after receiving an order, and items are customized to meet individual customer requirements.
- **Make-to-Stock (MTS)**: Products are manufactured in anticipation of demand. These items are produced and stocked based on forecasts or historical demand patterns.
- **Engineer-to-Order (ETO)**: Manufacturing is designed and produced according to unique customer specifications. Each order requires engineering design and customization.

## MANUFACTURING PROCESSES AND ADVANCEMENTS IN ECC PRODUCTION

Engineered Cementitious Composites (ECC) are created through specialized methods to optimize fiber dispersion and material properties (Lu & Zhong, 2022). The Figure 1 depicts the advancements in ECC production and the manufacturing processes involved. ECC production advancements involve exploring innovative fibers, optimizing mix designs, and utilizing advanced manufacturing technologies, with research focusing on fiber-matrix interactions, self-healing capabilities, and tailored ECC variants.

- **Fiber Dispersion:** Achieving a homogeneous dispersion of high-volume discrete fibers within the cementitious matrix is critical. Advanced mixing techniques, such as high-shear mixing or sequential addition methods, are employed to evenly distribute fibers throughout the mixture. This ensures the reinforcement is effectively distributed, enhancing the material's performance (Boopathi, Balasubramani, et al., 2023; Sampath et al., 2022; Sathish et al., 2023).
- **Optimization of Matrix Composition:** Fine-tuning the composition of cementitious materials, aggregates, and admixtures is essential to complement the reinforcing fibers (Karthik et al., 2023). Balancing these constituents improves the workability, flowability, and mechanical properties of ECC.

*Figure 1. Manufacturing processes and advancements in ECC production*

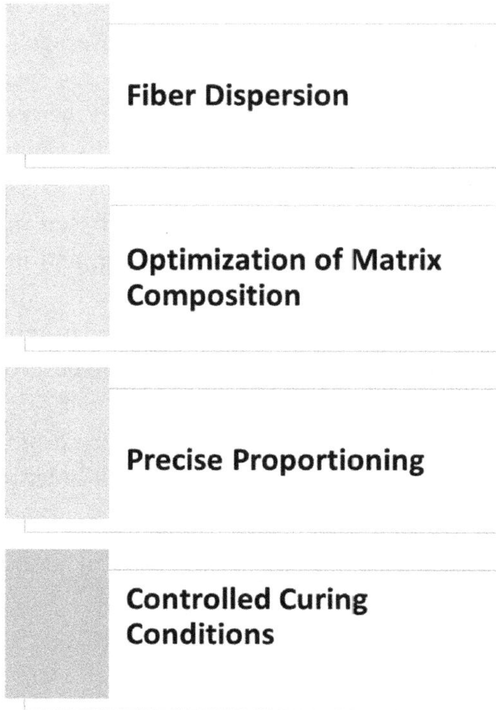


- **Precise Proportioning:** Accurate proportioning of materials is crucial for ECC production. Consistency in material ratios, fiber content, and mixing times is essential to ensure the desired properties and performance of the final product.
- **Controlled Curing Conditions:** Proper curing conditions, such as maintaining adequate moisture and temperature levels, are vital for ECC. Controlled curing helps in achieving optimal hydration of cementitious materials and promotes the development of desired mechanical properties.

## Advantages and Potential Applications of ECC in Construction

ECC's distinctive properties offer numerous benefits, indicating its potential use across various construction sectors (Li, 2003).

- ECC's superior tensile ductility and crack control capabilities enhance its durability, making it ideal for dynamic loads in structures like earthquake-resistant buildings, bridges, and pavements. Its self-healing properties reduce maintenance needs and prolong structure service life.

- ECC's exceptional deformation resistance enhances its resistance to impact, seismic events, and harsh environmental conditions. Its unique mechanical behavior enables thinner sections, complex shapes, and innovative structural designs, allowing architects and engineers to explore new construction possibilities.
- ECC, a sustainable construction material, can be used for repair and retrofitting of deteriorated structures, reducing maintenance and repair needs, and aligning with sustainability goals in construction.

ECC's superior ductility, self-healing capabilities, and durability make it a versatile material suitable for seismic zones, infrastructure rehabilitation, and sustainable construction practices, offering a promising future for resilient and long-lasting infrastructure.

## Alternative Materials in Construction

The use of alternative materials in construction is a significant shift towards sustainable and eco-friendly practices, with geopolymers and recycled aggregates being key alternatives for reducing environmental impact and improving material efficiency (Anderson et al., 2020).

**Geopolymers:** Geopolymers are inorganic, cementitious materials typically synthesized from industrial by-products or naturally occurring materials, like fly ash, slag, metakaolin, or volcanic ash. They are formed through a chemical process that involves activating aluminosilicate materials with alkali solutions, resulting in a binder with properties comparable to conventional cement (Palaniappan et al., 2023; Senthil et al., 2023).

## Advantages of Geopolymers

- **Lower Carbon Footprint:** Geopolymers often utilize industrial waste materials, reducing reliance on traditional cement production, which is a significant source of $CO_2$ emissions.
- **Enhanced Durability:** They exhibit excellent mechanical properties, including high compressive strength, low permeability, and resistance to chemical corrosion, making them suitable for various construction applications.

- **Thermal Insulation:** Geopolymer-based materials can offer improved thermal insulation properties compared to conventional concrete, contributing to energy efficiency in buildings.

## Challenges

- **Standardization:** Geopolymer technology is still evolving, and standardization of production methods and quality control measures is necessary for widespread adoption.
- **Material Consistency:** Achieving consistent material properties can be challenging due to variations in raw materials and curing conditions.

**Recycled Aggregates:** Recycled aggregates are derived from crushed concrete, asphalt, or industrial by-products, which are processed and reused as aggregates in new construction materials, reducing the demand for virgin aggregates (Ding et al., 2022c).

## Advantages of Recycled Aggregates

- **Waste Reduction:** Recycling construction and demolition waste into aggregates diverts materials from landfills, promoting a circular economy in construction.
- **Resource Conservation:** Using recycled aggregates reduces the need for extracting and processing natural resources, conserving energy and minimizing environmental impact.
- **Cost-Effectiveness:** In some cases, recycled aggregates can be more cost-effective than virgin materials, contributing to overall project savings.

## Challenges

- **Quality Control:** Variability in the quality of recycled aggregates due to inconsistent composition and potential contaminants can affect material performance.
- **Regulatory Hurdles:** Regulations and standards governing the use of recycled aggregates vary across regions and may hinder their widespread adoption without proper guidelines.

The integration of alternative materials like geopolymers and recycled aggregates into mainstream construction methodologies requires research, standardization, and technological advancements to ensure sustainable and resource-efficient practices.

## COMPARATIVE ANALYSIS OF PERFORMANCE, DURABILITY, AND SUSTAINABILITY

Comparing the performance, durability, and sustainability aspects of different construction materials, such as traditional concrete, Engineered Cementitious Composites (ECC), geopolymers, and recycled aggregates, provides valuable insights into their suitability for various applications (Li, 2008):

### Performance

- **Traditional Concrete:** While concrete offers good compressive strength, it tends to be brittle under tension, making it susceptible to cracking. Its performance under dynamic loads or harsh environmental conditions might be limited.
- **ECC:** Engineered Cementitious Composites exhibit superior tensile ductility, strain-hardening behavior, and crack control capabilities compared to traditional concrete. ECC's ability to deform significantly before failure and its self-healing properties enhance its performance under various loading conditions.
- **Geopolymers:** Geopolymers generally possess excellent mechanical properties, such as high compressive strength and durability. They offer comparable or superior performance to traditional concrete in terms of strength and resistance to chemical corrosion.
- **Recycled Aggregates:** The performance of concrete using recycled aggregates can vary based on the quality and properties of the recycled materials. While it may slightly compromise initial strength, proper processing and quality control can ensure acceptable performance levels.

### Durability

- **Traditional Concrete:** Concrete's durability depends on factors like water permeability, resistance to chemical attack, and susceptibility to cracking. It may suffer from issues like cracking due to shrinkage and external factors (Da Costa et al., 2019).

- **ECC:** ECC's exceptional crack control and self-healing properties contribute significantly to its durability. Its ability to mitigate crack propagation enhances resistance to environmental degradation and increases service life.
- **Geopolymers:** Geopolymers often exhibit low permeability and high resistance to chemical corrosion, offering good durability compared to traditional concrete. However, long-term performance data may be limited due to their relatively recent commercial use.
- **Recycled Aggregates:** When used in concrete, the durability of recycled aggregates can be influenced by the quality of the source material and the presence of contaminants. Proper processing and quality control are crucial to ensuring satisfactory durability.

### Sustainability

- **Traditional Concrete:** Production of conventional concrete involves high energy consumption, substantial CO2 emissions, and depletion of natural resources like aggregates and water (Z. Zhang et al., 2021).
- **ECC:** ECC's use of high volumes of fine fibers, which can often be recycled or made from sustainable sources, contributes to its sustainability. Additionally, its longer service life and reduced maintenance needs align with sustainability goals.
- **Geopolymers:** Geopolymers show promise in sustainability due to their ability to utilize industrial by-products or waste materials in production, reducing reliance on traditional cement and lowering carbon emissions.
- **Recycled Aggregates:** Using recycled aggregates helps reduce waste going into landfills and conserves natural resources, making it an environmentally sustainable practice.

Selecting construction materials based on project requirements, environmental factors, and long-term performance expectations requires a thorough evaluation of performance, durability, and sustainability.

## APPLICATIONS OF ALTERNATIVE MATERIALS

Figure 2 depicts the various applications of different alternative materials (Karthik et al., 2023; Kavitha, Malini, et al., 2023).

- **Geopolymer Concrete in Infrastructure:** Geopolymer concrete has been successfully used in infrastructure projects, including bridges, pavements, and buildings. For instance, in Australia, the Swan Street Bridge in Melbourne was repaired using geopolymer concrete, showcasing its durability and suitability for structural rehabilitation.

*Figure 2. Applications of various alternative materials*

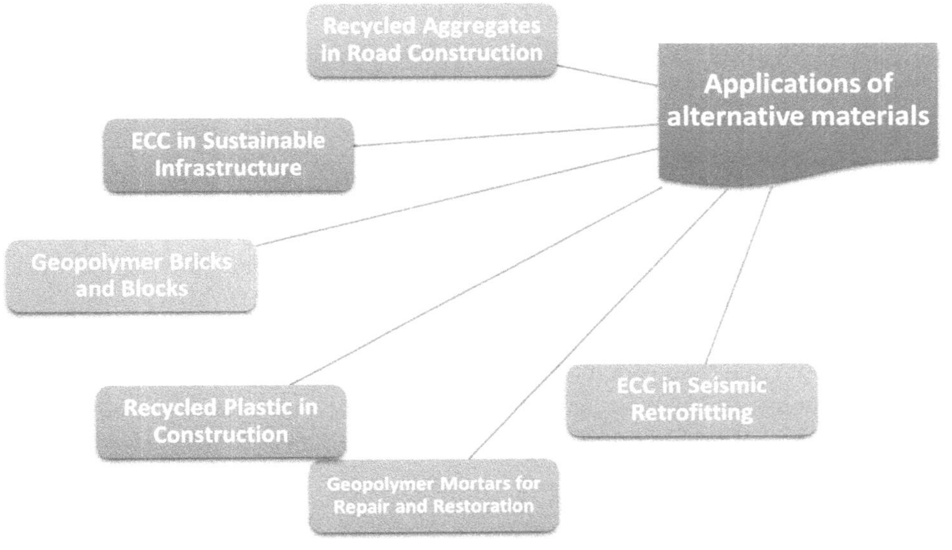


- **Recycled Aggregates in Road Construction:** Recycled aggregates derived from demolished concrete have been extensively used in road construction projects globally. Roads in various cities, such as Los Angeles and Barcelona, have incorporated recycled aggregates to reduce the environmental impact and conserve natural resources.
- **Geopolymer Bricks and Blocks:** Geopolymer-based bricks and blocks have found applications in sustainable construction. These materials offer improved thermal insulation and reduced environmental impact compared to conventional fired bricks. Projects in India and Southeast Asia have utilized geopolymer bricks for housing and commercial structures.
- **ECC in Seismic Retrofitting:** Engineered Cementitious Composites have been employed in seismic retrofitting projects to enhance the resilience of structures against earthquakes. Buildings in seismic-prone areas, such as

Japan and California, have undergone retrofitting using ECC overlays to improve their ability to withstand seismic forces.

- **Geopolymer Mortars for Repair and Restoration:** Geopolymer-based mortars have been utilized for repair and restoration work in heritage structures and historical buildings. These materials offer excellent adhesion and compatibility with old masonry, contributing to the preservation of architectural heritage.
- **Recycled Plastic in Construction:** Innovative applications of recycled plastics, such as plastic lumber, roofing tiles, and insulation materials, have been seen in construction. Projects in developing countries have used recycled plastic materials for low-cost housing solutions, showcasing their potential in affordable and sustainable building practices.
- **ECC in Sustainable Infrastructure:** Engineered Cementitious Composites have been instrumental in creating sustainable infrastructure, including environmentally friendly pavements and durable, crack-resistant precast elements. These applications have been seen in various parts of the world, promoting sustainable construction practices.

Successful applications demonstrate the versatility and potential of alternative materials in construction, showcasing their performance, durability, and sustainability advantages in real-world projects.

## INNOVATIONS IN CONSTRUCTION METHODS

### Reinventing Structural Design and Construction Methodologies With ECC

Reinventing structural design and construction methodologies with Engineered Cementitious Composites (ECC) involves leveraging the unique properties of ECC to enable innovative approaches and enhance the performance of structures (Bao & Li, 2020):

- **Design Flexibility:** ECC's exceptional tensile ductility and crack control capabilities enable engineers to rethink structural design. It allows for slimmer sections, innovative shapes, and complex geometries, offering architects greater design freedom without compromising structural integrity.
- **Improved Resilience:** ECC's ability to undergo significant deformation before failure makes it highly suitable for structures in seismic zones. It allows

buildings to better withstand seismic forces by absorbing and dissipating energy, enhancing the overall resilience of the structure.

- **Enhanced Durability:** ECC's self-healing properties and crack control capabilities contribute to increased durability. Structures built with ECC are less prone to cracking and have improved resistance to environmental factors, leading to longer service life and reduced maintenance needs.
- **Retrofitting and Rehabilitation:** ECC overlays and retrofitting techniques are used to strengthen and rehabilitate deteriorated structures. By applying ECC as a thin layer on existing surfaces, it reinforces the structure and extends its lifespan, offering a sustainable solution for infrastructure maintenance.
- **Construction in Challenging Environments:** ECC's properties make it suitable for construction in harsh environments or extreme climates. Its resistance to cracking and enhanced durability make it valuable for infrastructure projects in regions prone to temperature fluctuations, moisture, or chemical exposure (Boopathi, 2022a, 2023; Janardhana et al., 2023).
- **Prefabrication and Rapid Construction:** ECC's workability and unique properties facilitate prefabrication of elements off-site, allowing for faster and more efficient construction. Prefabricated ECC elements can be transported and assembled, reducing on-site labor and construction time.
- **Sustainable Construction Practices:** ECC's potential for utilizing recycled materials as components aligns with sustainable construction practices. Incorporating recycled fibers or other sustainable additives contributes to reducing environmental impact and promoting circularity in construction.

The integration of Energy Conservation Concrete (ECC) in structural design and construction is a significant shift, promoting resilience, durability, and sustainability. As engineers and architects continue to explore its capabilities, it is expected to expand.

## Integration of Alternative Materials in Conventional Construction Practices

The integration of alternative materials like geopolymers, recycled aggregates, and Engineered Cementitious Composites into conventional construction methods necessitates careful consideration and adaptation of established methodologies (Barhemat et al., 2022).

- **Material Testing and Standardization:** Before integration, alternative materials undergo rigorous testing to assess their properties and behavior in various conditions. Standardization of testing protocols ensures compatibility with existing construction standards and regulations.

- **Design Adaptation:** Engineers and architects adapt design methodologies to accommodate the unique properties of alternative materials. This involves recalculating structural parameters, such as load-bearing capacities, considering the materials' mechanical characteristics.
- **Quality Assurance and Control:** Implementing robust quality assurance processes ensures consistency and reliability in material properties. This includes monitoring sourcing, processing, and application methods to maintain quality throughout the construction process.
- **Compatibility and Mix Design:** Developing compatible mix designs involves understanding how alternative materials interact with traditional components like cement, aggregates, and admixtures. Optimizing proportions and formulations ensures cohesive and durable construction materials.
- **Construction Techniques and Handling:** Training construction teams in handling and application techniques specific to alternative materials is crucial. This includes proper handling of recycled aggregates, specialized curing requirements for geopolymers, and techniques for fiber dispersion in ECC.
- **Adoption of Guidelines and Codes:** Updating building codes and guidelines to incorporate the use of alternative materials helps legitimize their usage and ensures compliance with safety and performance standards.
- **Educational Initiatives:** Educating stakeholders, including engineers, contractors, and clients, about the benefits, limitations, and best practices for using alternative materials fosters confidence and acceptance in their integration.
- **Pilot Projects and Case Studies:** Implementing pilot projects and documenting successful case studies demonstrating the performance and advantages of alternative materials encourages wider adoption within the industry.
- **Collaboration and Research:** Collaborative efforts among academia, industry, and regulatory bodies drive ongoing research and development, addressing challenges and refining methodologies for integrating alternative materials.
- **Economic Viability Assessment:** Evaluating the cost-effectiveness and long-term benefits of using alternative materials ensures their practicality and viability in construction projects (Rahamathunnisa et al., 2024; Ravisankar et al., 2023; Vijayakumar et al., 2024).

The integration of alternative materials into conventional construction practices necessitates a multifaceted approach that includes technical adaptation, regulatory alignment, skill development, and collaboration, aiming for sustainable, resilient, and efficient practices.

## Environmental and Economic Benefits of Adopting These Innovations

The use of innovative materials like Engineered Cementitious Composites (ECC), geopolymers, and recycled aggregates in construction provides substantial environmental and economic advantages (Broome & Hayes, 1997). Figure 3 illustrates the environmental and economic advantages of incorporating innovative materials.

## Environmental Benefits

- **Reduced Carbon Footprint:** Alternative materials often utilize industrial by-products or recycled components, reducing reliance on virgin resources. This lowers the carbon emissions associated with traditional material production processes, contributing to mitigating climate change.
- **Waste Reduction and Circular Economy:** Incorporating recycled aggregates and utilizing waste materials in geopolymers reduces the amount of construction and demolition waste sent to landfills. It promotes a circular economy by reusing materials that would otherwise be discarded (Selvakumar, Adithe, et al., 2023; Selvakumar, Shankar, et al., 2023).
- **Conservation of Natural Resources:** Using alternative materials lessens the demand for natural resources like aggregates and reduces the need for quarrying, preserving natural landscapes and habitats.
- **Energy Efficiency:** Some alternative materials, like ECC and geopolymers, require lower curing temperatures or use less energy during production compared to traditional materials, leading to overall energy savings (Boopathi, 2024b; Venkateswaran et al., 2023).
- **Improved Indoor Air Quality:** Geopolymer-based materials, for example, emit fewer volatile organic compounds (VOCs) than some conventional materials, contributing to better indoor air quality in buildings.

*Figure 3. Environmental and economic benefits of adopting innovative materials*

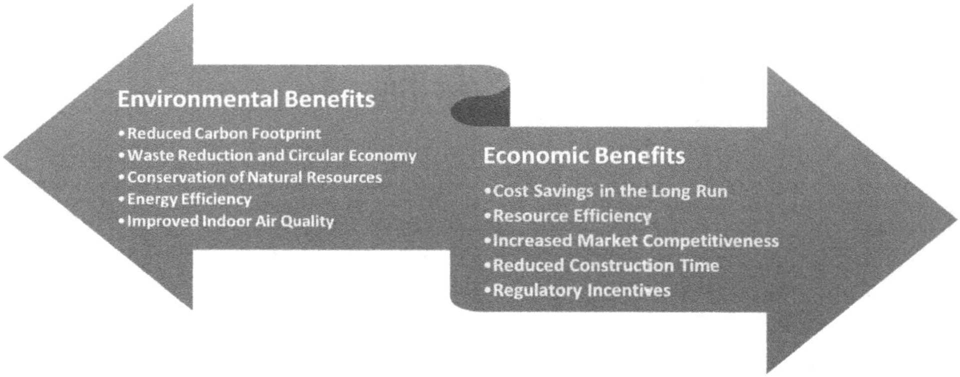


## Economic Benefits

- **Cost Savings in the Long Run:** While some alternative materials may have higher upfront costs, their enhanced durability and reduced maintenance needs over their lifecycle can result in significant cost savings in the long term.
- **Resource Efficiency:** Using recycled aggregates or waste materials in construction can lead to cost savings by reducing the need to purchase virgin materials, thus lowering material procurement costs.
- **Increased Market Competitiveness:** Companies adopting innovative materials can gain a competitive edge by showcasing their commitment to sustainability, attracting environmentally conscious clients and investors.
- **Reduced Construction Time:** Some alternative materials, such as prefabricated ECC elements, enable faster construction, reducing labor costs and project timelines.
- **Regulatory Incentives:** Some regions offer incentives or tax benefits for using sustainable or innovative materials in construction, providing economic advantages for adopting these materials.

The construction industry should embrace innovative materials for a sustainable and economically viable future, as they align with environmental stewardship, offer cost efficiencies, differentiate in the market, and comply with sustainability regulations.

## CHALLENGES AND FUTURE DIRECTIONS

### Addressing Hurdles in the Widespread Adoption of ECC and Alternative Materials

The widespread use of Engineered Cementitious Composites (ECC) and alternative materials faces several challenges that necessitate targeted strategies to overcome (Ullah et al., 2023).

- **Cost Considerations:** Initial costs for alternative materials like ECC can be higher than traditional materials. Strategies to address this hurdle include promoting lifecycle cost analyses that demonstrate long-term savings due to reduced maintenance and enhanced durability.
- **Limited Awareness and Education:** Many in the construction industry may be unfamiliar with alternative materials, their properties, and their applications. Educating stakeholders through training programs, workshops, and case studies can help build confidence in using these materials (Das et al., 2024; Durairaj et al., 2023).
- **Standardization and Codes:** The lack of standardized testing methods and codes specific to alternative materials poses challenges in their widespread adoption. Collaborative efforts among industry stakeholders to develop and update standards can help legitimize these materials and ensure compliance with safety regulations.
- **Supply Chain Challenges:** Availability and consistent quality of raw materials for alternative materials can be a concern. Strengthening supply chains, exploring multiple sourcing options, and investing in research to develop alternative material variants can address this issue (Mohanty, Venkateswaran, et al., 2023; Verma et al., 2024).
- **Technological Advancements:** Continuous research and development are crucial for improving the performance and scalability of alternative materials. Advancements in manufacturing techniques, such as automated mixing processes for ECC, can enhance efficiency and reduce production costs.
- **Skills and Expertise:** Specialized knowledge and skills may be required for handling and utilizing alternative materials effectively. Training programs and knowledge-sharing platforms can bridge this skill gap among construction professionals.
- **Market Acceptance and Perception:** Building trust and confidence in the performance and reliability of alternative materials is essential. Demonstrating successful applications through pilot projects and showcasing their long-term benefits can change perceptions within the industry and among clients.

- **Regulatory Support:** Collaborating with policymakers to incorporate alternative materials into building codes, incentivizing their use, and creating favorable regulations can accelerate their adoption.
- **Demonstrating Performance in Real-World Scenarios:** Conducting comprehensive performance evaluations and long-term studies in various environments can provide empirical evidence of the benefits of alternative materials, increasing their acceptance.

The widespread adoption of ECC and alternative materials, facilitated by industry collaboration, research, education, policy advocacy, and strategic investments, will lead to more sustainable, resilient, and innovative construction practices.

## Potential Solutions and Research Avenues to Enhance Material Properties

The study aims to improve the properties of construction materials like Engineered Cementitious Composites (ECC) through various research and solutions (Istiaque Ahmed et al., 2021).

- **Fiber Development and Optimization:** Research focused on developing advanced fibers with superior mechanical properties can enhance the performance of ECC. Exploring new types of fibers, optimizing their size, shape, and composition, and understanding their interaction within the matrix can improve tensile strength and ductility.
- **Matrix Composition and Additives:** Investigating alternative binders, additives, and admixtures that complement ECC can lead to optimized matrix compositions. Innovative materials, such as nano-materials or organic additives, may improve bonding, workability, and durability.
- **Self-Healing Properties:** Advancing research in self-healing mechanisms, such as microcapsules or bio-inspired materials, can enhance ECC's ability to autonomously repair micro-cracks, extending its service life and reducing maintenance requirements.
- **Nano-engineered Materials:** Exploring nanotechnology to manipulate material properties at the molecular level can enhance mechanical strength, durability, and chemical resistance. Nano-reinforcements within the matrix can improve ECC's overall performance (Boopathi, Umareddy, et al., 2023; Boopathi & Davim, 2023b, 2023a).
- **Performance Modeling and Simulation:** Developing advanced computational models and simulations can predict material behavior under different conditions, aiding in the design and optimization of ECC structures.

- **Life Cycle Assessment (LCA):** Conducting comprehensive life cycle assessments to evaluate the environmental impact of alternative materials. This involves considering the entire life cycle, from raw material extraction to disposal, to identify areas for improvement.
- **Scaling Up Production Techniques:** Research aimed at refining large-scale manufacturing processes for alternative materials can improve efficiency and reduce production costs. Automated mixing, curing technologies, and quality control measures can streamline production (Nanda et al., 2024).
- **Durability Studies in Real-world Environments:** Conducting long-term studies and field trials in diverse environmental conditions can validate material performance, providing empirical data to support their widespread adoption.
- **Multi-Disciplinary Collaboration:** Collaborative efforts between material scientists, engineers, architects, and industry stakeholders can foster innovation and drive research towards practical applications (Boopathi et al., 2021; Paul et al., 2024; Sathish et al., 2023).
- **Standardization and Testing Protocols:** Developing standardized testing methods and protocols specific to alternative materials can establish benchmarks and facilitate their integration into construction codes and practices.

By investing in these research avenues and implementing innovative solutions, the properties of ECC and alternative materials can be enhanced, unlocking new possibilities for sustainable, durable, and high-performance construction materials.

## Future Prospects and Trends in Construction Materials and Technologies

The future of construction materials and technologies is poised for transformative advancements driven by sustainability, digitalization, and innovation. Several key prospects and trends are anticipated to shape the industry (Bhushan et al., 2020):

- The focus is on sustainability and circular economy, promoting resource efficiency and eco-friendly construction through recycled materials, bio-based alternatives, and low-impact production methods. The integration of nanotechnology in construction materials will enhance strength, durability, and functionality, revolutionizing the industry with improved properties and innovative functionalities (Sharma et al., 2024).
- Smart materials, incorporating sensors, actuators, and self-monitoring, will improve structures' adaptability, enabling self-repair, energy harvesting,

and real-time performance monitoring. The growth of 3D printing and prefabrication techniques will facilitate rapid, cost-effective construction of complex structures, minimizing material waste and onsite labor (Maheswari et al., 2023; Syamala et al., 2023).

- The use of biophilic design principles and biocompatible materials can promote healthier living spaces and improved well-being. Digitalization and Building Information Modeling (BIM) can optimize design, construction, and maintenance processes, enhancing collaboration, efficiency, and project outcomes by integrating augmented reality, virtual reality, and other digital technologies (Kavitha, Varalatchoumy, et al., 2023; Sengeni et al., 2023).
- The need for resilient materials and the increasing use of robotics and automation in construction processes, such as autonomous vehicles and drones, is becoming increasingly important, particularly in vulnerable regions, to enhance safety, efficiency, and precision on job sites (Mohanty, Jothi, et al., 2023; Pramila et al., 2023).
- The focus is shifting towards carbon-neutral and low-carbon materials in construction, utilizing carbon capture, utilization, and storage technologies to combat climate change. These materials are also being developed with adaptive properties that enable structures to self-adjust and optimize performance.

The industry's commitment to sustainability, efficiency, and innovation is evident in current trends, driven by collaboration among researchers, manufacturers, policymakers, and construction professionals, resulting in resilient, sustainable, and intelligent built environments (Ghose et al., 2018).

*Figure 4. Real-world examples: ECC and alternative materials*

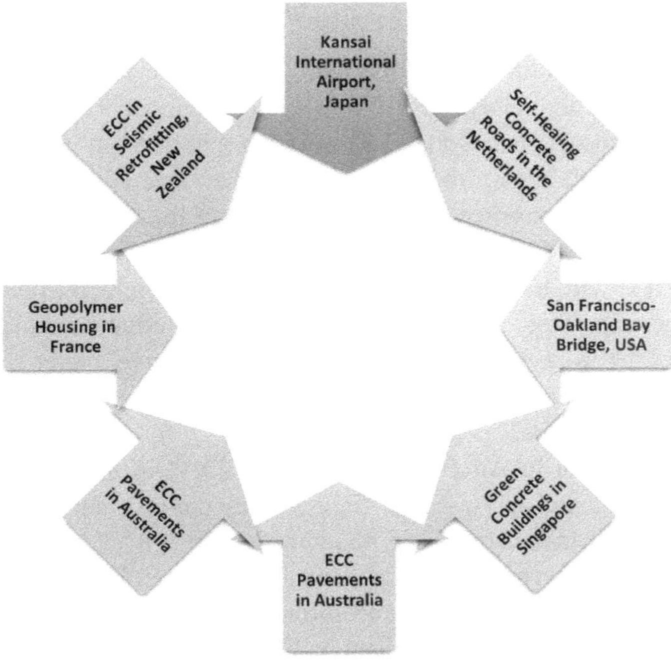


## CASE STUDIES AND APPLICATIONS

### Real-World Examples: ECC and Alternative Materials

Several successful projects globally have showcased the efficacy and versatility of Engineered Cementitious Composites (ECC) and alternative materials, setting benchmarks for innovative construction practices (Lim et al., 2023): Figure 4 illustrates real-world examples of ECC and alternative materials.

- **Kansai International Airport, Japan:** ECC was utilized in the seismic retrofitting of the Kansai International Airport's terminal building. The airport, located on a man-made island in a seismic zone, required enhanced resilience. ECC overlays were applied to strengthen critical structural elements, significantly improving the building's capacity to withstand seismic forces.
- **Self-Healing Concrete Roads in the Netherlands:** A project in the Netherlands implemented self-healing concrete made with bacteria-activated healing agents. When cracks form in the road surface, the bacteria within

the concrete become active, producing limestone to fill the cracks. This innovation significantly reduces maintenance needs and prolongs the lifespan of road infrastructure.

- **San Francisco-Oakland Bay Bridge, USA:** The seismic retrofitting of this critical infrastructure used ECC in various components, including bridge decks and support structures. ECC's exceptional ductility and crack control capabilities made it instrumental in enhancing the bridge's ability to withstand seismic events, ensuring safety and continuity of transportation.
- **Green Concrete Buildings in Singapore:** Singapore has embraced green construction with buildings using eco-friendly materials like green concrete, which incorporates recycled materials and alternative binders. These structures showcase the sustainability of alternative materials, reducing environmental impact while maintaining structural integrity (Boopathi, 2023; Janardhana et al., 2023).
- **ECC Pavements in Australia:** The use of ECC in pavements and roads has gained traction in Australia. ECC pavements offer increased durability and crack resistance, reducing maintenance costs and traffic disruptions. Projects in Melbourne and Sydney have implemented ECC overlays for road rehabilitation, demonstrating its potential in extending pavement lifespan.
- **Geopolymer Housing in France:** In France, geopolymer-based materials were used in the construction of eco-friendly social housing. These buildings showcased the feasibility of using geopolymers in construction, providing energy-efficient structures with reduced environmental impact (Boopathi, Khare, et al., 2023; Palaniappan et al., 2023).
- **ECC in Seismic Retrofitting, New Zealand:** In earthquake-prone regions like New Zealand, ECC has been applied in retrofitting historical buildings to enhance their resilience against seismic activities. Its ability to mitigate cracking and improve structural performance has been crucial in preserving architectural heritage.

These examples highlight the successful integration of ECC and alternative materials in diverse construction projects worldwide. They demonstrate the materials' versatility, durability, and ability to address specific challenges in seismic resilience, sustainability, and infrastructure maintenance, setting a precedent for their wider adoption in the construction industry.

## Detailed Analysis of the Performance, Durability, and Cost-Effectiveness

### Performance

- **ECC in Infrastructure:** In seismic-prone areas, ECC demonstrates exceptional performance due to its high tensile ductility and crack control. Its ability to undergo significant deformation without failure makes it suitable for bridges, pavements, and buildings, improving resilience against seismic events (Gu et al., 2021).
- **Geopolymers in Building Construction:** Geopolymers exhibit high compressive strength and resistance to chemical corrosion, making them suitable for building construction. Their durability and fire resistance qualities make them valuable for facades, structural elements, and insulation materials.
- **Recycled Aggregates in Road Construction:** Recycled aggregates in road construction offer good performance, especially in non-structural layers like sub-bases and base courses. Their use reduces the demand for natural aggregates while providing satisfactory load-bearing capacity (Selvakumar, Adithe, et al., 2023).

### Durability

- **ECC's Longevity:** ECC's self-healing properties and crack control contribute to enhanced durability. Its ability to limit crack propagation and autonomously heal micro-cracks increases service life, reducing maintenance needs and lifecycle costs (Tan et al., 2022).
- **Geopolymers' Chemical Resistance:** Geopolymers showcase excellent resistance to harsh chemicals and aggressive environments. This durability feature makes them suitable for wastewater treatment plants, chemical storage facilities, and marine structures.
- **Recycled Aggregates' Performance:** The durability of concrete incorporating recycled aggregates depends on the quality of the recycled material. Proper processing and quality control ensure acceptable performance in various construction applications, though sometimes initial strength may be slightly compromised.

## Cost-Effectiveness

- **ECC's Lifecycle Cost:** While ECC may have higher initial costs due to specialized materials and mix designs, its enhanced durability and reduced maintenance lead to long-term cost savings. The extended service life and reduced need for repairs offset the initial investment.
- **Geopolymers vs. Conventional Concrete:** Geopolymer-based materials may have slightly higher upfront costs due to specialized production, but their durability and energy-saving properties can contribute to cost-effectiveness over the structure's life cycle.
- **Recycled Aggregates' Economic Benefits:** Incorporating recycled aggregates can reduce material procurement costs. However, factors like sorting, processing, and ensuring consistent quality might affect the overall cost-effectiveness in some cases (Babu et al., 2022; Mohanty, Venkateswaran, et al., 2023; Ravisankar et al., 2023).

The cost-effectiveness of these materials varies based on factors like local availability, project specifications, and long-term performance. While some alternatives may have higher upfront costs, their superior performance and reduced maintenance needs often translate to overall economic benefits over time. The suitability of these materials depends on the specific construction context and project objectives (Bhushan et al., 2020; Istiaque Ahmed et al., 2021).

# CONCLUSION

The study on Engineered Cementitious Composites (ECC) and alternative materials in construction highlights their superior performance, durability, seismic resilience, and self-healing capabilities, indicating their potential for creating more robust and long-lasting structures compared to traditional materials. The use of alternative materials like geopolymers, recycled aggregates, and ECC promotes sustainability by reducing carbon footprint, minimizing waste, and conserving natural resources. However, widespread adoption faces challenges like initial cost considerations, standardization issues, and increased industry awareness and education, despite their numerous advantages.

Research and innovation are crucial for improving the properties, scalability, and cost-effectiveness of alternative materials. Advancements in fiber development, manufacturing techniques, and understanding their long-term performance will drive their adoption. These materials have diverse applications in construction, from seismic retrofitting to infrastructure. Despite higher upfront costs, their long-term benefits, including extended service life, reduced maintenance, and cost savings, make them economically viable options.

The future of construction materials will be shaped by sustainability, technological innovation, and resilience. Smart materials, nanotechnology, digitalization, and manufacturing advancements will create efficient, durable, and environmentally conscious built environments. ECC and alternative materials offer a promising path for sustainable, resilient, and innovative practices. Collaborative efforts, research-driven innovations, and a shift towards sustainable methodologies will drive their wider adoption.

## REFERENCES

Al-Fakih, A., Mohammed, B. S., & Liew, M. (2021a). On rubberized engineered cementitious composites (R-ECC): A review of the constituent material. *Case Studies in Construction Materials*, *14*, e00536. doi:10.1016/j.cscm.2021.e00536

Al-Fakih, A., Mohammed, B. S., & Liew, M. (2021b). On rubberized engineered cementitious composites (R-ECC): A review of the constituent material. *Case Studies in Construction Materials*, *14*, e00536. doi:10.1016/j.cscm.2021.e00536

Anderson, M., Bogus, S. M., Arce, G., Hungria, R., & Hassan, M. (2020). Advantageous Construction Techniques for ECC Overlays. In *Tran-SET 2020* (pp. 247–257). American Society of Civil Engineers Reston.

Babu, B. S., Kamalakannan, J., Meenatchi, N., Karthik, S., & Boopathi, S. (2022). Economic impacts and reliability evaluation of battery by adopting Electric Vehicle. *IEEE Explore*, 1–6.

Bao, Y., & Li, V. C. (2020). Feasibility study of lego-inspired construction with bendable concrete. *Automation in Construction*, *113*, 103161. doi:10.1016/j.autcon.2020.103161

Barhemat, R., Mahjoubi, S., Li, V. C., & Bao, Y. (2022). Lego-inspired reconfigurable modular blocks for automated construction of engineering structures. *Automation in Construction*, *139*, 104323. doi:10.1016/j.autcon.2022.104323

Bhushan, B., Khamparia, A., Sagayam, K. M., Sharma, S. K., Ahad, M. A., & Debnath, N. C. (2020). Blockchain for smart cities: A review of architectures, integration trends and future research directions. *Sustainable Cities and Society*, *61*, 102360. doi:10.1016/j.scs.2020.102360

Boopathi, S. (2022a). An investigation on gas emission concentration and relative emission rate of the near-dry wire-cut electrical discharge machining process. *Environmental Science and Pollution Research International*, *29*(57), 86237–86246. doi:10.1007/s11356-021-17658-1 PMID:34837614

Boopathi, S. (2022b). Performance Improvement of Eco-Friendly Near-Dry wire-Cut Electrical Discharge Machining Process Using Coconut Oil-Mist Dielectric Fluid. *World Scientific: Journal of Advanced Manufacturing Systems*.

Boopathi, S. (2023). Impact analysis of ceramic tile powder aggregates on self-compacting concrete. *Engineering Research Express*.

Boopathi, S. (2024a). Advancements in Machine Learning and AI for Intelligent Systems in Drone Applications for Smart City Developments. In *Futuristic e-Governance Security With Deep Learning Applications* (pp. 15–45). IGI Global. doi:10.4018/978-1-6684-9596-4.ch002

Boopathi, S. (2024b). Energy Cascade Conversion System and Energy-Efficient Infrastructure. In Sustainable Development in AI, Blockchain, and E-Governance Applications (pp. 47–71). IGI Global.

Boopathi, S., Balasubramani, V., & Kumar, R. S. (2023). Influences of various natural fibers on the mechanical and drilling characteristics of coir-fiber-based hybrid epoxy composites. *Engineering Research Express*, *5*(1), 015002. doi:10.1088/2631-8695/acb132

Boopathi, S., & Davim, J. P. (2023a). Applications of Nanoparticles in Various Manufacturing Processes. In *Sustainable Utilization of Nanoparticles and Nanofluids in Engineering Applications* (pp. 1–31). IGI Global. doi:10.4018/978-1-6684-9135-5.ch001

Boopathi, S., & Davim, J. P. (2023b). *Sustainable Utilization of Nanoparticles and Nanofluids in Engineering Applications*. IGI Global. doi:10.4018/978-1-6684-9135-5

Boopathi, S., Khare, R., KG, J. C., Muni, T. V., & Khare, S. (2023). Additive Manufacturing Developments in the Medical Engineering Field. In Development, Properties, and Industrial Applications of 3D Printed Polymer Composites (pp. 86–106). IGI Global.

Boopathi, S., Myilsamy, S., & Sukkasamy, S. (2021). *Experimental Investigation and Multi-Objective Optimization of Cryogenically Cooled Near-Dry Wire-Cut EDM Using TOPSIS Technique*. IJAMT.

Boopathi, S., Umareddy, M., & Elangovan, M. (2023). Applications of Nano-Cutting Fluids in Advanced Machining Processes. In *Sustainable Utilization of Nanoparticles and Nanofluids in Engineering Applications* (pp. 211–234). IGI Global. doi:10.4018/978-1-6684-9135-5.ch009

Boopathi, S., Venkatesan, G., & Anton Savio Lewise, K. (2023). Mechanical Properties Analysis of Kenaf–Grewia–Hair Fiber-Reinforced Composite. In *Springer: Recent Advances in Mechanical Engineering* (pp. 101–110). Springer. doi:10.1007/978-981-16-9057-0_11

Broome, J. C., & Hayes, R. W. (1997). A comparison of the clarity of traditional construction contracts and of the New Engineering Contract. *International Journal of Project Management*, *15*(4), 255–261. doi:10.1016/S0263-7863(96)00078-6

Cho, B. H., Chung, W., & Nam, B. H. (2020). Molecular dynamics simulation of calcium-silicate-hydrate for nano-engineered cement composites—A review. *Nanomaterials (Basel, Switzerland)*, *10*(11), 2158. doi:10.3390/nano10112158 PMID:33138107

Da Costa, F., Righi, D., Graeff, A., & da Silva Filho, L. (2019). Experimental study of some durability properties of ECC with a more environmentally sustainable rice husk ash and high tenacity polypropylene fibers. *Construction & Building Materials*, *213*, 505–513. doi:10.1016/j.conbuildmat.2019.04.092

Das, S., Lekhya, G., Shreya, K., Shekinah, K. L., Babu, K. K., & Boopathi, S. (2024). Fostering Sustainability Education Through Cross-Disciplinary Collaborations and Research Partnerships: Interdisciplinary Synergy. In Facilitating Global Collaboration and Knowledge Sharing in Higher Education With Generative AI (pp. 60–88). IGI Global.

Ding, Y., Yu, K., & Li, M. (2022). A review on high-strength engineered cementitious composites (HS-ECC): Design, mechanical property and structural application. *Structures*, *35*, 903–921. doi:10.1016/j.istruc.2021.10.036

Durairaj, M., Jayakumar, S., Karpagavalli, V., Maheswari, B. U., Boopathi, S., & ... (2023). Utilization of Digital Tools in the Indian Higher Education System During Health Crises. In *Multidisciplinary Approaches to Organizational Governance During Health Crises* (pp. 1–21). IGI Global. doi:10.4018/978-1-7998-9213-7.ch001

Ghose, S., Hsieh, K., Boroumand, A., Ausavarungnirun, R., & Mutlu, O. (2018). Enabling the adoption of processing-in-memory: Challenges, mechanisms, future research directions. *arXiv Preprint arXiv:1802.00320*.

Gowri, N. V., Dwivedi, J. N., Krishnaveni, K., Boopathi, S., Palaniappan, M., & Medikondu, N. R. (2023). Experimental investigation and multi-objective optimization of eco-friendly near-dry electrical discharge machining of shape memory alloy using Cu/SiC/Gr composite electrode. *Environmental Science and Pollution Research International*, *30*(49), 1–19. doi:10.1007/s11356-023-26983-6 PMID:37126160

Gu, D., Zhao, W., Xie, Y., Wang, X., Su, K., & Zolotarev, O. V. (2021). A personalized medical decision support system based on explainable machine learning algorithms and ecc features: Data from the real world. *Diagnostics (Basel)*, *11*(9), 1677. doi:10.3390/diagnostics11091677 PMID:34574018

Ingle, R. B., Swathi, S., Mahendran, G., Senthil, T., Muralidharan, N., & Boopathi, S. (2023). Sustainability and Optimization of Green and Lean Manufacturing Processes Using Machine Learning Techniques. In *Circular Economy Implementation for Sustainability in the Built Environment* (pp. 261–285). IGI Global. doi:10.4018/978-1-6684-8238-4.ch012

Istiaque Ahmed, K., Tahir, M., Hadi Habaebi, M., Lun Lau, S., & Ahad, A. (2021). Machine learning for authentication and authorization in iot: Taxonomy, challenges and future research direction. *Sensors (Basel)*, *21*(15), 5122. doi:10.3390/s21155122 PMID:34372360

Janardhana, K., Singh, V., Singh, S. N., Babu, T. R., Bano, S., & Boopathi, S. (2023). Utilization Process for Electronic Waste in Eco-Friendly Concrete: Experimental Study. In Sustainable Approaches and Strategies for E-Waste Management and Utilization (pp. 204–223). IGI Global.

Karthik, K., Teferi, A. B., Sathish, R., Gandhi, A. M., Padhi, S., Boopathi, S., & Sasikala, G. (2023). Analysis of delamination and its effect on polymer matrix composites. *Materials Today: Proceedings*. Advance online publication. doi:10.1016/j.matpr.2023.07.199

Kavitha, C., Malini, P. G., Kantumuchu, V. C., Kumar, N. M., Verma, A., & Boopathi, S. (2023). An experimental study on the hardness and wear rate of carbonitride coated stainless steel. *Materials Today: Proceedings*, *74*, 595–601. doi:10.1016/j.matpr.2022.09.524

Kavitha, C., Varalatchoumy, M., Mithuna, H., Bharathi, K., Geethalakshmi, N., & Boopathi, S. (2023). Energy Monitoring and Control in the Smart Grid: Integrated Intelligent IoT and ANFIS. In Applications of Synthetic Biology in Health, Energy, and Environment (pp. 290–316). IGI Global.

Li, V. C. (2003). On engineered cementitious composites (ECC) a review of the material and its applications. *Journal of Advanced Concrete Technology*, *1*(3), 215–230. doi:10.3151/jact.1.215

Li, V. C. (2008). *Engineered cementitious composites (ECC) material, structural, and durability performance*. Academic Press.

Li, V. C., Bos, F. P., Yu, K., McGee, W., Ng, T. Y., Figueiredo, S. C., Nefs, K., Mechtcherine, V., Nerella, V. N., Pan, J., van Zijl, G. P. A. G., & Kruger, P. J. (2020). On the emergence of 3D printable engineered, strain hardening cementitious composites (ECC/SHCC). *Cement and Concrete Research*, *132*, 106038. doi:10.1016/j.cemconres.2020.106038

Lim, T. H., Ellis, B. R., & Skerlos, S. J. (2023). Life cycle assessment of railway ties fabricated with ductile cementitious composites and carbonation curing. *Journal of Cleaner Production*, *413*, 137276. doi:10.1016/j.jclepro.2023.137276

Lu, D., & Zhong, J. (2022). Carbon-based nanomaterials engineered cement composites: A review. *Journal of Infrastructure Preservation and Resilience*, *3*(1), 1–20. doi:10.1186/s43065-021-00045-y

Ma, H., Yi, C., & Wu, C. (2021). Review and outlook on durability of engineered cementitious composite (ECC). *Construction & Building Materials*, *287*, 122719. doi:10.1016/j.conbuildmat.2021.122719

Maheswari, B. U., Imambi, S. S., Hasan, D., Meenakshi, S., Pratheep, V., & Boopathi, S. (2023). Internet of things and machine learning-integrated smart robotics. In Global Perspectives on Robotics and Autonomous Systems: Development and Applications (pp. 240–258). IGI Global. doi:10.4018/978-1-6684-7791-5.ch010

Malathi, J., Kusha, K., Isaac, S., Ramesh, A., Rajendiran, M., & Boopathi, S. (2024). IoT-Enabled Remote Patient Monitoring for Chronic Disease Management and Cost Savings: Transforming Healthcare. In Advances in Explainable AI Applications for Smart Cities (pp. 371–388). IGI Global.

Mohanty, A., Jothi, B., Jeyasudha, J., Ranjit, P., Isaac, J. S., & Boopathi, S. (2023). Additive Manufacturing Using Robotic Programming. In *AI-Enabled Social Robotics in Human Care Services* (pp. 259–282). IGI Global. doi:10.4018/978-1-6684-8171-4.ch010

Mohanty, A., Venkateswaran, N., Ranjit, P., Tripathi, M. A., & Boopathi, S. (2023). Innovative Strategy for Profitable Automobile Industries: Working Capital Management. In Handbook of Research on Designing Sustainable Supply Chains to Achieve a Circular Economy (pp. 412–428). IGI Global.

Murali, B., Padhi, S., Patil, C. K., Kumar, P. S., Santhanakrishnan, M., & Boopathi, S. (2023). Investigation on hardness and tensile strength of friction stir processing of Al6061/TiN surface composite. *Materials Today: Proceedings*.

Nanda, A. K., Sharma, A., Augustine, P. J., Cyril, B. R., Kiran, V., & Sampath, B. (2024). Securing Cloud Infrastructure in IaaS and PaaS Environments. In Improving Security, Privacy, and Trust in Cloud Computing (pp. 1–33). IGI Global. doi:10.4018/979-8-3693-1431-9.ch001

Palaniappan, M., Tirlangi, S., Mohamed, M. J. S., Moorthy, R. S., Valeti, S. V., & Boopathi, S. (2023). Fused Deposition Modelling of Polylactic Acid (PLA)-Based Polymer Composites: A Case Study. In Development, Properties, and Industrial Applications of 3D Printed Polymer Composites (pp. 66–85). IGI Global.

Paul, A., Thilagham, K., KG, J., Reddy, P. R., Sathyamurthy, R., & Boopathi, S. (2024). Multi-criteria Optimization on Friction Stir Welding of Aluminum Composite (AA5052-H32/B4C) using Titanium Nitride Coated Tool. Engineering Research Express.

Pramila, P., Amudha, S., Saravanan, T., Sankar, S. R., Poongothai, E., & Boopathi, S. (2023). Design and Development of Robots for Medical Assistance: An Architectural Approach. In Contemporary Applications of Data Fusion for Advanced Healthcare Informatics (pp. 260–282). IGI Global.

Rahamathunnisa, U., Sudhakar, K., Padhi, S., Bhattacharya, S., Shashibhushan, G., & Boopathi, S. (2024). Sustainable Energy Generation From Waste Water: IoT Integrated Technologies. In Adoption and Use of Technology Tools and Services by Economically Disadvantaged Communities: Implications for Growth and Sustainability (pp. 225–256). IGI Global.

Ravisankar, A., Sampath, B., & Asif, M. M. (2023). Economic Studies on Automobile Management: Working Capital and Investment Analysis. In Multidisciplinary Approaches to Organizational Governance During Health Crises (pp. 169–198). IGI Global.

Sampath, B., Naveenkumar, N., Sampathkumar, P., Silambarasan, P., Venkadesh, A., & Sakthivel, M. (2022). Experimental comparative study of banana fiber composite with glass fiber composite material using Taguchi method. *Materials Today: Proceedings*, *49*, 1475–1480. doi:10.1016/j.matpr.2021.07.232

Sathish, T., Sunagar, P., Singh, V., Boopathi, S., Al-Enizi, A. M., Pandit, B., Gupta, M., & Sehgal, S. S. (2023). Characteristics estimation of natural fibre reinforced plastic composites using deep multi-layer perceptron (MLP) technique. *Chemosphere*, *337*, 139346. doi:10.1016/j.chemosphere.2023.139346 PMID:37379988

Selvakumar, S., Adithe, S., Isaac, J. S., Pradhan, R., Venkatesh, V., & Sampath, B. (2023). A Study of the Printed Circuit Board (PCB) E-Waste Recycling Process. In Sustainable Approaches and Strategies for E-Waste Management and Utilization (pp. 159–184). IGI Global.

Selvakumar, S., Shankar, R., Ranjit, P., Bhattacharya, S., Gupta, A. S. G., & Boopathi, S. (2023). E-Waste Recovery and Utilization Processes for Mobile Phone Waste. In *Handbook of Research on Safe Disposal Methods of Municipal Solid Wastes for a Sustainable Environment* (pp. 222–240). IGI Global. doi:10.4018/978-1-6684-8117-2.ch016

Sengeni, D., Padmapriya, G., Imambi, S. S., Suganthi, D., Suri, A., & Boopathi, S. (2023). Biomedical waste handling method using artificial intelligence techniques. In *Handbook of Research on Safe Disposal Methods of Municipal Solid Wastes for a Sustainable Environment* (pp. 306–323). IGI Global. doi:10.4018/978-1-6684-8117-2.ch022

Senthil, T., Puviyarasan, M., Babu, S. R., Surakasi, R., Sampath, B., & Associates. (2023). Industrial Robot-Integrated Fused Deposition Modelling for the 3D Printing Process. In Development, Properties, and Industrial Applications of 3D Printed Polymer Composites (pp. 188–210). IGI Global.

Sharma, M., Sharma, M., Sharma, N., & Boopathi, S. (2024). Building Sustainable Smart Cities Through Cloud and Intelligent Parking System. In *Handbook of Research on AI and ML for Intelligent Machines and Systems* (pp. 195–222). IGI Global.

Shoji, D., He, Z., Zhang, D., & Li, V. C. (2022). The greening of engineered cementitious composites (ECC): A review. *Construction & Building Materials*, *327*, 126701. doi:10.1016/j.conbuildmat.2022.126701

Subha, S., Inbamalar, T., Komala, C., Suresh, L. R., Boopathi, S., & Alaskar, K. (2023). A Remote Health Care Monitoring system using internet of medical things (IoMT). *IEEE Explore*, 1–6.

Syamala, M., Komala, C., Pramila, P., Dash, S., Meenakshi, S., & Boopathi, S. (2023). Machine Learning-Integrated IoT-Based Smart Home Energy Management System. In *Handbook of Research on Deep Learning Techniques for Cloud-Based Industrial IoT* (pp. 219–235). IGI Global. doi:10.4018/978-1-6684-8098-4.ch013

Tan, T. G., Szalachowski, P., & Zhou, J. (2022). Challenges of post-quantum digital signing in real-world applications: A survey. *International Journal of Information Security*, *21*(4), 937–952. doi:10.1007/s10207-022-00587-6

Ullah, S., Zheng, J., Din, N., Hussain, M. T., Ullah, F., & Yousaf, M. (2023). Elliptic Curve Cryptography; Applications, challenges, recent advances, and future trends: A comprehensive survey. *Computer Science Review*, *47*, 100530. doi:10.1016/j.cosrev.2022.100530

Venkateswaran, N., Vidhya, K., Ayyannan, M., Chavan, S. M., Sekar, K., & Boopathi, S. (2023). A Study on Smart Energy Management Framework Using Cloud Computing. In 5G, Artificial Intelligence, and Next Generation Internet of Things: Digital Innovation for Green and Sustainable Economies (pp. 189–212). IGI Global. doi:10.4018/978-1-6684-8634-4.ch009

Verma, R., Christiana, M. B. V., Maheswari, M., Srinivasan, V., Patro, P., Dari, S. S., & Boopathi, S. (2024). Intelligent Physarum Solver for Profit Maximization in Oligopolistic Supply Chain Networks. In *AI and Machine Learning Impacts in Intelligent Supply Chain* (pp. 156–179). IGI Global. doi:10.4018/979-8-3693-1347-3.ch011

Vijayakumar, G. N. S., Domakonda, V. K., Farooq, S., Kumar, B. S., Pradeep, N., & Boopathi, S. (2024). Sustainable Developments in Nano-Fluid Synthesis for Various Industrial Applications. In Adoption and Use of Technology Tools and Services by Economically Disadvantaged Communities: Implications for Growth and Sustainability (pp. 48–81). IGI Global.

Zhang, D., Yu, J., Wu, H., Jaworska, B., Ellis, B. R., & Li, V. C. (2020). Discontinuous micro-fibers as intrinsic reinforcement for ductile Engineered Cementitious Composites (ECC). *Composites. Part B, Engineering*, *184*, 107741. doi:10.1016/j.compositesb.2020.107741

Zhang, Z., Liu, S., Yang, F., Weng, Y., & Qian, S. (2021). Sustainable high strength, high ductility engineered cementitious composites (ECC) with substitution of cement by rice husk ash. *Journal of Cleaner Production*, *317*, 128379. doi:10.1016/j.jclepro.2021.128379

Zhu, J.-X., Xu, L.-Y., Huang, B.-T., Weng, K.-F., & Dai, J.-G. (2022). Recent developments in Engineered/Strain-Hardening Cementitious Composites (ECC/SHCC) with high and ultra-high strength. *Construction & Building Materials*, *342*, 127956. doi:10.1016/j.conbuildmat.2022.127956

# Chapter 4
# Harden and Self-Sensing Properties of Engineered Cementitious Composite Reinforced With Nano-Carbon

**Divya Sabapathi**
*PSG College of Technology, India*

**N. Shanmugasundaram**
https://orcid.org/0000-0002-2312-5442
*PSG College of Technology, India*

**S. Praveenkumar**
https://orcid.org/0000-0002-3931-6104
*PSG College of Technology, India*

**K. Gayathiri**
*PSG College of Technology, India*

## ABSTRACT

*Engineering cementitious composites (ECC) were discovered as a better alternative to conventional concrete due to their ability to achieve higher tensile strength. In subsequent years, additional studies were conducted to modify the materials in ECC to achieve a more sustainable and environmentally friendly composition. Incorporating nano-carbons such as fullerenes, carbon nanotubes, carbon nanoparticles, and graphene into ECC makes the advanced material highly advantageous. This chapter investigates the behavior of ECC reinforced with nano-carbons. The performance of ECC blended with nano-carbon is analyzed by comparing the fresh, mechanical and durability properties. In addition, the self-sensing and self-healing functionalities of ECC blended with nano-carbon are observed. The crack behaviour and healing performance of ECC are monitored due to the conductive ability of nano-carbon. Adding graphene in the range of 0.01 to 0.08% in ECC improves the young's modulus compared to conventional ECC.*

DOI: 10.4018/978-1-6684-8182-0.ch004

## 1. INTRODUCTION

In recent years, the construction industry requires a novel concrete composite for better infrastructure development, and hence research was carried out to meet construction needs. The development of engineered cementitious composites (ECC) was one such quest for enhanced composite (Costa et al. 2019; Shah, Hu, and Yao 2022). Engineered cementitious composite (ECC) is a unique, ultra-ductile fiber-reinforced concretes created during the past 20 years to improve the brittle behavior of traditional concrete (Altwair and Abu Jarir 2018). Compared to other types of fiber-reinforced cementitious composites, engineered cementitious composites (ECC) are distinguished by their high ductility with medium fiber content. In the beginning, ECC was conceptualized theoretically using a micromechanics method in order to endure the harsh environment and also formulated to have higher mechanical and toughness capabilities (Guo et al. 2021). In addition to their distinct characteristics, ECC is highly capable of self-healing attributes even subjected to different environmental conditions. ECC is also known as bendable concrete, which is a type of high-performance fiber-reinforced concrete with a 3–7% strain capacity (Li et al. 2022; Xu, Huang, and Dai 2023). In order to create sustainable ECC, supplemental cementitious materials (SCM) such as fly ash, granulated powdered blast furnace slag, and silica fumes are used in large quantities (N. S. and S. Praveenkumar 2022). On the list, nanomaterials have become the key component for improving the characteristics of ECC.

According to Yildirim et al. (Yıldırım et al. 2020), conventional ECC is regarded as isolative and resistant to damage because its individual components are not conductive, and because the mixing of water becomes less effective with time as a result of advancements in hydration and pozzolanic activity. However, changes in specific parameters such as water availability, chemistry and ionic state, and continuing pozzolanic reactions can all have a substantial impact on the electrical and conductivity properties of ECC (Peng et al. 2022). A common practice is to add electrically conductive materials to cement-based systems, including ECC, to eliminate these variations and ensure improved stability in electrical measurements. Nonetheless, the integration of conductive materials can disrupt the electrically conductive network of cementitious systems by covering the conductive materials or directly obstructing the channel by densing the microstructure (Ajith, Shanmugasundaram, and Praveenkumar 2021). To sustain the potency of self-sensing in ECC systems, it is necessary to confirm that the electrical characteristics remain constant or at appropriate levels even at its later ages.

On the other hand, carbon nanoparticles have piqued the interest of investigators due to their impressive mechanical, durability, and conductivity characteristics, which provide cementitious composites enhanced mechanical, durability and other defining performance (Divya, Praveenkumar, and Tayeh 2022). Carbon nanofibre, carbon nanotube and graphene are the most commonly used carbon-nanomaterials for the application of concrete (Van Dac Ho, Ching-Tai Ng, Campbell J. Coghlan, Andy Goodwin and Togay Ozbakkaloglu 2020). According to the results, the properties of cementitious materials reinforced with graphene are better than carbon nanofibers and carbon nanotubes. Tyson et al. (Chuah et al. 2014) observed that mortar blended with carbon nanofibers shows increase in displacement up to 150% compared to plain cement mortar. Furthermore, the study indicates that efficient interaction between modified nano-carbon and PVA fiber in ECC can result in superior characteristics. Lin et al. (Lin and Du 2020) suggested that adding graphene creates planes for the reactivity of cement hydration and strengthens the covalent bond. Similarly, Zheng et al. (Jing et al. 2017), observes that graphene improves the structural interface and elevates the effectiveness of cement-based composites because of the high specific energy, nano-carbons are rapidly agglomerated. Reliable dispersion procedures result in the homogeneous dispersion of nano-carbons into cementitious system, allowing them to exercise their effect in increasing performance fully.

In this study, the fresh, hardened, and self-sensing potential of a nano-carbon reinforced ECC mixture was reviewed and analyzed. The investigation of the self-sensing and self-healing qualities of nano-carbon mixed ECC drew special attention. Therefore, the inclusion of nano-carbons has the potential of minimizing the challenges related to the full-scale utilization of an ECC. The porosity in ITZ decreased due to the introduction of GO in ECC; meanwhile, the compressive strength of the ECC samples was improved. The addition of GO also narrowed the gap between the porosity of ITZ, and therefore, the entire microstructure of ECC became more homogenous. The mechanism of bond between ECC and nano-carbon is shown in figure 1.

*Figure 1. Mechanism of ECC without and with modified nano-carbons*

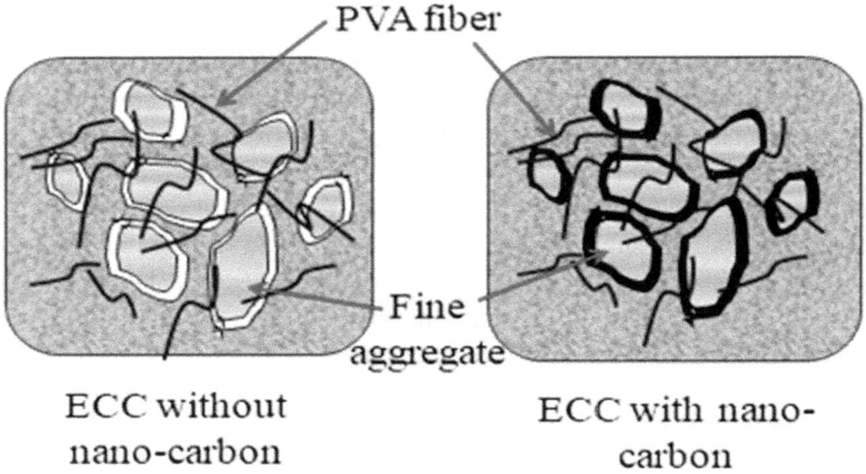


## 1.1 Research Significance

The characteristics of Engineering Cementitious Composite are determined by their fresh and hardened properties. However, many studies focused on mechanical properties, but only limited research was carried out on conductivity properties and the effect of SCM in ECC. The novelty of this review article is the influence of nano-carbon on the workability, mechanical, and conductivity properties of ECC.

This study mainly focuses on understanding the following criteria:

i. Workability of graphene blended Engineering Cementitious Composite.
ii. Mechanical behavior of nano-carbon modified Engineering Cementitious Composite.
iii. Development of Smart Engineered Cementitious Composite using nano-carbons.

## 2. WORKABILITY OF NANO-TAILORED ECC

An expertly integrated ECC mix should have the consistency of cream and an even finish. The fresh attribute of ECC must be tested before casting to ensure optimal self-compacting tendency. A concrete slump cone test might be employed to access the fresh properties of ECC. The consistency of the fresh Engineered Cement Composite is evaluated using the slump cone test. Isyaka Abdulkadir et al (Abdulkadir et al. 2022) noticed that adding 0.08% graphene to ECC increases viscosity by 92.6% when compared with conventional ECC. The profound specific surface area of nano-carbons has a significant impact on the workability of ECC (Gayathiri and Praveenkumar 2022). Similarly, negatively charged graphene agglomerate with cement particles due to the electrostatic force and increases the viscosity of the composite. The free flow capacity decreases by increasing the graphene dosage because less gravitational shearing force is needed to triumph over the yield stress of matrix. The ECC matrix become more cohesive with the highly reactive graphene, which additionally lends it a higher viscosity and slower flow rate than mixes without graphene. The workability of nano material blended ECC composite is shown in figure 2.

Farhad Aslani et al (Aslani, Wang, and Zheng 2019), studied the behavior of carbon fiber in ECC. The specific surface area of carbon fiber in ECC necessitate for a little increase in superplasticizers (SP) to ensure adequate workability. A small quantity of carbon fiber may augment the packing density of ECC, but the water demand increases, therefore increasing the SP dosage lubricates the cement particle and benefits towards improved flowability. However, increasing the concentration of SP to a certain point, carbon fiber adsorption effect causes flowability to adverse impact. From the previous studies, it is observed that good workability and self-compaction properties of ECC with nano-carbon are ensured approximately at 0.75% of superplasticizers.

*Figure 2. Workability of nano-carbon blended ECC*

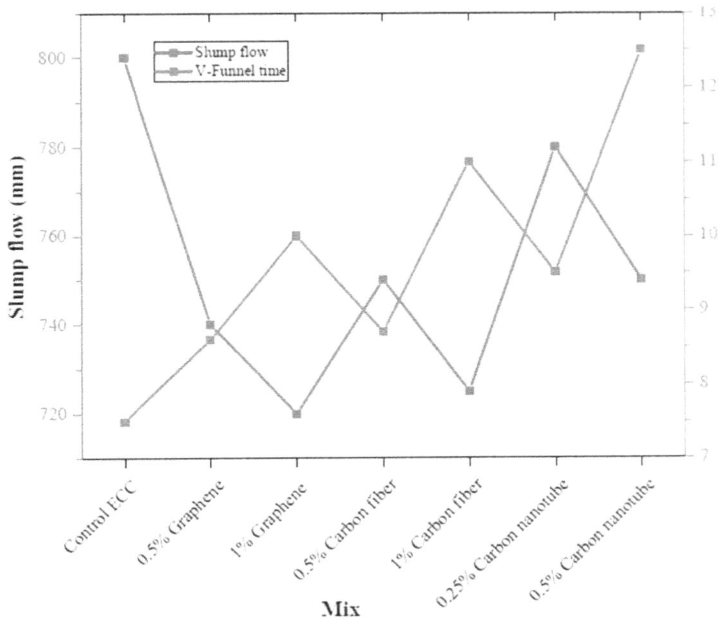

## 3. MECHANICAL PROPERTIES OF NANO ECC

This was expected due to advancements in the strength-developing reactions with time which resulted in less overall porosity and highly tortuous pore system, continuously increasing compressive strength results.

G.Yildirim et al (Yıldırım et al. 2020). investigated ECC reinforced with carbon fibers (CF), carbon nanotubes (CNT) and black carbon (CB) to ascertain its mechanical characteristics. On the one hand, compressive strength outcomes tend to be enhanced by the inclusion of carbon-based components. On the other hand, the rates of strength increment over a period differed markedly. The mechanical performance of ECC was especially apparent in its early ages. The 7-day compressive strength of the control ECC matrixes was 27.5 MPa, whereas the incorporation of CF, CNT, and CB improved the strength values to 39, 48, and 42.5 MPa, respectively (Mazzoli et al. 2018). After 7 days, there were no longer any appreciable differences between the compressive capacity results for mixes containing and excluding carbon-based components, indicating that the early phases of hydration are when adding carbon-based is most useful for developing compressive strength. The substantially packed matrices of the samples, which became more substantial with the inclusion of carbon-based materials owing to its notable specific surface areas of those materials, may be pinned for the smaller amounts of enhancement in the compression strength outcomes of ECC blends composed of various carbon-based materials. Moreover, matrix densification is expected to slow the progression of improvement in strength over time by delaying the pozzolanic activity and filler impact, decreasing the space obtainable for the consolidation of hydration and pozzolanic product reactions, and other effects (Yan Feng, Jakob Kero, Qixing Yang, Qiusong Chen and Qi 2019). Every carbon-based components employed in the current investigation are inactive, indicating that they don't play a direct participation in the hydration reactions cementitious composite. Therefore, it makes perceived to assert their roles involve offering more nucleation sites for the consolidation of hydration products, instead of promoting the pozzolanic ability of cementitious components. Table 2 presents the compressive strength enhancement from various authors and figure 3 shows the comparison graph of compressive strength of ECC.

*Figure 3. Compressive strength of ECC with nano-carbons*

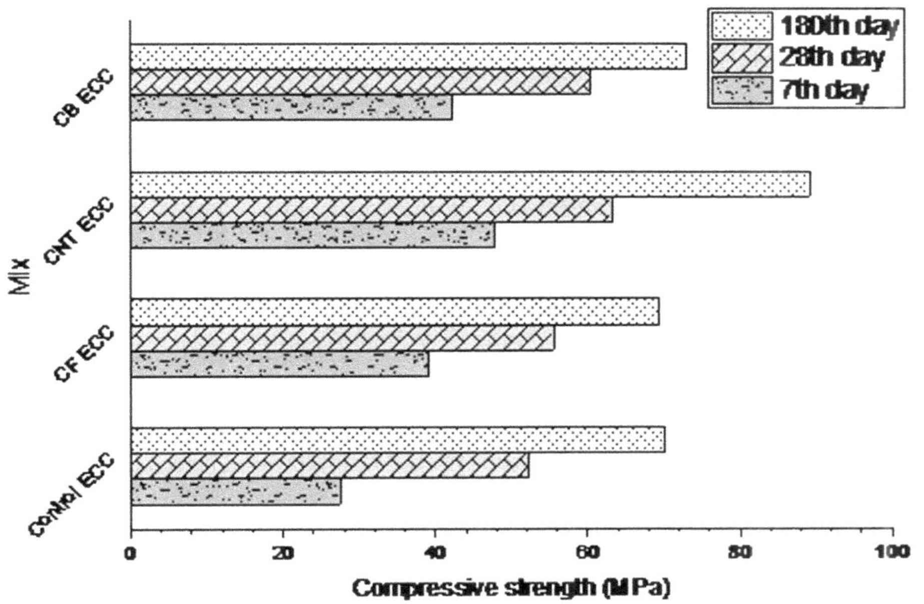


*Table 1. Compressive strength enhancement of ECC with Nano-carbons*

| S.No | Nano-Carbon | Dosage (%) | Compressive Strength Improvement (%) | Remarks | Reference |
|---|---|---|---|---|---|
| 1 | Carbon fiber | 0.5 | 14.7 | Increasing the fiber quantity improves the strength. | (Siad et al. 2018) |
| | | 1 | 21.4 | | |
| 2 | Carbon-Nanotube | 0.25 | 19.9 | Carbon fibers acts as a inert material to enhance the strength | (Yıldırım et al. 2020) |
| | | 0.5 | 30 | | |
| 3 | Graphene Oxide | 0.25 | 13.3 | Uniform dispersion of graphene increases the compressive strength. | (Sakulich and Li 2011) |
| | | 0.5 | 25.2 | | |
| | | 0.75 | 34.6 | | |
| | | 1 | 38.5 | | |
| 4 | Carbon fiber | 0.5 | 42.5 | Density of the composite increases in the addition of nano-carbons. | (Abdulkadir et al. 2022) |
| | Carbon-Nanotube | 0.5 | 73.45 | | |
| | Carbon black | 0.5 | 54.18 | | |

Adding 1% carbon fiber to ECC produced the best results when compared to control ECC, yielding 47% improvements. The considerably greater tensile strength and elastic modulus of carbon fibers in comparison with PVA fibers might suggest why carbon fibers blend ECC have larger improved flexural strength than control ECC. The increased fiber quantity of carbon fiber in ECC certainly drew more $Ca^+$ and $OH^-$ ions, leading to better production of hydration products surrounding PVA and carbon fibres (Hong et al. 2020). Also, it could additionally point to greater bonding capacity due to raised fiber quantity, which may restrict the emergence of cracks, particularly at the beginning phases of microcrack development (Sabapathy et al. 2020). The increased stiffness, first fracture strengths, and lower deflection values of carbon fiber incorporated ECC indicate an improved bond among carbon fiber, PVA fiber and the composite (Shanmugasundaram et al. 2022). The deformation rates o CNT blended ECC were analogous to the control ECC, with a minor enhancement of 3% with 0.25% of carbon nanotubes and a modest decline of -6% was observed in ECC with 0.5% of CNT (Aslani, Wang, and Zheng 2019). The results obtained suggest that a denser structure with larger nano-sized carbon nanotubes, determining that compressive strength advancement had a lesser impact on flexural capabilities. This outcome is consistent with several research in the literature that indicates the flexural behavior and young's modulus of ECCs is more dependent on bonding characteristics between fibers and the matrix. Figure 4 shows the flexural strength of ECC blended with modified nano-carbons.

*Figure 4. Flexural strength behavior of ECC with various nano-carbons*

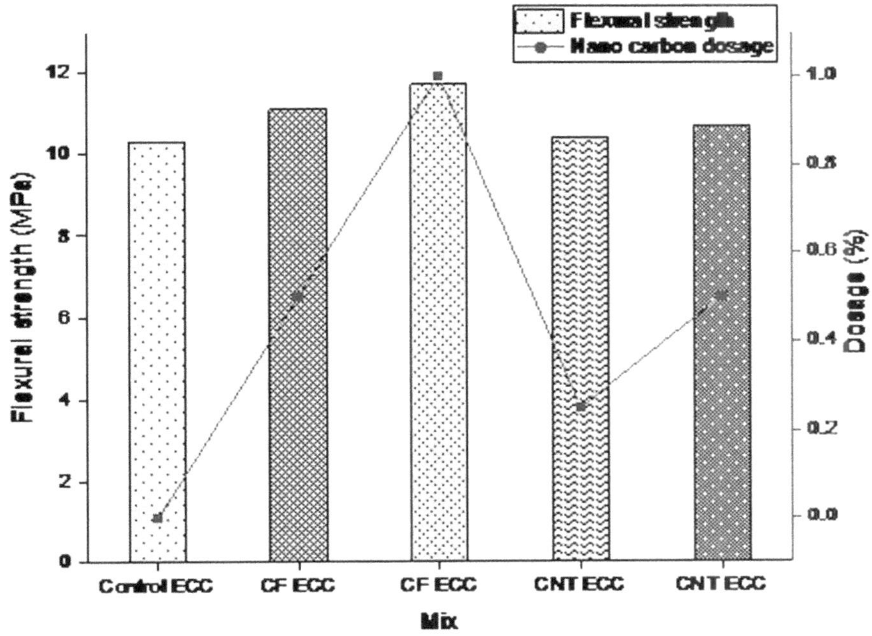


## 4. DEVELOPMENT OF SMART ENGINEERED CEMENTITIOUS COMPOSITE USING NANO-CARBONS

Smart concrete manufacturing provides a different way to track the health of reinforced concrete structures. It allows for the detection of strain or stress in concrete structures earlier than they collapse. Smart concrete technology has been extensively tested in laboratories but has still reached the industry (Mehmet Serkan Kırgız 2019; K. G. and S. Praveenkumar 2023). The test involved to determine the smartness of the concrete is electrical resistivity, thermal conductivity and electromagnetic shielding effect.

### 4.1 Electrical Resistivity

The electrical resistivity experiment was shown in studies to be a valid measure of the transport ability of concrete components, particularly considering the chemical and physical characteristics of the porous network involving porosity and tortuosity. Figure 5 shows the electrical conductivity of nano carbon blended ECC.

Through increased carbon finer and carbon nanotube content, resistance to electricity at various ages and moisture contents revealed notable decreases as compared to control ECC, nevertheless the usage of carbon nanotube, particularly at decreased phases, showed much slower reduction rates in electrical resistivity rather than carbon fibers (Abdulhussein and Katalin Kopecskó 2012). In this regard, in comparison to control ECC, the average electrical resistivity measured in saturated samples decreased by 85% to 90% after 28 days of healing (Gao et al. 2018). The results obtained demonstrate the effectiveness of adding carbon fibers and carbon nanotubes to ECC mixtures in order to increase conductance and, consequently, their potential for self-sensing. The recent research on the electrical characteristics of carbon fibers and carbon nanotube-based cementitious materials correspond with the increased piezoresistive behavior of carbon fiber induced cementitious composites compared to those with carbon nanotube.

*Figure 5. Electrical conductivity of ECC with nano carbons*

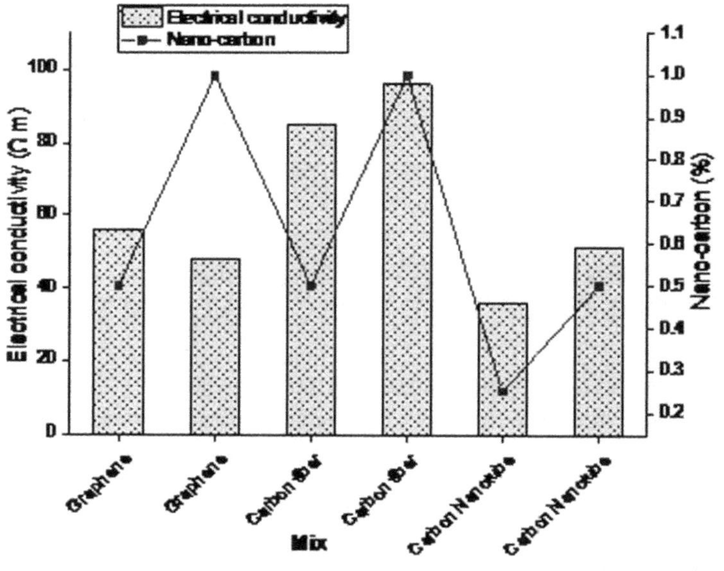


## 4.2 Thermal Conductivity

Thermal conductivity was decreased by 4.65% and 20.5%, to be precise, with the inclusion of 0.5% and 1% graphene and raised by 23.72% with the introduction of 0.5% carbon fiber (Ma, Qian, and Li 2019). In accordance with various studies published in the scientific journals, outcomes resulting from injecting carbon nanotube dispersion tend to be less substantial than anticipated (Ezzatollah Shamsaei 2018). For instance, the application of carbon fibers showed to be superior to the inclusion of graphene in improving thermal conductivity, despite being utilized at an excessive amount of 2% by the quantity of ECC (Shanmugasundaram and Praveenkumar 2021).

## 4.3 Electromagnetic Shielding Effect

In the case of electromagnetic shielding, incorporating graphene into the cement base increases the material's efficient anisotropy and interfacial polarization, each of which increases scattering and results in an increased degree of shielding efficiency than the control sample. The mixture functions like a 3D structure with two distinct electric conductivity areas whereby electromagnetic radiation is dispersed and barely ingested, depending on the microwave frequency (Ozturk et al. 2020; Shengtian Zhai, Bo Pang 2021). Both the transmission and reflection factor of the concrete mixture is directly impacted by the superposition of dispersed and multi-reflected electromagnetic radiation, which makes the shielding effect wavelength dependant (S.Divya et al. 2024). Instead, the inclusion of electrically conductive graphene in the mixture results in the formation of an array of evenly distributed conductive fillers, which raises electromagnetic declines in the areas where electromagnetic waves are strewn. This reduces the variations in the shielding effect caused by localized impacts in favor of uniform recognition of the electromagnetic field, and that in turn helps to boost the shielding effect throughout the full frequency spectrum (Mazzoli et al. 2018; Siad et al. 2018). The shielding effect is measured by the equation 1 and figure 7 shows setup to determine the electromagnetic shielding effect. Figure 6 depicts the thermal conductivity of ECC with various derivatives of nano-carbon. Increasing graphene dosage increases the resistance to thermal property due to its filler effect. Overall, thermal conductivity of carbon fiber is high compared to other derivatives.

*Figure 6. Thermal conductivity of nano-carbon blended ECC*

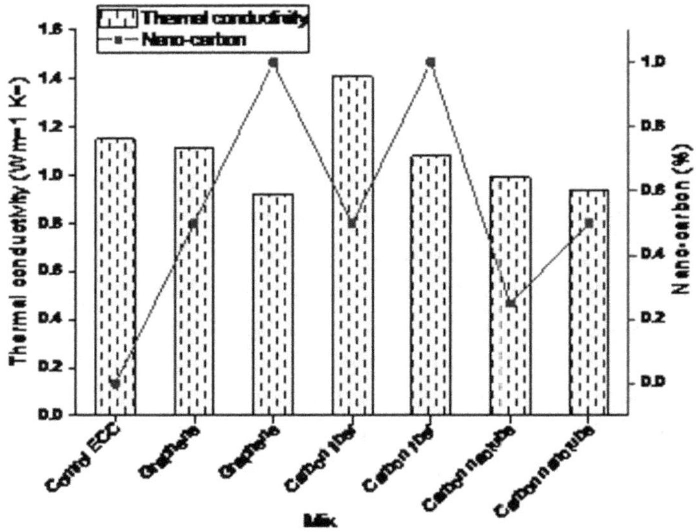


$$SE = 10\log_{10}\left\{\frac{\left\langle\left|S_{41,ns}\right|^2\right\rangle\left\langle\left|S_{21,s}\right|^2\right\rangle\left\langle\left|S_{43,s}\right|^2\right\rangle}{\left\langle\left|S_{41,s}\right|^2\right\rangle\left\langle\left|S_{21,ns}\right|^2\right\rangle\left\langle\left|S_{43,ns}\right|^2\right\rangle}\right\} \tag{1}$$

Where,
Sij = Parameter for scattering between i and J {.}
S= Sample
Ns = no sample in the aperture

*Figure 7. Setup to determine the electromagnetic shielding effect (Mazzoli et al., 2018)*

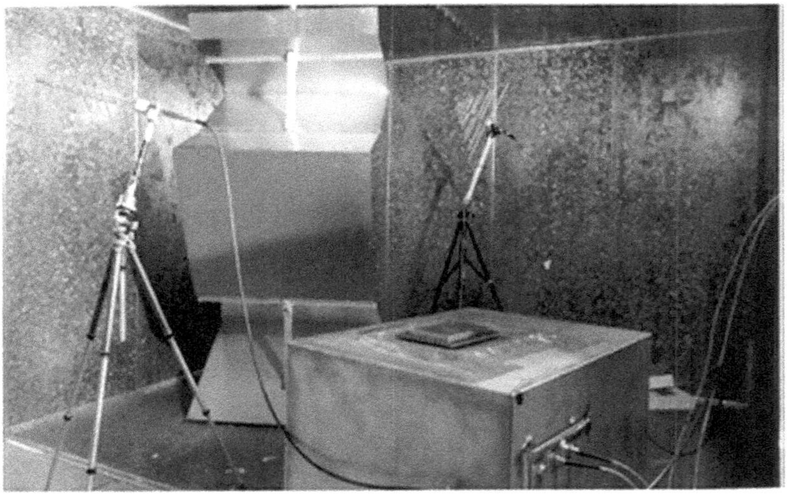

## 5. MICROSTRUCTURE ANALYSIS OF ECC IN NANO SCALE

The use of scanning electron microscopy aided in the understanding of the microstructure and morphological structure of ECC blends. Carbon fibers are immediately visible and identifiable on the specimen's interface (Divya and Shanmugam 2023). Carbon fibers may additionally be observed embedded within the cement matrix and hydration products, indicating that strong interaction happens between the carbon fibers and the cement matrix. Strong adhesion means that stress is passed throughout the matrix and carbon fibers, resulting in increased compressive strength.

Carbon nanotubes are simple to identify and plentiful in microscopic cracks surrounding PVA fibers that had been pulled away from the matrix. Multiple nanotubes were found in every single rupture in the region near PVA fiber (Peng Zhang et al 2021). Nanotubes were merely infrequently observed in similar-size fissures elsewhere in the other region of the ECC microstructure. This observation suggests that the nanotubes consolidate around the PVA fibres, even though it is not definitive. This consolidation could enhance the strength of the material generally by spanning fractures and reinforcing microcracks that spread outwards from the PVA fibers (Altayeb, Wang, and Musa 2021). Multiple nanotubes, involving certain linking nanotubes, were visible sticking out from both sides of the crack face at minimal magnification. The elevated magnification allowed for clearer observation of these nanotubes and the confirmation of their crack-bridging properties. Figure 8 presents the morphological analysis of nano-carbon blended ECC.

*Figure 8. SEM image of carbon nanotube and carbon fiber reinforced ECC (Siad et al. 2018; Yıldırım et al. 2020)*

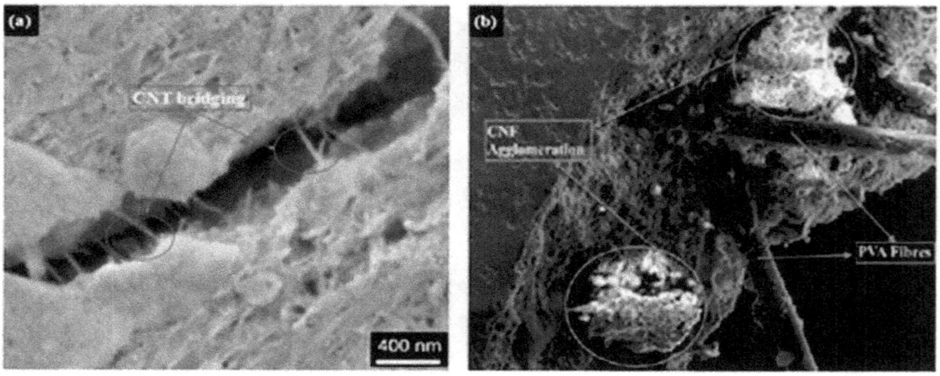


## 6. CONCLUSION

This work looked into the creation of a versatile ECC combining self-sensing efficiency with strong mechanical and ductility characteristics. Performance was evaluated on the basis of compressive and flexural strengths, cracking behavior, electrical resistivity, thermal conductivity, electromagnetic effect, and morphological study for carbon fibers, carbon nanotubes, and graphene-integrated ECC compositions. From the findings, the subsequent inferences were made:

- The compressive strength enhancement was observed by adding nano carbons to ECC mixes, particularly at earlier healing period, where the graphene combination showed gains of up to 70% over the control sample.
- The inclusion of carbon-based components to ECC greatly enhanced the self-sensing functioning, while the degree of the benefits depended on the loading circumstances and the age of the samples.
- The thermal conductivity of ECC improved with the inclusion of graphene, which was added in percentages of 0.5% and 1% by cement volume. The enhancement in thermal characteristics may be partially explained by the non-uniform dispersion of the GO microparticles during the production of the blend and their comparatively large diameters.

- Engineered cementitious composites with the inclusion of graphene are regarded as a promising novel category of electromagnetic shielding materials, capable of maintaining significant effectiveness over time and with potential use in the range of frequencies for digital communication.

## Credit Authorship Contribution Statement

S. Divya: Conceptualization, Investigation, Writing – original draft. S. Praveenkumar: Writing - review & editing, Supervision. N. Shanmugasundaram: Writing – review, Investigation. K.Gayathiri: Writing – review, Investigation.

## ACKNOWLEDGMENT

The authors wish to thank Dr. K. Prakasan, Principal, PSG College of Technology, Coimbatore for the facilities and support provided in carrying out this research work at Advanced Concrete Research Laboratory

## REFERENCES

Abdulhussein, Z. A., & Kopecskó, K. (2012). Effect of Supplementary Cementitious Materials on Properties of Concrete. *ACI Materials Journal*, *109*(1), 71–80.

Abdulkadir, I., Mohammed, B. S., Montasir, O. A. A., & Liew, M. S. (2022). Effects of Graphene Oxide and Crumb Rubber on the Fresh Properties of Self-Compacting Engineered Cementitious Composite Using Response Surface Methodology. *Materials (Basel)*, *15*(7), 2519. doi:10.3390/ma15072519 PMID:35407853

Ajith, G., Shanmugasundaram, N., & Praveenkumar, S. (2021). Effect of Mineral Admixtures and Manufactured Sand on Compressive Strength of Engineered Cementitious Composite. *Journal of Building Pathology and Rehabilitation*, *6*(1), 1–9. doi:10.1007/s41024-021-00137-y

Altayeb, M., Wang, X., & Musa, T. H. (2021). An Ensemble Method for Predicting the Mechanical Properties of Strain Hardening Cementitious Composites. *Construction & Building Materials*, *286*, 122807. doi:10.1016/j.conbuildmat.2021.122807

Altwair, N. M., & Abu Jarir, S. J. (2018).. . *Influence of Surface Roughness on Adhesion Between the Existing and New Plain Concretes.*, (September), 581–591.

Aslani, F., Wang, L., & Zheng, M. (2019). The Effect of Carbon Nanofibers on Fresh and Mechanical Properties of Lightweight Engineered Cementitious Composite Using Hollow Glass Microspheres. *Journal of Composite Materials*, *53*(17), 2447–2464. doi:10.1177/0021998319827078

Chuah, S., Pan, Z., Sanjayan, J. G., Wang, C. M., & Duan, W. H. (2014). Nano Reinforced Cement and Concrete Composites and New Perspective from Graphene Oxide. *Construction & Building Materials*, *73*, 113–124. doi:10.1016/j.conbuildmat.2014.09.040

Costa, F., Righi, D. P., Graeff, A. G., & Silva Filho, L. C. P. (2019). Experimental Study of Some Durability Properties of ECC with a More Environmentally Sustainable Rice Husk Ash and High Tenacity Polypropylene Fibers. *Construction & Building Materials*, *213*, 505–513. doi:10.1016/j.conbuildmat.2019.04.092

Divya, S. (2023). Nucleation Effect of Graphene in Mechanical and Microstructure Properties of Mortar with Copper Slag and Quarry Dust. *Fullerenes, Nanotubes, and Carbon Nanostructures*, *0*(0), 1–20. doi:10.1080/1536383X.2023.2255319

Divya, S., Praveenkumar, S., & Bassam, A. (2022). Performance of Modified Nano Carbon Blended with Supplementary Materials in Cement Composite – An Interpretive Review. *Construction & Building Materials*, *346*(June), 128452. doi:10.1016/j.conbuildmat.2022.128452

Divya, S., Saranya, R., & Praveenkumar, S. (2023, September). 2024. "Pozzolanic Attributes of Hydraulic Cement Paste Hybridized with Agricultural By-Product and Nano-Carbon.". *Construction & Building Materials*, *411*, 134119. doi:10.1016/j.conbuildmat.2023.134119

Feng, Y., Kero, J., Yang, Q., Chen, Q., Engström, F., Samuelsson, C., & Qi, C. (2019). Mechanical Activation of Granulated Copper Slag and Its Influence on Hydration Heat and Compressive Strength of Blended Cement. *Materials (Basel)*, *12*(5), 772. doi:10.3390/ma12050772 PMID:30845730

Gao, S., Wang, Z., Wang, W., & Qiu, H. (2018). Effect of Shrinkage-Reducing Admixture and Expansive Agent on Mechanical Properties and Drying Shrinkage of Engineered Cementitious Composite (ECC). *Construction & Building Materials*, *179*, 172–185. doi:10.1016/j.conbuildmat.2018.05.203

Gayathiri, K., & Praveenkumar, S. (2022). *Silicon Influence of Nano Silica on Fresh and Hardened Properties of Cement-Based Materials – A Review*. Springer Netherlands. doi:10.1007/s12633-021-01598-z

Guo, P., Meng, W., Xu, M., Li, V. C., & Bao, Y. (2021). Predicting Mechanical Properties of High-Performance Fiber-Reinforced Cementitious Composites by Integrating Micromechanics and Machine Learning. *Materials (Basel)*, *14*(12), 3143. doi:10.3390/ma14123143 PMID:34201068

Ho, V. D., Ng, C.-T., Coghlan, C. J., Goodwin, A., Mc Guckin, C., & Dusan, L. T. O. (2020). Electrochemically Produced Graphene with Ultra Large Particles Enhances Mechanical Properties of Portland Cement Mortar. *Construction & Building Materials*, *234*, 117403. doi:10.1016/j.conbuildmat.2019.117403

Hong, D. L. H. (2020). Deformation Properties of Rubberized Ecc Incorporating Nano Graphene Using Response Surface Methodology. *Materials (Basel)*, *13*(12), 1–14. doi:10.3390/ma13122831 PMID:32545443

Jing, G., Ye, Z., Lu, X., & Hou, P. (2017). Effect of Graphene Nanoplatelets on Hydration Behaviour of Portland Cement by Thermal Analysis. *Advances in Cement Research*, *29*(2), 63–70. doi:10.1680/jadcr.16.00087

Li, P., Liu, J., Suh, H., Zal Nezhad, E., & Bae, S. (2022). Understanding the Role of Graphene Oxide Nanoribbons–Functionalized Carbon Nanotubes–Graphene Oxide (GNFG) Complex in Enhancing the Fire Resistance of Cementitious Composites. *Construction & Building Materials*, *348*(July), 128637. doi:10.1016/j.conbuildmat.2022.128637

Lin, Y., & Du, H. (2020). Graphene Reinforced Cement Composites: A Review. *Construction & Building Materials*, *265*, 120312. doi:10.1016/j.conbuildmat.2020.120312

Ma, H., Qian, S., & Li, V. C. (2019). Tailoring Engineered Cementitious Composite with Emulsified Asphalt for High Damping. *Construction & Building Materials*, *201*, 631–640. doi:10.1016/j.conbuildmat.2018.12.222

Mazzoli, A., Corinaldesi, V., Donnini, J., Di Perna, C., Micheli, D., Vricella, A., Pastore, R., Bastianelli, L., Moglie, F., & Mariani Primiani, V. (2017, February). 2018. "Effect of Graphene Oxide and Metallic Fibers on the Electromagnetic Shielding Effect of Engineered Cementitious Composites.". *Journal of Building Engineering*, *18*, 33–39. doi:10.1016/j.jobe.2018.02.019

Mehmet Serkan Kırgız. (2019). *Smart Nanoconcretes and Cement-Based Materials In Nano Size Particle Packing for Nanoconcretes and Cement Based Materials: Mathematical Models*. Theory, and Technology.

Ozturk, M., Karaaslan, M., Akgol, O., & Korkut, U. (2020). Mechanical and Electromagnetic Performance of Cement Based Composites Containing Different Replacement Levels of Ground Granulated Blast Furnace Slag, Fly Ash, Silica Fume and Rice Husk Ash. *Cement and Concrete Research*, *136*(January), 106177. doi:10.1016/j.cemconres.2020.106177

Peng, K. D., Zeng, J.-J., Huang, B.-T., Huang, J.-Q., Zhuge, Y., & Dai, J.-G. (2022). Bond Performance of FRP Bars in Plain and Fiber-Reinforced Geopolymer under Pull-out Loading. *Journal of Building Engineering*, *57*(June), 104893. doi:10.1016/j.jobe.2022.104893

Praveenkumar, K. (2023). Retaining the Particle Packing Approach and Its Application in Developing the Cement Composites towards Sustainability. *Journal of Building Pathology and Rehabilitation*, *8*(1), 26. Advance online publication. doi:10.1007/s41024-023-00271-9

Praveenkumar, N. (2022). Mechanical Properties of Engineered Cementitious Composites (ECC) Incorporating Different Mineral Admixtures and Fibre: A Review. *Journal of Building Pathology and Rehabilitation*, *7*(1), 40. Advance online publication. doi:10.1007/s41024-022-00182-1

Sabapathy, L., Mohammed, B. S., Al-Fakih, A., Wahab, M. M. A., Liew, M. S., & Amran, Y. H. M. (2020). Acid and Sulphate Attacks on a Rubberized Engineered Cementitious Composite Containing Graphene Oxide. *Materials (Basel)*, *13*(14), 3125. doi:10.3390/ma13143125 PMID:32668788

Sakulich, A. R., & Li, V. C. (2011). Nanoscale Characterization of Engineered Cementitious Composites (ECC). *Cement and Concrete Research*, *41*(2), 169–175. doi:10.1016/j.cemconres.2010.11.001

Shah, Y. I., Hu, Z., & Yao, P. (2022). Interface Bond Strength Performance of Overlap Joints within the Corrugated Pipes Used in Prefabricated Bridges. *Advances in Structural Engineering*, *25*(6), 1240–1253. doi:10.1177/13694332211072324

Shamsaei, E., & Basquiroto de Souza, F. (2018). Graphene-Based Nanosheets for Stronger and More Durable Concrete: A Review. *Construction & Building Materials*, *183*, 642–660. doi:10.1016/j.conbuildmat.2018.06.201

Shanmugasundaram, N., & Praveenkumar, S. (2021). Influence of Supplementary Cementitious Materials, Curing Conditions and Mixing Ratios on Fresh and Mechanical Properties of Engineered Cementitious Composites – A Review. *Construction & Building Materials*, *309*(October), 125038. doi:10.1016/j.conbuildmat.2021.125038

Shanmugasundaram, Praveenkumar, Gayathiri, & Divya. (2022). Prediction on Compressive Strength of Engineered Cementitious Composites Using Machine Learning Approach. *Construction and Building Materials, 342*(PA), 127933. . doi:10.1016/j.conbuildmat.2022.127933

Siad, H., Lachemi, M., Sahmaran, M., Mesbah, H. A., & Hossain, K. A. (2018). Advanced Engineered Cementitious Composites with Combined Self-Sensing and Self-Healing Functionalities. *Construction & Building Materials*, *176*, 313–322. doi:10.1016/j.conbuildmat.2018.05.026

Xu, Huang, & Dai. (2023). *Utilization of Artificial Geopolymer Aggregates in High-Strength Engineered Cementitious Composites (HS-ECC).* Academic Press.

Yıldırım, G. (2020). Self-Sensing Capability of Engineered Cementitious Composites: Effects of Aging and Loading Conditions. *Construction & Building Materials*, 231.

Zhai, S., Pang, B., Liu, G., Zhang, Y., Xu, K., She, W., & Zhang, Y. (2021). Investigation on Preparation and Multifunctionality of Reduced Graphene Oxide Cement Mortar. *Construction & Building Materials*, *275*, 122119. doi:10.1016/j.conbuildmat.2020.122119

Zhang, P., Wang, K., Wang, J., Guo, J., & Ling, Y. (2021). Macroscopic and Microscopic Analyses on Mechanical Performance of Metakaolin/Fly Ash Based Geopolymer Mortar. *Journal of Cleaner Production*, *294*, 126193. doi:10.1016/j.jclepro.2021.126193

# Chapter 5
# Development of Models for Mechanical Properties of Engineered Cementitious Composites

**P. Jagadesh**
https://orcid.org/0000-0002-9803-748X
*Coimbatore Institute of Technology, India*

**K. S. Amirtavarshini**
*Coimbatore Institute of Technology, India*

**Haytham F. Isleem**
https://orcid.org/0000-0002-4826-7420
*Qujing Normal University, China*

**Athanasia K. Thomoglou**
*Democritus University of Thrace, Greece*

**Maristella E. Voutetaki**
*Democritus University of Thrace, Greece*

## ABSTRACT

*Engineering cement composites (ECC) are a type of fiber cement composite material and possess high mechanical and durability properties. The main objective of this study is to develop a model for the relationship between compressive strength and split tensile strength. To develop these relationships, mechanical properties that are available in the literature for various types of ECC are collected in which 75% data is used for training and 25% for testing purpose. The study is further extended to check*

DOI: 10.4018/978-1-6684-8182-0.ch005

*the relationships or models available in the standards, codes, and in literature for various mechanical properties are also evaluated. Sensitive assessment is performed to evaluate the new model developed for ECC based on mechanical properties. These relationships are required for designers, engineers, researchers, and academicians when one of the mechanical properties are not available for design purpose.*

## 1. INTRODUCTION

ECC (Engineered Cementitious Composites) extensively known as concrete with bendable characteristics, which was initially proposed and developed by Professor Victor Li, Professor at the University of Michigan, are fibre reinforced concrete aiming to attain more tensile strength of the concrete. ECCs are generally also recognized as Strain Hardening Cementitious Composites (SHCC) due to their property of strain hardening by the process of micromechanics where mortar-based composites are reinforced with suitable fibres, commonly polymer fibres to improve the ductile property of the concrete. These compounds also possess a self-healing property which makes it applicable to water or sewage conduits. These fibres not only provide the flexibility property but also acts as reinforcement in the concrete. Hence, the amount of reinforcement steel provided along with concrete can be reduced and thus reducing the cost of the construction project. ECCs can also be precast.

ECCs generally look similar to Ordinary Portland Cement (OPC), except the ductile nature of the compound. When compared to OPC, the emission of harmful gases is lesser from ECCs and are also comparatively lighter than OPC. ECC have a tensile strength of 3-7% more when compared to Ordinary Portland Cement (OPC), conventional concrete or cement mortar (Bentur and Mindess., 2007). Therefore, the ductile nature of the compound is more than its brittleness and thus eradicates the disadvantage of the low tensile strength of conventional concrete when replaced. Even several ECC has compressive strength greater than 80 MPa and even more, which is called as high strength ECC and ECC having compressive strength lesser than 80 MPa, is called as conventional strength ECC.

Considering a structure or elements of structure, mechanical performance (strength and durability properties) can be determined. Once the mechanical and functional performance requirements are converted into properties of the material (strength, strain, modulus, fracture energy, etc.), the models from micromechanics, which acts a vital tool for selection of materials. The material selection includes the selection of fibre ratio, dimensions and type of fibres, choice, type and shape of aggregates, cement types, etc. Apart from this, for design purposes, several models related to properties of composites are required to satisfy the performance requirements of a structure. These models not only reduce the cost of materials involved during the study of material properties or composites characterization but also saves the time involved in it. The effects of curing conditions, mixing ratios, experimental procedures, etc., which are specified in the various codes becomes significant in attaining the required ECC properties. ECC is built on the theory of micromechanics to tailor the fibre to achieve substantial tensile ductility (Engineered composites) of ECC that is reinforced to cultivate microscopic synergistic interactions. ECC covers a broad range of material design covering all kinds of properties at different levels like macro-, micro- and nano- scales.

The properties of ECC also include tiny crack width, absence of anisotropically weak planes, use of a small amount of fibres (around 2% fibre ratio) and also ease of processing when compared to conventional materials (Yao et al., 2021). Various properties of ECC like ductility, tensile strength, strain, cracking strength, porosity, cracking width, pore volume, pore size distribution, etc., may vary due to the type, dimensions and aspect ratio. The usage of cement, fly ash, silica sand, polyvinyl fibrers, etc., commonly increase the tensile strength of ECC (Shanmugasundaram and Praveen Kumar., 2022). Considering the cost of ECC, locally available and suitable fibres such as ground granulated blast-furnace slag, rice husk ash, silica fume, bagasse ash, etc., can be used for an economical compound.

One of the advantages of ECC, is that it possess lightweight because due to the non-usage of coarse aggregates which also reduce the young's modulus of ECC. On comparing the attaining the compressive strength duration of ECC with OPC, it requires less time. The main factor behind the usage of ECCs are is that it can be made sustainable by incorporating materials that are locally available as the reinforced fibre. For example, usage of industry and agricultural by-products for the fibre that is reinforced into ECC. Another advantage of using industrial and agricultural wastes not only reduce the land pollution but also reduce the air pollution (Shanmugasundaram and Praveen Kumar., 2022). ECCs can also be used as repair materials (Yildirum et al., 2018). Due to infrastructure development across the globe, there is increasing in usage of cement and concrete across the construction sector is noted. And also, concrete is most consumable material next to water. Therefore, increasing in utilization of concrete, leads to increasing utilization of coarse aggregates results in demolition of natural resources. Achieving the target strength has to carefully designed as each ingredient and their properties and quantities affect it. Feasibility of the materials used as reinforcement should be analysed and, in some cases, the initial cost may be high.

Applications of ECC are broadening with developments in research and implementations globally in areas like eco environmental constructions infrastructure in different countries, marine and coastal construction, structural and non-structural applications construction at hot arid climates, bridge, roadways, and airport pavements and runways, earthquake resistant structures, chemical resistant structures, impact resistance, high rise buildings, fire resistance, high durability structures, hydraulic structures, repair and restoration works, prestress or precast concrete construction, load carrying or high pressure pipelines, etc. (Yildirum et al., 2018).

## 1.1 Novelty of the Study

One of the governing factors for composite materials for design and application in the Civil Engineering field is the durable and mechanical properties. Mechanical properties are influenced by material properties, mixing ratio, curing conditions, etc. Studies related to influence of these factors on mechanical properties are needed for researchers, academicians, engineers, contractors, etc. It is also helpful to select the suitable materials based on mechanical properties of ECC in various Civil Engineering and Concrete Construction applications. For practical implementation, it is impossible to conduct tests for all the mechanical properties like compressive strength and tensile strength of ECC during casting or at erection site. And also, the cost involved in materials and the man power consumption need to be reduced. Hence, it is necessary for engineers or designers to develop a relationship between the mechanical properties for an easy way to understand the other properties. Majorly, most of studies proposed to develop relationship with respect to compressive strength, because it is one of the major experimental tests available now a days.

## 1.2 Objective of the Present Study

The present study aims for compiling an experimental database for ECC with different dosages of supplementary or alternative cementitious materials and fibres for their mechanical properties. The core objective of the current research is to develop a new model for mechanical properties from the experimental results available in literatures for ECC. In particular, the relationship between compressive strength and split tensile strength of ECC is focused. The study is further extended to check the relationships of these mechanical properties available in codes, Indian and International standards and literature for conventional concrete.

# 2. EXPERIMENTAL DATABASE

Researchers have experimentally determined the various properties of ECC incorporated concrete based on the fibres aspect ratio, fibres type, the ratio of fibres used, water to binder ratio, etc., Mechanical properties like strength properties (compressive strength and split tensile strength) from various research papers along with the material used as ECC, their respective fibre ratio and water to binder ratio are tabulated in Table 1.

*Table 1. Fibre ratio, water to binder ratio, tensile strength, and compressive strength properties of ECC available in literature*

| S No | Author | Year | Material Used | Compressive Strength (MPa) | Direct Tensile Strength (MPa) | Fibre Ratio (%) | Water to Binder Ratio |
|---|---|---|---|---|---|---|---|
| 1 | Zhang et al | 2020 | PVA fibres | 54.77 | 4.09 | 2.00 | |
| 2 | Zhang et al | 2020 | Fly ash | 54.85 | 6.52 | | 0.25 |
| 3 | Zhang et al | 2020 | Combination of SCM | 80.44 | 9.98 | | 0.42 |
| 4 | Zhang et al | 2020 | Combination of SCM | 108.57 | 13.07 | | 0.42 |
| 5 | Yu et al | 2020 | PVA fibres | 67.38 | 5.23 | 2.00 | |
| 6 | Yu et al | 2020 | PVA fibres | 67.38 | 7.79 | 2.00 | |
| 7 | Yu et al | 2020 | Fly ash | 67.97 | 5.18 | | 0.42 |
| 8 | Yu et al | 2020 | LS | 64.85 | 5.18 | | 0.41 |
| 9 | Yu et al | 2020 | FSS | 68.19 | 5.22 | | 0.41 |
| 10 | Guan et al | 2019 | GGBS | 93.57 | 7.58 | | 0.28 |
| 11 | Singh et al | 2020 | FSS | 70.66 | 7.52 | | 0.55 |
| 12 | Soe et al | 2013 | PE fibres | 78.92 | 6.64 | 2.00 | |
| 13 | Soe et al | 2013 | Combination of SCM | 80.44 | 6.38 | | 0.55 |
| 14 | Zhou et al | 2021 | PVA fibres | 45.43 | 5.23 | 2.00 | |
| 15 | Ali et al | 2017 | Combination of SCM | 56.88 | 5.36 | | 0.28 |
| 16 | Siad et al | 2018 | PVA fibres | 45.70 | 4.45 | 1.40 | |
| 17 | Siad et al | 2018 | Fly ash | 45.74 | 4.40 | | 0.31 |
| 18 | Meng et al | 2017 | PVA fibres | 60.64 | 5.76 | 2.20 | |
| 19 | Meng et al | 2017 | Fly ash | 60.61 | 5.87 | | 0.29 |
| 20 | Yu et al | 2018 | PE fibres | 114.63 | 17.36 | 2.00 | |
| 21 | Yu et al | 2018 | Combination of SCM | 114.66 | 17.47 | | 0.16 |
| 22 | Zhu et al | 2020 | Fly ash | 52.72 | 4.72 | | 0.30 |
| 23 | Zhu et al | 2020 | Combination of SCM | 47.50 | 3.79 | | 0.33 |
| 24 | Huang et al | 2020 | SF | 136.23 | 8.00 | | 0.21 |
| 25 | Mohammed et al | 2018 | PVA fibres | 101.35 | 5.29 | 2.00 | |
| 26 | Mohammed et al | 2018 | Fly ash | 101.94 | 5.18 | | 0.13 |
| 27 | Mohammed et al | 2018 | Combination of SCM | 93.57 | 4.03 | | 0.17 |
| 28 | Wang et al | 2020 | Fly ash | 46.65 | 3.18 | | 4.21 |
| 29 | Adesina et al | 2020 | Fly ash | 55.76 | 4.35 | | 2.02 |
| 30 | Li et al | 2019 | Hybrid | 44.70 | 5.50 | 2.00 | |
| 31 | Li et al | 2019 | Fly ash | 45.14 | 5.23 | | 0.31 |
| 32 | Ma et al | 2020 | PVA fibres | 46.30 | 5.08 | 2.00 | |

*Continued on following page*

*Table 1. Continued*

| S No | Author | Year | Material Used | Compressive Strength (MPa) | Direct Tensile Strength (MPa) | Fibre Ratio (%) | Water to Binder Ratio |
|---|---|---|---|---|---|---|---|
| 33 | Ma et al | 2020 | Fly ash | 46.35 | 5.00 | | 0.26 |
| 34 | Yu et al | 2018 | PE fibres | 148.48 | 18.93 | 2.00 | |
| 35 | Yu et al | 2018 | Combination of SCM | 148.42 | 18.90 | | 0.17 |
| 36 | Ding | 2018 | PE fibres | 119.10 | 16.47 | 2.00 | |
| 37 | Qin et al | 2020 | PVA fibres | 78.53 | 10.25 | 2.00 | |
| 38 | Qin et al | 2020 | Fly ash | 78.28 | 10.03 | | 0.30 |
| 39 | Xu et al | 2021 | PVA fibres | 59.78 | 7.06 | 2.00 | |
| 40 | Xu et al | 2021 | Fly ash | 59.40 | 6.98 | | 0.26 |
| 41 | Bo-TaoHuanget et al | 2021 | PE fibres | 138.56 | 8.11 | 2.00 | |
| 42 | Bo-TaoHuanget et al | 2021 | SF | 138.58 | 8.09 | | 0.19 |
| 43 | Himasha et al | 2020 | PET | 52.64 | 3.98 | 2.00 | |
| 44 | Ma et al | 2018 | PVA fibres | 49.83 | 4.87 | 2.00 | |
| 45 | Ma et al | 2018 | Fly ash | 49.38 | 4.72 | | 0.73 |
| 46 | Wang et al | 2019 | PVA fibres | 55.37 | 5.08 | 2.00 | |
| 47 | Wang et al | 2019 | BFS | 57.82 | 4.76 | | 0.33 |
| 48 | Ratnayake et al | 2020 | PVA fibres | 79.66 | 4.92 | 2.00 | |
| 49 | Ratnayake et al | 2020 | Fly ash | 80.40 | 4.81 | | 0.27 |
| 50 | Lei et al | 2019 | PE fibres | 166.34 | 6.44 | 2.00 | |
| 51 | Lei et al | 2019 | SF | 167.76 | 6.43 | | 0.27 |
| 52 | Lei et al | 2019 | Combination of SCM | 120.29 | 6.38 | | 0.28 |
| 53 | Liu et al | 2017 | PVA fibres | 49.50 | 5.76 | 2.00 | |
| 54 | Zhang et al | 2019 | PE fibres | 86.36 | 10.98 | 2.00 | |
| 55 | Zhang et al | 2019 | Combination of SCM | 87.00 | 10.90 | | 0.42 |
| 56 | Ye et al | 2019 | UHMWPE | 118.60 | 11.29 | 2.00 | |
| 57 | Ye et al | 2019 | Combination of SCM | 120.29 | 11.18 | | 0.14 |
| 58 | Ding et al | 2018 | PE fibres | 115.13 | 18.09 | 2.00 | |
| 59 | Ding et al | 2018 | Combination of SCM | 114.66 | 17.98 | | 0.34 |
| 60 | Ling et al | 2019 | Combination of SCM | 102.94 | 5.69 | | 0.34 |
| 61 | Zhou | 2010 | Combination of SCM | 55.94 | 4.35 | | 0.28 |
| 62 | Mohammed | 2018 | Hybrid | 85.87 | 5.34 | 1.25 | |
| 63 | Mohammed | 2018 | Fly ash | 88.29 | 5.27 | | 0.14 |
| 64 | Said et al | 2015 | PE fibres | 65.04 | 5.29 | 1.00 | |
| 65 | Said et al | 2015 | Fly ash | 65.16 | 5.23 | | 0.38 |
| 66 | Said et al | 2015 | FSS | 65.71 | 5.22 | | 0.38 |
| 67 | Shi et al | 2018 | PVA fibres | 63.85 | 4.56 | 1.50 | |
| 68 | Zhou et al | 2019 | PE fibres | 100.75 | 16.06 | 2.00 | |
| 69 | Zhou et al | 2019 | Combination of SCM | 92.63 | 16.04 | | 0.24 |
| 70 | Ding et al | 2020 | PE fibres | 115.62 | 16.11 | 2.00 | |

*Continued on following page*

*Table 1. Continued*

| S No | Author | Year | Material Used | Compressive Strength (MPa) | Direct Tensile Strength (MPa) | Fibre Ratio (%) | Water to Binder Ratio |
|---|---|---|---|---|---|---|---|
| 71 | Ding et al | 2020 | Combination of SCM | 115.13 | 16.04 | | 0.15 |
| 72 | Ding et al | 2020 | FSS | 115.58 | 16.00 | | 0.15 |
| 73 | Zhenbo et al | 2015 | Steel | 62.06 | 6.12 | 1.70 | |
| 74 | Zhenbo et al | 2015 | Hybrid | 46.19 | 4.71 | 3.20 | |
| 75 | Krishnaraja et al | 2018 | PVA fibres | 50.37 | 5.86 | 2.00 | |
| 76 | Krishnaraja et al | 2018 | Steel | 53.13 | 6.96 | 2.00 | |
| 77 | Krishnaraja et al | 2018 | Hybrid | 51.65 | 6.38 | 2.00 | |
| 78 | Krishnaraja et al | 2018 | Fly ash | 53.02 | 6.84 | | 0.36 |
| 79 | Sahmaran et al | 2009 | PVA fibres | 61.85 | 5.29 | 2.00 | |
| 80 | Sahmaran et al | 2009 | Fly ash | 62.13 | 5.13 | | 0.28 |
| 81 | Guo et al | 2020 | PE fibres | 63.55 | 8.11 | 2.00 | |
| 82 | Guo et al | 2020 | Fly ash | 64.25 | 8.04 | | 0.26 |
| 83 | Guo et al | 2020 | USS | 60.77 | 7.99 | | 0.25 |
| 84 | Gürbüz et al | 2020 | PE fibres | 57.60 | 7.22 | 2.00 | |
| 85 | Gürbüz et al | 2020 | Combination of SCM | 57.82 | 7.16 | | 0.22 |
| 86 | Deng et al | 2018 | PVA fibres | 81.47 | 4.40 | 2.00 | |
| 87 | Deng et al | 2018 | Fly ash | 57.88 | 4.49 | | 0.27 |
| 88 | Deng et al | 2018 | GGBS | 81.38 | 4.35 | | 0.28 |
| 89 | Deng et al | 2018 | Combination of SCM | 67.66 | 4.12 | | 0.27 |

*Figure 1. Fibres used in ECC*

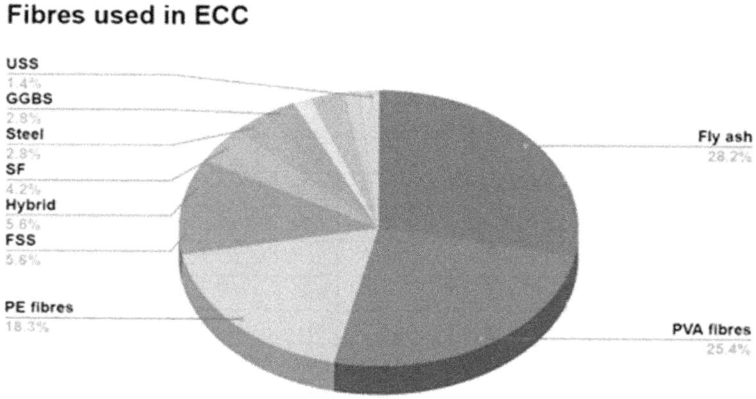

As it is represented in Figure 1, from the data used in this paper it can be noted that in terms of the fibres used in ECC, the most commonly used is fly ash fibres then are the PVA fibres. Usage of PE fibres are significantly used after the above-mentioned fibres with a percentage close to 15%. Usage of GGBS, Steel, Silica fumes are comparatively in lesser percentages.

*Figure 2. Number of journals per year in ECC*

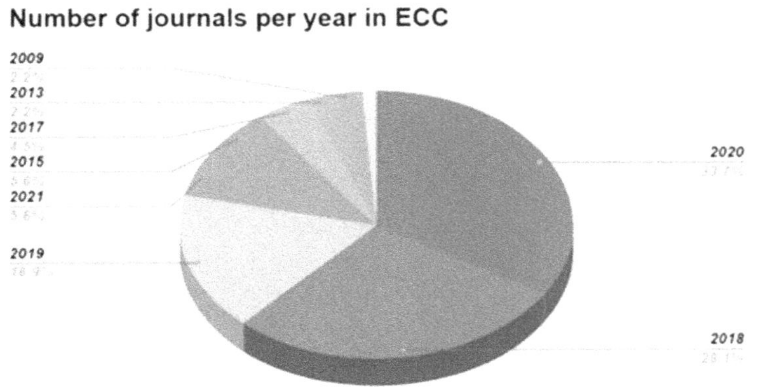


Figure 2 shows the number of journals published each year on ECC incorporated concrete. Various of these journals discuss the mechanical properties associated with the ECC concrete. The experimental data from these journal papers is used for analysis and to formulate a association between the compressive and split tensile strength.

From these experimental database results (Table 1), compressive and split tensile strength are focused for further study in this paper. Figure 3 and Figure 4 represent the histogram of compressive and split tensile strength obtained from various research papers, respectively. From Figure 3 and Figure 4, the frequency and the cumulative appearances of the value of compressive and tensile strength are depicted. It can be understood that in terms of compressive strength, the value of 50MPa has the highest frequency of about 12. Similarly, in terms of split tensile strength the value of 5.5MPa has the highest frequency of about 19. The cumulative percentages for compressive and split tensile strength are also depicted in the below figures (Figure 3 and Figure 4). All the values that are used to formulate the below graphs are obtained from literature database (Table 1).

*Figure 3. Histogram of compressive strength of ECC (from literature)*

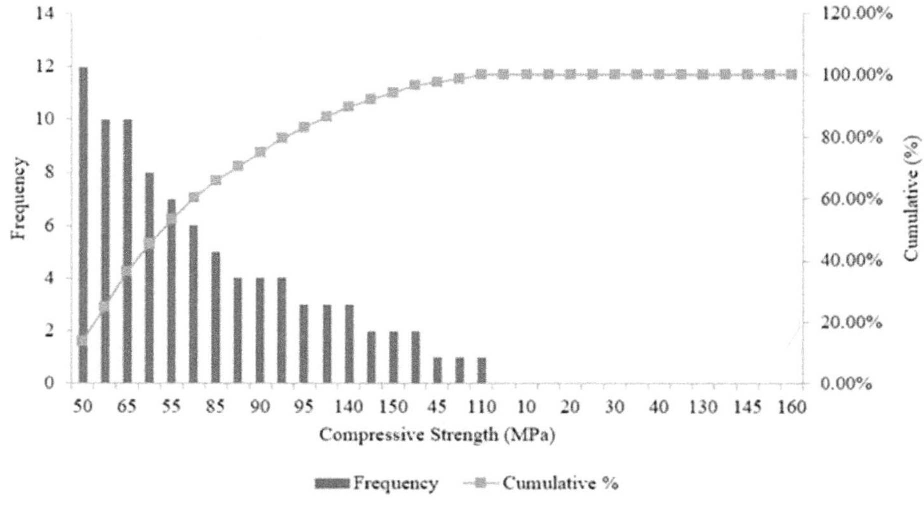


*Figure 4. Histogram of split tensile strength of ECC (from literature)*

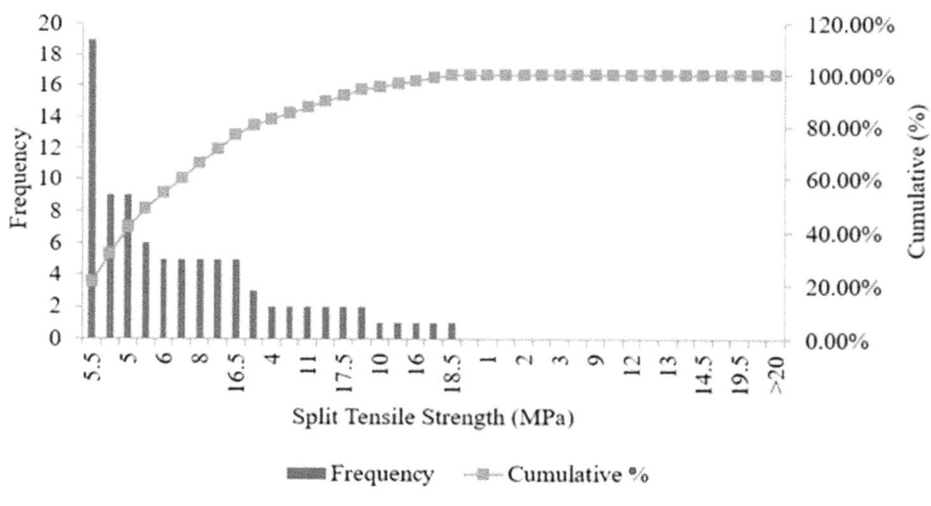

## 3. METHODS

Primarily, to develop the connection between the compressive and split tensile strength, results from various international codes and various literatures with respect to conventional concrete are observed and tabulated below. From literature, it can be noted that, the association between the compressive and split tensile strength is expressed in equation (1),

$$f_s = A(f_c)^v \tag{1}$$

Secondly, the reliability of these relationships from various literature are assessed in order to arrive at the best possible relationship. To execute this, the Root Mean Square Error (RMSE), Mean Absolute Error (MAE) and Integrated Absolute Error (IAE) and are taken into account. For this assessment, the above criteria are analysed with the lowest value of RMSE, IAE and MAE in order to determine the best hold relationship.

From all the relationships observed from various international standards and research papers, linear regression lines were plotted using least square method to find the best fit experimental relationship between compressive strength and split tensile strength. Further, the confidence and predication interval were plotted with suitable best fit line. From these two intervals it can be observed that reliability of these relationships mentioned.

### 3.1 Compressive and Split Tensile Strength Relationship Models

The value of ***A*** and ***V*** varies through a range in different international standards and literatures. From various international standards, it is observed (Table 2) that the value if ***A*** ranges from 0.13 to 0.59 with the average value of ***A*** around 0.37. Similarly, the value of ***V*** ranges from 0.5 to 0.85 with the average value of ***V*** as 0.6 and the median as 0.5.

*Table 2. Relationship between compressive and split tensile strength for concrete available in various international standards*

| | |
|---|---|
| ACI 318-2008 | $f_s = 0.56\sqrt{f_c}$ |
| CEB-FIB-1997 | $f_s = 0.30(f_c)^{2/3}$ |
| ACI 363R-1992 | $f_s = 0.59\sqrt{f_c}$ |
| AS3600-2009 | $f_s = 0.4\sqrt{f_c}$ |
| JCI., 2011 | $f_s = 0.13(f_c)^{0.85}$ |
| JSCE., 2002 | $f_s = 0.44\sqrt{f_c}$ |
| AIJ., 2018 | $f_s = 0.18(f_c)^{0.75}$ |
| BS8007:1987 | $f_s = 0.12(f_c)^{0.7}$ |

In the case of various research papers, the association between compressive strength and split tensile strength can be found in the same format as mentioned above. From the literature it can be observed that when the value of ***V*** is fixed as 0.5 (Table 3), the values of ***A*** ranges from 0.45 to 0.6 with median around 0.56.

*Table 3. Relationship between compressive strength and split tensile strength for concrete with power as 0.5 in various literatures*

| | |
|---|---|
| Gunasekaran et al., 2017 | $f_s = 0.45\sqrt{f_c}$ |
| Smadi et al., 1991 | $f_s = 0.46\sqrt{f_c}$ |
| Wafa and Ashour., 1992 | $f_s = 0.58\sqrt{f_c}$ |
| Huste et al., 2004 | $f_s = 0.55\sqrt{f_c}$ |
| Thomas and Ramasamy., 2007 | $f_s = 0.57\sqrt{f_c}$ |
| Choi and Yuan., 2005 | $f_s = 0.6\sqrt{f_c}$ |

It can also be further observed from literature on conventional concrete (Table 4) that the value of $\boldsymbol{A}$ ranges from 0.106 to 0.47 with the median value of 0.294. With these range of values for $\boldsymbol{A}$, the value of $\boldsymbol{v}$ ranges from -0.37 to 0.95 with the median value at 0.67.

*Table 4. Relationship between compressive and split tensile strength for concrete with various power in literatures*

| | |
|---|---|
| Raphael et al., 1984 | $f_s = 0.313(f_c)^{0.667}$ |
| Gardner., 1990 | $f_s = 0.34(f_c)^{0.66}$ |
| Gardner et al., 1988 | $f_s = 0.47(f_c)^{0.59}$ |
| Arioglu et al., 2016 | $f_s = 0.321(f_c)^{0.661}$ |
| Oluokon et al., 1991 | $f_s = 0.294(f_c)^{0.69}$ |
| Carnio and Lew., 1982 | $f_s = 0.272(f_c)^{0.71}$ |
| Ahamed et al., | $f_s = 0.462(f_c)^{0.55}$ |
| Neville., 1995 | $f_s = 0.249(f_c)^{0.67}$ |
| Arioglu et al., 1985 | $f_s = 0.387f_c(f_c)^{-0.37}$ |
| Selim et al., 2008 | $f_s = 0.106(f_c)^{0.948}$ |
| Lavanya et al., 2015 | $f_s = 0.249(f_c)^{0.772}$ |
| Rashid et al., 1992 | $f_s = 0.47(f_c)^{0.56}$ |
| Xu and Shi., 2009 | $f_s = 0.21(f_c)^{0.83}$ |
| Ramados., 2014 | $f_s = 0.12(f_c)^{0.95}$ |
| Permual., 2014 | $f_s = 0.188(f_c)^{0.84}$ |

## 3.2 Sensitive Assessments

Sensitivity analysis is used to estimate the impact of various values of an autonomous variable on a specific non-autonomous variable under a pre-defined norm. It can also be defined as sensitivity analyses study on several causes of uncertainty in a model that contributes to the mathematical model's complete uncertainty. This method is usually used within exact limitations. These limitations may depend on various types of input variables.

In order to check the confidence of the finest associations in literatures, the MAE (Lavanya and Jegan., 2016) (Equation 2) must be closer to zero, IAE (Arioglu et al., 2016) (Equation 3) must be less than 10% and RMSE (Yan et al., 2013) (Equation 4) must be closer to zero. Finest associations between the strength properties, the lowest value of MAE, IAE and RMSE is considered.

$$MAE = \frac{1}{n}\sum_{i=1}^{i=n}\left|P_i - O_i\right| \quad (2)$$

$$IAE = \sum_{i=0}^{i=n}\frac{\{(O_i - P_i)^2\}^{1/2}}{\Sigma O_i}x100 \quad (3)$$

$$\mathrm{RMSE} = \sqrt{\frac{1}{n}\sum_{i=1}^{i=n}\left[\{O_i - P_i\}^2\right]} \quad (4)$$

Where, $P_i$ is forecast value from relationships in literatures and $O_i$ is experiential value from experiment.

## 3.3 Confidence and Prediction Intervals

In earlier days, research on various types of concrete based on their ingredients have led to various relations between its strength properties. The development of these associations led to proper method of testing of concrete properties including their mechanical properties and determination of new materials that can be used in concrete. Hence, it is vital to develop these relationships between the mechanical properties. This study intentions to estimate the relationship between compressive and split tensile strength in order to predict the application of various concrete with ECC.

Firstly, the impact of water to binder ratio and fibre ratio on the compressive strength of the material was determined (Figures 5 and 6). Similarly, such relationships were formed with respect to split tensile strength for concrete with ECC (Figures 7 and 8). Subsequently, from these relationships, using the least square method the linear egression lines were plotted in order to estimate the best fit experimental data. Further, investigation of residual and adjusted $R^2$ values was done for some regression models in literatures.

The confidence interval and predication interval bands for each of the regression lines were determined and they were plotted together with the most appropriate best fit line. From the confidence intervals, models for the prediction intervals are estimated. Nevertheless, from the specific data points are estimated based on the prediction intervals (Gunasekara and Setunge., 2017). Confidence interval is defined as that there is a 95% probability that the accurate linear regression line of the population will lie inside the confidence interval of the regression line designed from the sample data (Equation 5) (Gunasekara and Setunge., 2017). Predication interval is well-defined as that for a precise value $X_0$, there is a 95% probability that the actual value of Y agreeing to $X_0$ is situated within this interval (Equation 6) (Gunasekara and Setunge., 2017).

$$Y_{Pred} \pm t_{0.05} * S_{yx} * \sqrt{\left[ \left\{ \frac{1}{n} \right\} + \left\{ \frac{\left(X - \bar{X}\right)^2}{SS_x} \right\} \right]} \quad (5)$$

$$Y_{Pred} \pm t_{0.05} * S_{yx} * \sqrt{1 + \left[ \left\{ \frac{1}{n} \right\} + \left\{ \frac{\left(X - \bar{X}\right)^2}{SS_x} \right\} \right]} \quad (6)$$

where,
$Y_{Pred}$ is forecast of Y values;
$t_{0.05}$ is the t critical value for 95% probability interval;
$n$ is number of data points used to establish the regression line;
$\bar{X}$ is the mean value of sample data points;
$SS_x$ is the sum of the squares of standard error of X values and
$S_{yx}$ is standard error in the estimate.

## 4. RESULTS AND DISCUSSIONS

In order to carry out the steps of the methodology as explained above, a detailed examination of the experimental data, observations and formulation of the association between the compressive strength and the split tensile strength is discussed further. Firstly, from the entire data set specified and detailed analysis is conducted based on the impact of ECC on different mechanical (strength) properties of the concrete.

### 4.1 Compressive Strength

#### 4.1.1 Effect of Water to Binder Ratio on Compressive Strength

Generally, in concrete the role of water is to bind all the constituents of concrete in order to arrive at a mixture that ideally performs uniformly. The water in the concrete also tends to widen the space between the material in the concrete and it indirectly affects the compaction (Panda et al., 2020). An optimum water to binder ratio must be used for a cement mixture as this highly affects the compressive strength of the concrete (Chan et al., 2018). When this ratio is lesser than required, it may result in shrinkage cracks and it makes the concrete difficult to work with and it also lowers the compressive strength. On the contrary, when this ratio is increased, it may lead to larger water pores and also bleeding of concrete. This will also reduce the compressive strength of the concrete (www.wecivilengineers.wordpress.com). From Figure 5, it is noted that for increasing water to binder ratio there is lesser compressive strength is noted.

*Figure 5. Influence of water to binder ratio on the compressive strength of ECC*

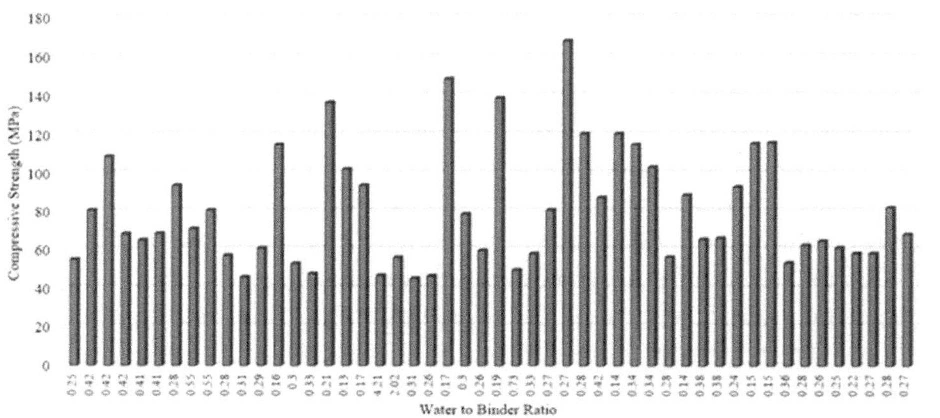


Impact of water to binder ratio on the compressive strength of the ECC can be understood from Figure 5. The water to binder ratio used in various research papers vary from 0.13 to 4.21. It can also be noted that the maximum compressive strength of about 160MPa was achieved when a water to binder ratio of 0.27 was used. Similarly, a low compressive strength of about 50MPa was obtained when the water to binder ratio was kept at 0.31. The most commonly used water to binder ratio across all research papers lies in the range between 0.25 and 0.3. When there is an increase in the water to binder ratio beyond 0.3, a decrease in compressive strength can be observed due to higher water pore penetration. When the water to binder ratio is optimised the pozzolanic reactions due to the hydration of cement takes place which in turn actively increases the compressive strength with the formation of hydrated products.

## 4.1.2 Effect of Fibre Ratio on Compressive Strength

The percentage of fibres used in concrete directly affects the compressive strength and hence for each type of fibre used in ECC, an optimum ratio of fibre must be used in order to attain a higher strength possible. From the various types of fibres used, each type varies based on the properties and influence on compressive and tensile strength of the concrete. From the fibres used in this paper, most commonly adopted ratio of fibres is 2%. If this ratio increases beyond 2%, there is a reduction in the compressive strength (Shanmugasundaram and Praveen Kumar., 2022). This fibre ratio not only affects the compressive strength but also affects the water content to be used and the bonding nature of cement. These factors indirectly affect the compressive strength. From analysis various such results, a fibre ratio of 2% can be recommended when using PVA, PE, GGBS, SF and Fly ash.

*Figure 6. Influence of fibre ratio of compressive strength of ECC*

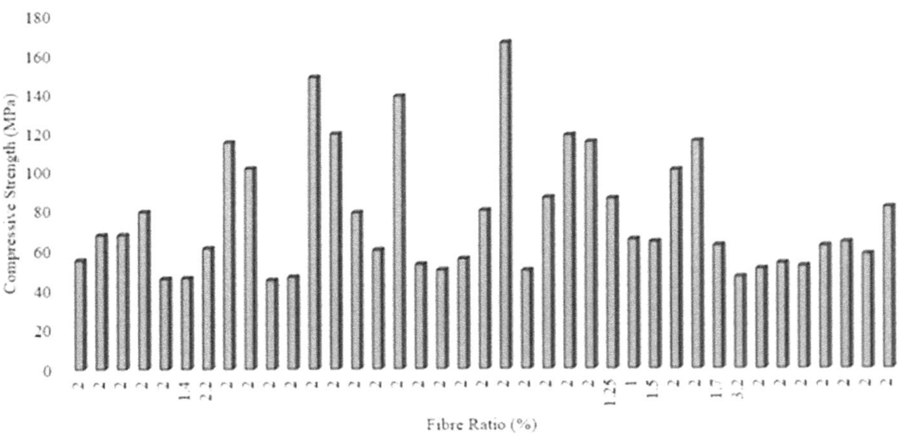


Figure 6 shows the influence of fibre ratio on the compressive strength of the concrete with ECC. The ratio of the fibres used in concrete varies from 1% to 3.2% with the maximum compressive strength of about 160MPa achieved from a fibre ratio of 2%. Furthermore, while considering hybrid fibres i.e., more than one type of fibre used in ECC incorporated concrete the compressive strength varies with different fibre ratios. An optimum ratio of hybrid fibres could not be determined due to varying properties of different kinds of fibres. Several factors that contribute to increase in compressive strength for the cement composites are fibre aspect ratio, load application, types of fibre, characteristics of fibre, etc., One of the major factors that contribute to increase in the compressive strength of composites is the orientation of fibre in the cement composites. If the fibre is present in the composites in such manner that the length of the fibre is present as perpendicular to the load applied on the composites.

## 4.2 Split Tensile Strength

### 4.2.1 Effect of Fibre Ratio on Split Tensile Strength

As it is suggested in most of study that the inclusion of fibres results in increase in split tensile strength of composite, where it presents. One of the main factors that contribute to increase in split tensile strength is the orientation of the fibre in composites. If the fibre is oriented in composite as length of fibre as parallel to the loading applied, there is more chance of increase in split tensile strength. Diameter of the fibre and the tensile characteristics of the fibre plays a vital role for the improvement of split tensile strength. From Figure 7, it is noted that the optimum dosage of fibre ratio used for composites is 2%. This indicates that if there is increase in fibre ratio further, there is reduction in split tensile strength is noted. When fibre ratio is used as lower dosage below 2%, there is no such increase in split tensile strength is noted. This is due to lower dosage does not have influence on the split tensile strength, either due to non-availability of fibre for load transfer or different orientation of fibre.

*Figure 7. Influence of fibre ratio on the split tensile strength of ECC*

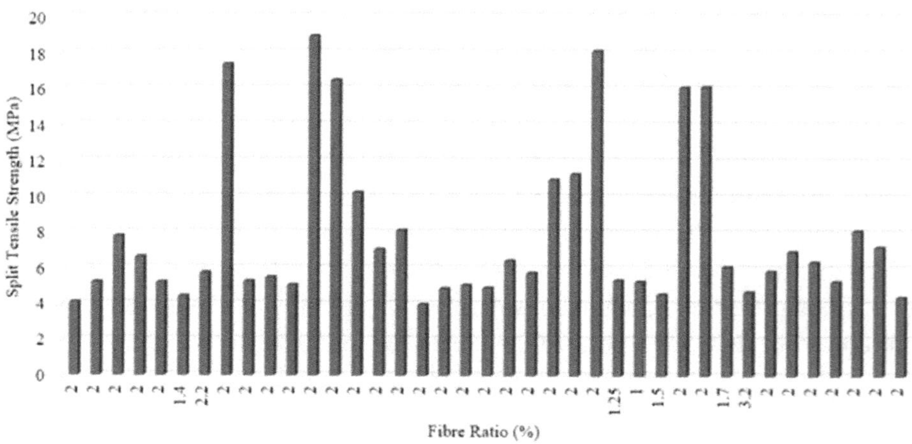

From Figure 7 it can be noted that for the determination of split tensile strength of ECC incorporated concrete the fibre ratio is varied between 1% and 3.2% in various research papers as similar to that of compressive strength. In terms of split tensile strength, the maximum strength of about 18MPa is obtained from a fibre ratio of 2%.

### 4.1.2 Effect of Water and Binder Ratio on Compressive Strength

From Figure 8 the influence of water to binder ratio on the split tensile strength is plotted against varying water to binder ratio from 0.13 to 0.73. The maximum split tensile strength was achieved when the water to binder ratio was kept at 0.17.

*Figure 8. Influence of water to binder ratio on the split tensile strength of ECC*

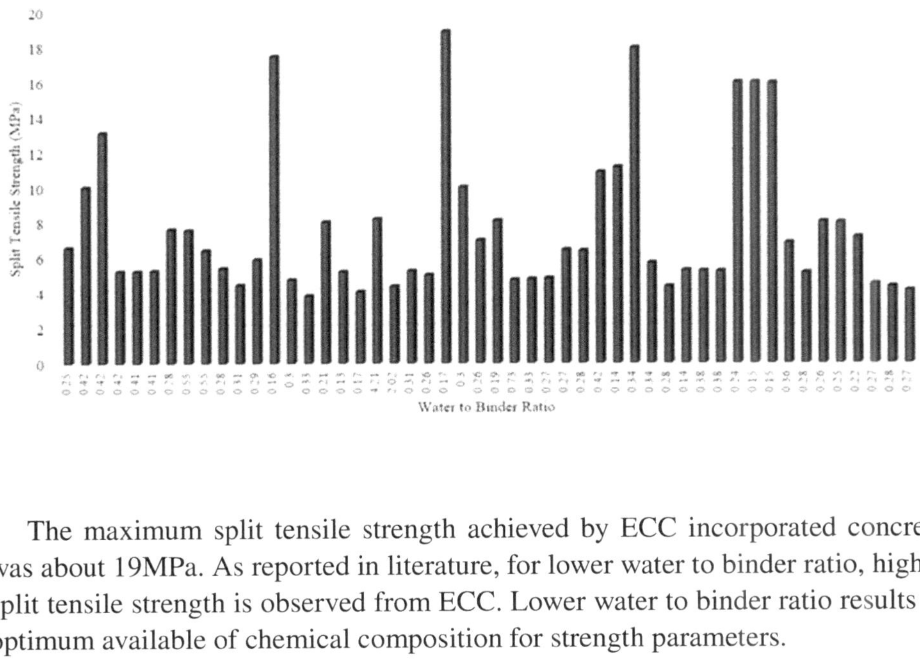


The maximum split tensile strength achieved by ECC incorporated concrete was about 19MPa. As reported in literature, for lower water to binder ratio, higher split tensile strength is observed from ECC. Lower water to binder ratio results in optimum available of chemical composition for strength parameters.

### 4.3 Sensitive Assessments

Relationship available in various international standards for compressive strength and split tensile strength is shown in Figure 9. Sensitive assessment of the various international standards like ACI, CEB-FIB, AS, JCI, JSCE, AIJ, BS, etc., among the compressive strength and split tensile strength indicates that if there is no need of determining all mechanical properties in order to reduce the cost, man power, etc.,

*Figure 9. Sensitive assessment for compressive strength and split tensile strength relationships available in various international standards*

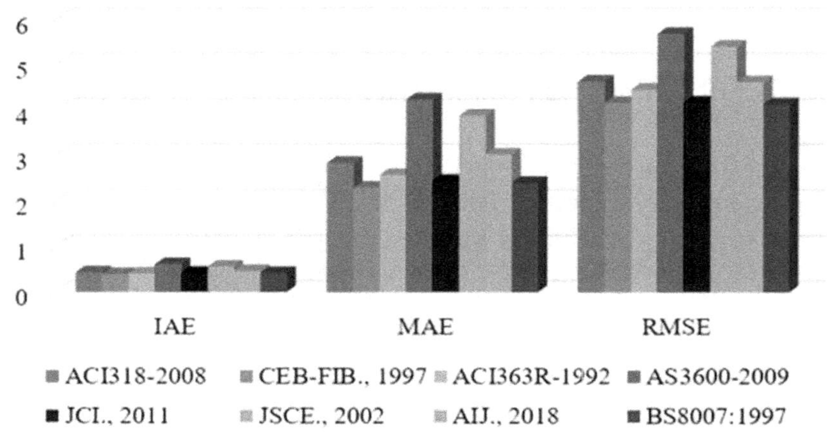


Lower the value of RMSE, IAE and MAE indicates that the model proposed is better. The order of international standards models is in the order as CEB-FIB., 1997<ACI363R-1992< BS8007:2007 < JCI., 2011 < ACI318-2008 < AIJ., 2018 < JSCE., 2002 < AS3600 – 2009 for IAE is noted. The order of international standards models is in the order as CEB-FIB., 1997 < BS8007:2007 < JCI., 2011 <ACI363R-1992 < ACI318-2008 < AIJ., 2018 < JSCE., 2002 < AS3600 – 2009 for MAE is noted. The order of international standards models is in the order as BS8007:2007 < CEB-FIB., 1997 < JCI., 2011 <ACI363R-1992 < AIJ., 2018 < ACI318-2008 < JSCE., 2002 < AS3600 – 2009 for MAE is noted.

Relationship between compressive and split tensile strength relationships with power as 0.5 available in literature are presented in Figure 10. Models from literature are analysed using sensitive assessment. Arrangement of 0.5 power relationship from literature as Choi and Yuan., 2005 < Wafa and Ashour., 1992 < Thomas and Ramasamy., 2007 < Huste et al., 2004 < Smadi et al., 1991 < Gunasekaran., 2017 for the IAE value is noted. For sensitive assessment MAE, the arrangement is Choi and Yuan., 2005 < Wafa and Ashour., 1992 < Thomas and Ramasamy., 2007 < Huste et al., 2004 < Smadi et al., 1991 < Gunasekaran., 2017 is noted. For sensitive assessment RMSE, the arrangement is Choi and Yuan., 2005 < Wafa and Ashour., 1992 < Thomas and Ramasamy., 2007 < Huste et al., 2004 < Smadi et al., 1991 < Gunasekaran., 2017 is noted.

*Figure 10. Sensitive assessment for the compressive and split tensile strength relationships with power as 0.5 available in literature*

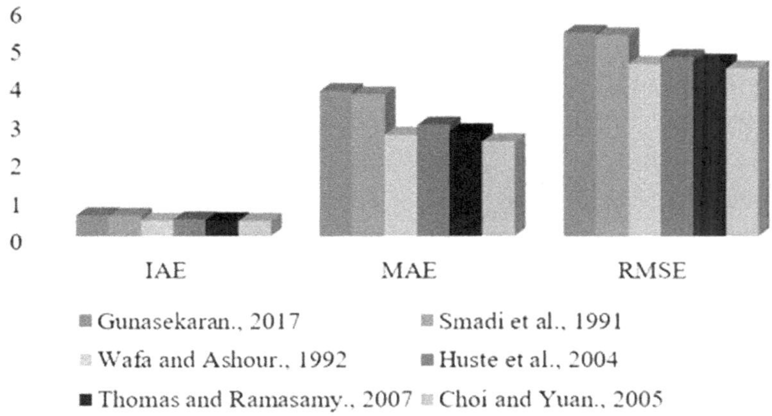


*Figure 11. Sensitive assessment for the compressive and split tensile strength relationships with various powers available in literature*

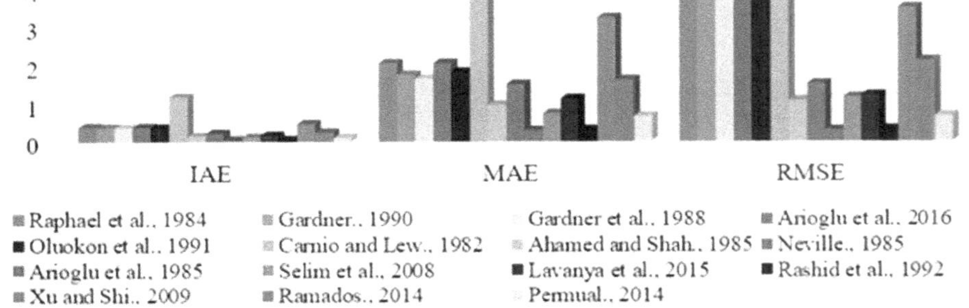

Sensitive assessment for the compressive and split tensile strength relationship with various powers available in literature are reported in Figure 11. Models from literature are analysed using sensitive assessment for various power from literatures are analysed using IAE, MAE and RMSE. Best arrangement of various powers from seven literature are in the order of Arioglu et al., 2016 < Rashid et al., 1992 < Permual., 2014 < Selim et al., 2008 < Ahamed and Shah., 1985 < Neville., 1995 < Ramados., 2014 for IAE. Best arrangement of various powers from seven literature are in the order of Arioglu et al., 2016 < Rashid et al., 1992 < Permual., 2014 < Selim et al., 2008 < Ahamed and Shah., 1985 < Neville., 1995 < Ramados., 2014 for MAE. Best arrangement of various powers from seven literature are in the order of Arioglu et al., 2016 < Rashid et al., 1992 < Permual., 2014 < Selim et al., 2008 < Ahamed and Shah., 1985 < Neville., 1995 < Ramados., 2014 for RMSE.

## 4.4 Confidence and Prediction Intervals

From Figure 12, it is noted that the most of models from international standards are lies with in upper and lower prediction. Relationships like BS 8007:1987, AS3600-2009 and JSCE., 2002 are lies outside the confidence intervals are noted. Even several points are out of the confidence intervals for the experimental values are noted. Relationships from international standards like ACI363R-1992, CEB-FIB-1997, AIJ., 2008, ACI 318-2008, JCI., 2011 are showing higher confidence relationship.

*Figure 12. Confidence and predicted intervals for the split tensile strength relationship available in various standards*

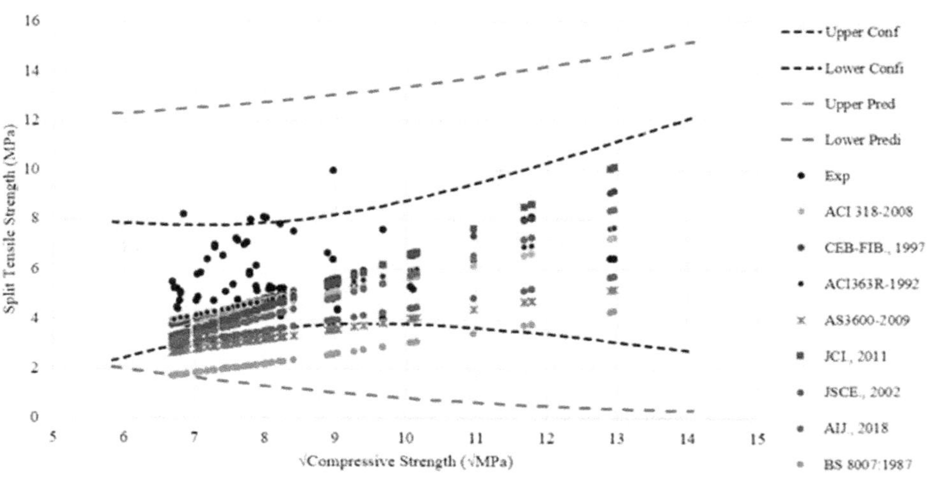

From Figure 13, it is noted that the split tensile and compressive strength with 0.5 power relationship shows better confidence relationship. Similar to Figure 12, experimental value is out of the confidence intervals are noted from Figure 13. Most of 0.5 power relationship between compressive and split tensile strength are lies between confidence intervals. From Figure 14, it is noted that the relationships from literature like Ramdos., 2014, Xu and Shi., 2009, Permual., 2014, Oluokon et al., 1991, Lavanya et al., 2015, etc., showing the relationship are lies outside of confidence interval and even Ramdos., 2014 are even lies outside the prediction interval is noted. Most of the models with various powers from literature are showing the strong relationship between confidence intervals.

*Figure 13. Confidence and predicted intervals for the split tensile strength with power as 0.5 relationship available in literature*

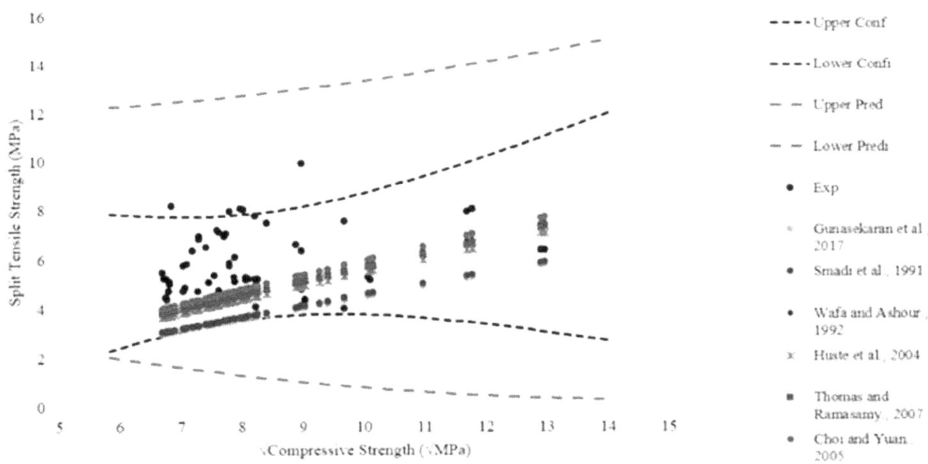

*Figure 14. Confidence and predicted intervals for the split tensile strength with various powers relationship available in literature*

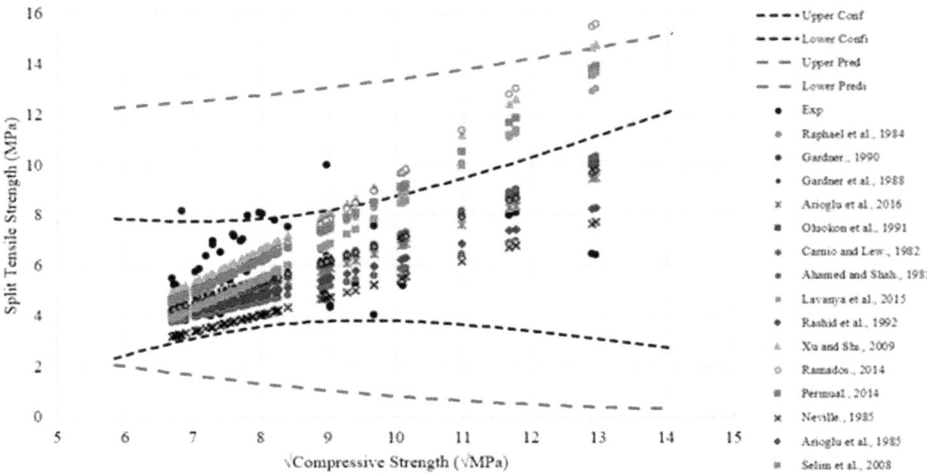


## 4.5 Proposed Models

Regression analysis are performed for the experimental database from literature to find out the relationship between the split tensile and compressive strength. Most of models from literatures shows power relationship, there are two different relationships developed with respect to constant power and different power. Several experimental data from literature are neglected for linear regression analysis. Linear regression analysis was performed for the 72 data from literature as shown in Figure 15.

*Figure 15. The linear relationship between the split tensile and compressive strength after reducing the data to 72*

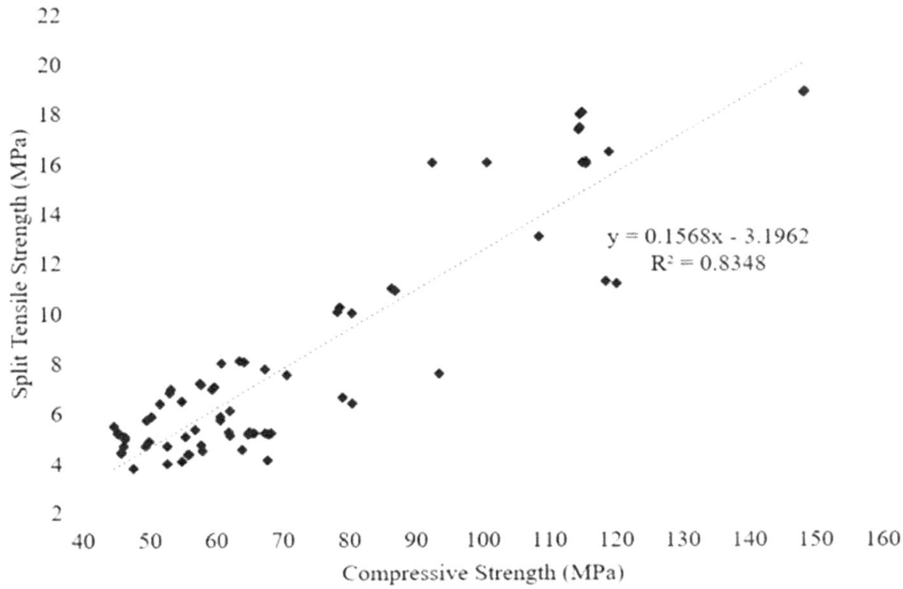


Linear relationship between the split tensile and compressive strength is proposed in Equation (7) with $R^2$ as 0.835.

$$f_{ck} = 0.1568 f_{sk} - 3.196 \tag{7}$$

*Figure 16. The power relationship between the split tensile and compressive strength after reducing the data to 72*

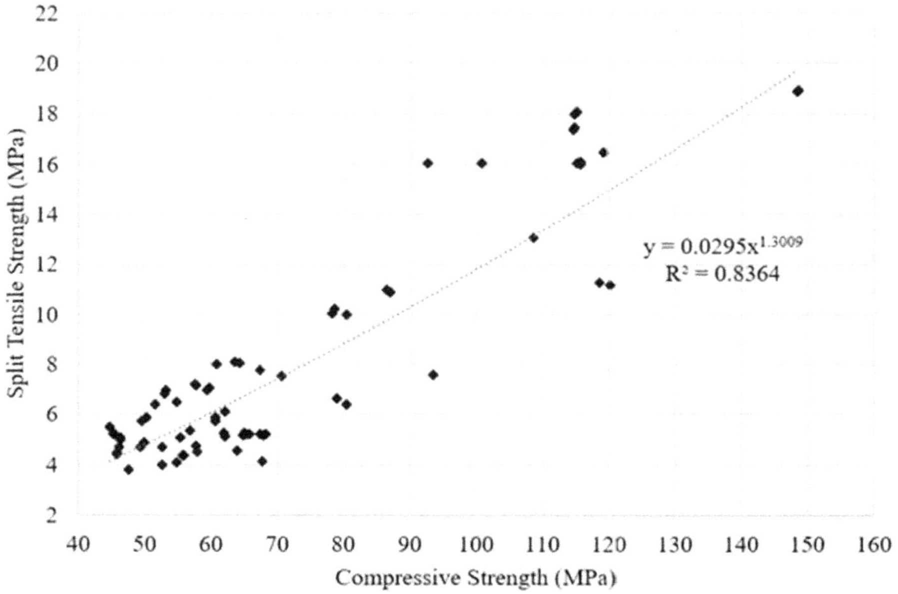


Power relationship between split tensile and compressive strength is developed from equation (8) for the 72 data from literature are shown from Figure 16 with $R^2$ as 0.836.

$$f_{sp} = 0.0295(f_c)^{1.3} \tag{8}$$

From Figure 17, it is noted that the best fit relationship is shown between split tensile and compressive strength is developed from Equation (9) for the experimental database with number of data as 62.

$$f_{sp} = 0.8391\sqrt{f_c} \tag{9}$$

*Figure 17. Best fit power relationship between the split tensile and compressive strength after reducing the data to 62*

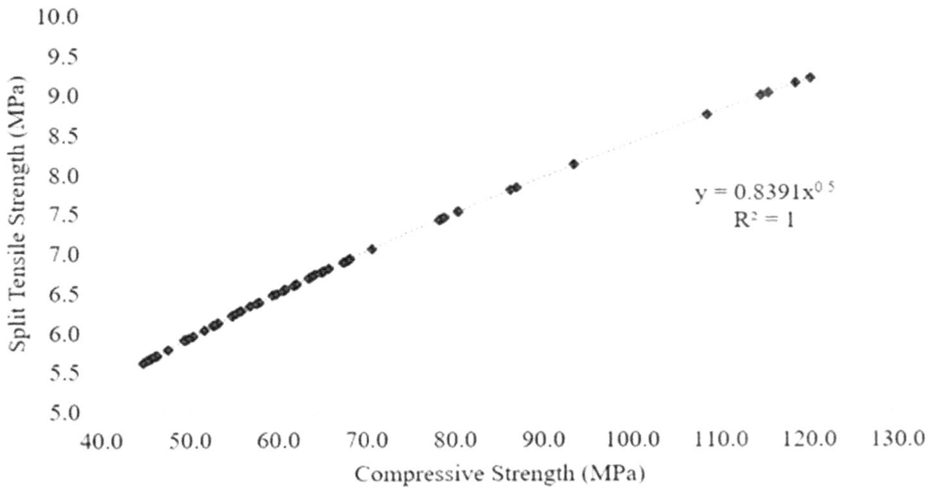


*Figure 18. Best fit linear relationship between the split tensile and compressive strength after reducing the data to 62*

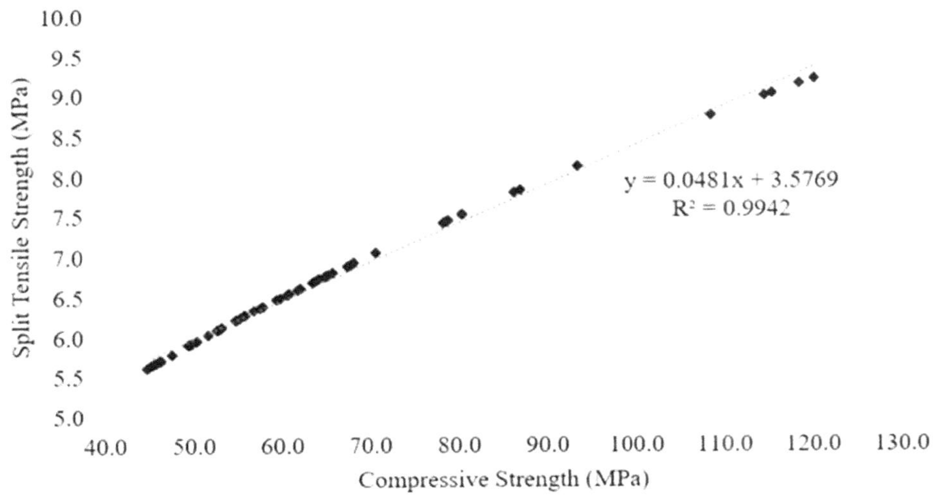

Whereas for the best linear regression relation between the compressive and split tensile strength after reducing the number of the data as 62 and the relationship proposed in the equation 10.

$$f_{ck} = 0.0481f_{sk} + 3.5769 \tag{10}$$

## 4.6 Error Assessments

Error value is estimated that the change between the predicted value and experimental value from literature. The error value for the international standards and proposed models are shown in Figure 19. Highest error value is noted for the equation (8), which indicates that it is worst proposed model. i.e., power relationship shows worst model.

*Figure 19. Error split tensile strength for international standard and proposed models*

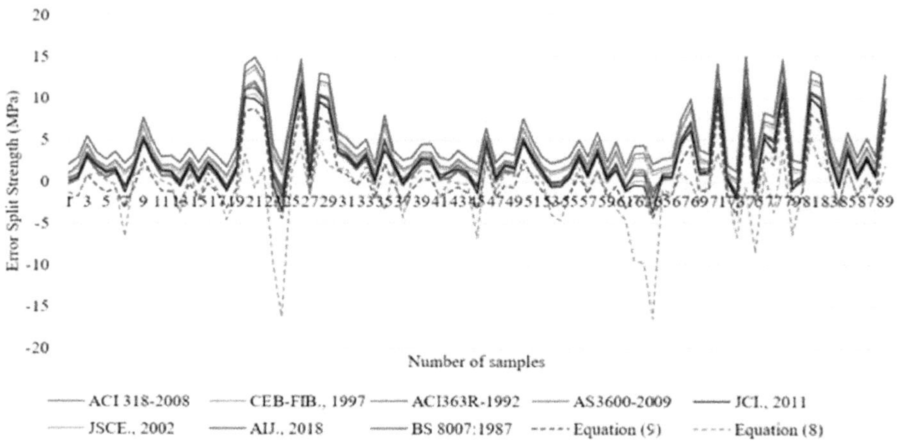

*Figure 20. Error split tensile strength for 0.5 power relationships from literature*

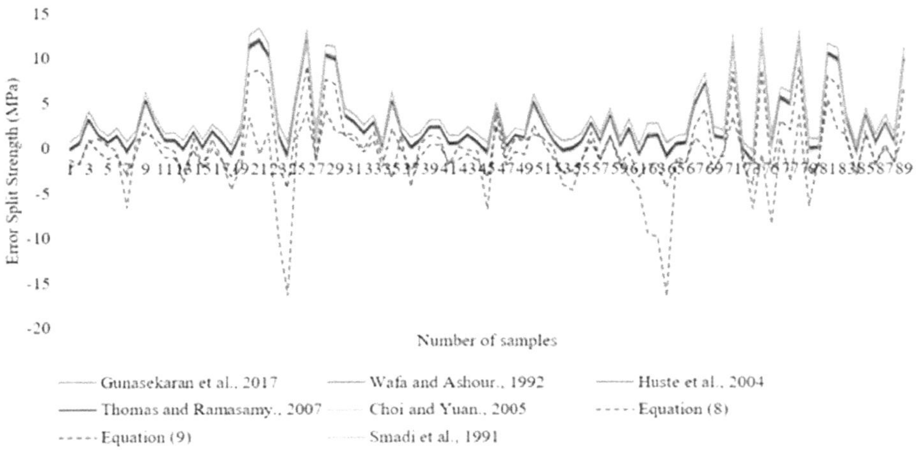


*Figure 21. Error split tensile strength for various power relationships from literature*

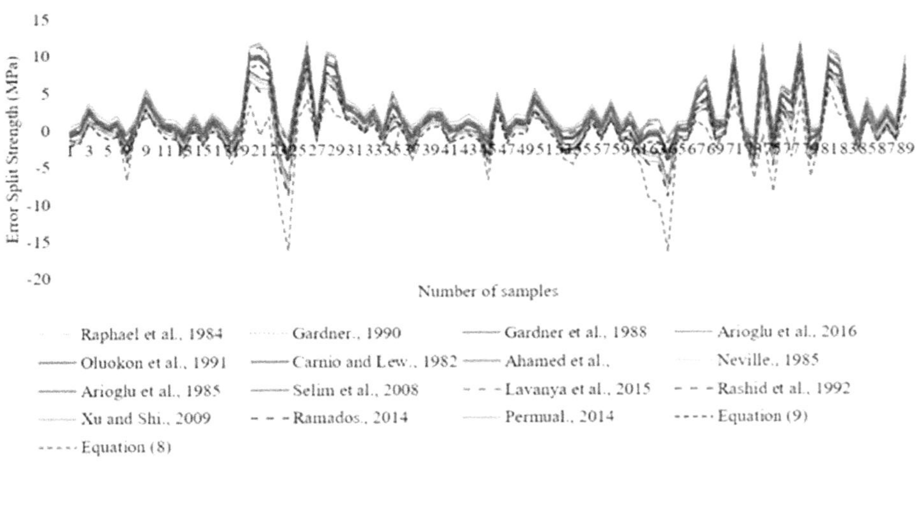


From Figure 20, it is noted that the equation 8 are showing worst poor power relationship. Figure 21, indicates that the equation 8 showing poor power relationship. From error assessment indicates that there is poor relationship is noted for the power equation.

## 5. CONCLUSION

One of the best material suitable to carry both compressive and split tensile strength is ECC. It is necessary to propose a model to develop relation between the compressive and split tensile strength for the ECC in order to avoid cost, material, time, etc., From the developed model of ECC between compressive and split tensile strength, following conclusion can be drawn:

- Strength of ECC is influenced by various factors like fibre aspect ratio, volume of aspect ratio, orientation of load, etc., From which it can be concluded that effect of fibre type and ratio on the strength.
- Strength of ECC is influenced by water to binder ratio, admixture, quality of water, etc. From which it evidently can be concluded that the effect of water to binder ratio on the strength.
- From three different relation between compressive and spilt tensile strength from cement composites were analysed. From this analysis, 0.5 power relationship shows better relationship from literatures and standards.
- Linear regression model and power regression model are proposed in this study. From ECC experimental database, it can be reported that the power regression model shows best model.

From this study, it can be concluded that the best power relationship between compressive and split tensile strength shows better model. From this study, it can be concluded that the cost, money, time, etc., are avoided for the engineers, contractors and scientists who are all involving in the ECC.

## 6. AUTHOR CONTRIBUTIONS

*Conceptualization:* P.Jagadesh *Investigation:* P.Jagadesh, K.S.Amirtavarshini, *Methodology:* P.Jagadesh, K.S.Amirtavarshini, Rebeca Martinez- Garcia *Project Administration:* P.Jagadesh, K.S.Amirtavarshini, Haytham F. Isleem, Rebeca Martinez- Garcia *Resources:* P.Jagadesh, K.S.Amirtavarshini, *Software:* P.Jagadesh, K.S.Amirtavarshini, *Supervision:* P.Jagadesh, *Validation:* P.Jagadesh, K.S.Amirtavarshini, *Visualization* P.Jagadesh, K.S.Amirtavarshini, Haytham F. Isleem, Rebeca Martinez- Garcia, *Writing* – Original Draft: P.Jagadesh, K.S.Amirtavarshini, Haytham F. Isleem, Rebeca Martinez- Garcia, *Writing – review & editing:* P.Jagadesh, K.S.Amirtavarshini, Haytham F. Isleem, Rebeca Martinez-Garcia

## 9. ACKNOWLEDGMENT

The authors acknowledge the Strength of Materials Laboratory and Structural Engineering Laboratory, Department of Civil Engineering, Coimbatore Institute of Technology, Coimbatore, Tamil Nadu, India.

## REFERENCES

ACI Committee 318. (2008). *Building Code Requirements for Structural Concrete (ACI 318 - 08) and Commentary*. American Concrete Institute.

Adesina, A., & Das, S. (2020). Mechanical performance of engineered cementitious composite incorporating glass as aggregates. *Journal of Cleaner Production*, *260*, 121113. doi:10.1016/j.jclepro.2020.121113

Ahmad, S. H., & Shah, S. P. (1985). Structural properties of high strength concrete and its implications for precast prestressed concrete. *PCI Journal*, *30*(6), 92–119. doi:10.15554/pcij.11011985.92.119

AIJ. (2008). Recommendations for Practice of Thermal Cracking Control of Massive Concrete in Building. *Architectural Institute of Japan, AIJ*, *2008*, 69.

Ali, M. A. E. M., Soliman, A. M., & Nehdi, M. L. (2017). Hybrid-fiber reinforced engineered cementitious composite under tensile and impact loading. *Materials & Design*, *117*, 139–149. doi:10.1016/j.matdes.2016.12.047

Arioglu. (1995). Discussion of relationship between splitting tensile strength and the compressive strength. *IMO Technical Journal, 4*, 1059-1062.

AS 3600, Handbook of Reinforced Concrete Deign, Cement Concrete & Aggregates Australia (2009)

BS8007:1987:British Standard Design of Concrete structures for retaining aqueous liquids", British Standards, 1987.

Bo-Tao. (2021). Tensile and compressive performance of high-strength engineered cementitious composites (ECC) with seawater and sea-sand. *RILEM Bookseries, 30*, 1034–1041. . doi:10.1007/978-3-030-58482-5

Carnio, N. J., & Lew, H. S. (1982). Re-Examination of the relation between splitting tensile and compressive strength of normal weight concrete. *ACI Journal Proceedings*, *79*(3), 214–219.

Chan, N., Rojanschi, C. Y., & Li, S. (2018). Effect of water-to-cement ratio and curing method on the strength, shrinkage and slump of the biosand filter concrete body. *Water Science and Technology*, *77*(6), 1744–1750. doi:10.2166/wst.2018.063 PMID:29595177

Choi, Y., & Yuan, R. L. (2005). Experimental relationship between splitting tensile strength and compressive strength of GFRC and PFRC. *Cement and Concrete Research*, *35*(8), 1587–1591. doi:10.1016/j.cemconres.2004.09.010

Deng, H., & Li, H. (2018). Assessment of self-sensing capability of Carbon Black Engineered Cementitious Composites. *Construction & Building Materials*, *173*, 1–9. doi:10.1016/j.conbuildmat.2018.04.031

Ding, Y., Yu, J., Yu, K.-Q., & Xu, S. (2018). Basic mechanical properties of ultra-high ductility cementitious composites: From 40 MPa to 120 MPa. *Composite Structures*, *185*, 634–645. doi:10.1016/j.compstruct.2017.11.034

Ding, Y., Yu, K., & Mao, W. (2020). Compressive performance of all-grade engineered cementitious composites: Experiment and theoretical model. *Construction & Building Materials*, *244*, 118357. doi:10.1016/j.conbuildmat.2020.118357

Ding, Y., Yu, K.-Q., Yu, J., & Xu, S. (2018). Structural behaviors of ultra-high performance engineered cementitious composites (UHP-ECC) beams subjected to bending-experimental study. *Construction & Building Materials*, *177*, 102–115. doi:10.1016/j.conbuildmat.2018.05.122

Fukuyama, H., Sato, Y., Li, V. C., Matsuzaki, Y., & Mihashi, H. (2000). Ductile Engineered Cementitious Composite Elements for Seismic Structural Application. WCEE, 12, 1–8.

Gardner, N. J. (1990). Effect of temperature on the early-age properties of type I, type III and type I/fly ash concretes. *ACI Materials Journal*, *87*(1), 68–78.

Gardner, N. J., Sau, P. L., & Cheung, M. S. (1988). Strength development and durability of concretes cast and cured at 0°C. *ACI Materials Journal*, *85*(6).

Guan, X., Li, Y., Liu, T., Zhang, C., Li, H., & Ou, J. (2019). An economical ultra-high ductile engineered cementitious composite with large amount of coarse river sand. *Construction & Building Materials*, *201*, 461–472. doi:10.1016/j.conbuildmat.2018.12.207

Gunasekara, C., Setunge, S., & Law, D. W. (2017). Correlations between mechanical properties of low-calcium fly ash geopolymer concretes. *Journal of Materials in Civil Engineering*, *29*(9), 1–9. doi:10.1061/(ASCE)MT.1943-5533.0001916

Guo, M., Zhong, Q., Zhou, Y., Hu, B., Huang, Z., & Yue, Y. (2020). Influence of flexural loading and chloride exposure on the fatigue behavior of high-performance lightweight engineered cementitious composites. *Construction & Building Materials, 249*, 118512. doi:10.1016/j.conbuildmat.2020.118512

Gürbüz, E., & Erdem, S. (2020). Development and thermo-mechanical analysis of high-performance hybrid fibre engineered cementitious composites with microencapsulated phase change materials. *Construction & Building Materials, 263*, 120139. doi:10.1016/j.conbuildmat.2020.120139

Himasha, M. W. C., Gamage, J. C. P. H., De Silva, G. I. P., & Attanayaka, V. (2020). Modification of Engineered Cementitious Composite Mortar to use as an adhesive for CFRP / Concrete bond. *11th Int. Conf. Sustain. Built Environ, 1*.

Huang, B. T., Wu, J. Q., Yu, J., Dai, J. G., Leung, C. K. Y., & Li, V. C. (2021). Seawater sea-sand engineered/strain-hardening cementitious composites (ECC/SHCC): Assessment and modeling of crack characteristics. *Cement and Concrete Research, 140*, 106292. Advance online publication. doi:10.1016/j.cemconres.2020.106292

Hueste, M. B. D., Chompreda, P., Trejo, D., Cline, D. B. H., & Keating, P. B. (2004). Mechanical properties of high-strength concrete for prestressed members. *Structural Journal., 101*, 457–465.

JSCE. (2002). Standard specifications for concrete structures-2002. Materials and Construction, Japan Society of Civil Engineers, JSCE.

Kewalramani, Mohamed, & Syed. (2017). Engineered cementitious composites for modern civil engineering structures in hot arid coastal climatic conditions. *Procedia Eng., 180*, 767–774. https://doi.org/.proeng.2017.04.237. doi:10.1016/j

Krishnaraja, A. R., & Kandasamy, S. (2018). Flexural performance of hybrid engineered cementitious composite layered reinforced concrete beams. *Periodica Polytechnica. Civil Engineering, 62*(4), 921–929. doi:10.3311/PPci.11748

Lavanya, G., & Jegan, J. (2015). Evaluation of relationship between split tensile strength and compressive strength for geopolymer concrete of varying grades and molarity. *International Journal of Applied Engineering Research: IJAER, 10*(15), 35523–35527.

Lei, D. Y., Guo, L. P., Chen, B., Curosu, I., & Mechtcherine, V. (2019). The connection between microscopic and macroscopic properties of ultra-high strength and ultra-high ductility cementitious composites (UHS-UHDCC). Compos. Part B Eng., 164, 144–157. doi:10.1016/j.compositesb.2018.11.062

Lepech, M. D., & Li, V. C. (2008). Large scale processing of Engineered Cementitious Composite. *ACI Materials Journal, 105*, 358–366.

Li, M., & Li, V. C. (2012). Rheology, Fiber Dispersion, and Robust Properties of Engineered Cementitious Composites. *Materials and Structures, 46*(3), 405–420. doi:10.1617/s11527-012-9909-z

Li, V. C. (1993). From mechanics to structural engineering - The design of cementitious composites for civil engineering applications. *Structural Engineering / Earthquake Engineering, 10*, 37s–48s.

Li, V. C., Stang, H., & Krenchel, H. (1993). Micromechanics of crack bridging in fiber reinforced concrete. *Materials and Structures, 26*(8), 486–494. doi:10.1007/BF02472808

Ling, Wang, Li, Shi, & Lu. (2019). Effect of slag on the mechanical properties and bond strength of fly ash-based engineered geopolymer composites. *Compos. Part B Eng., 164*, 747–757. https://doi.org/.compositesb.2019.01.092 doi:10.1016/j

Liu, H., Zhang, Q., Li, V., Su, H., & Gu, C. (2017). Durability study on engineered cementitious composites (ECC) under sulfate and chloride environment. *Construction & Building Materials, 133*, 171–181. doi:10.1016/j.conbuildmat.2016.12.074

Ma, H., & Zhang, Z. (2020). Paving an engineered cementitious composite (ECC) overlay on concrete airfield pavement for reflective cracking resistance. *Construction & Building Materials, 252*, 119048. doi:10.1016/j.conbuildmat.2020.119048

Ma, Zhang, Ding, & Tu. (2018). Investigation on the adhesive characteristics of Engineered Cementitious Composites (ECC) to steel bridge deck. *Constr. Build. Mater., 191*, 679–691. https://doi.org/.conbuildmat.2018.10.056 doi:10.1016/j

Meng, D., Huang, T., Zhang, Y. X., & Lee, C. K. (2017). Mechanical behaviour of a polyvinyl alcohol fibre reinforced engineered cementitious composite (PVA-ECC) using local ingredients. *Construction & Building Materials, 141*, 259–270. doi:10.1016/j.conbuildmat.2017.02.158

Minard, A/ (2009). Bendable Concrete Heals Itself -- Just Add Water. *National Geographic News*.

Mohammed, B. S., Achara, B. E., & Liew, M. S. (2018). The influence of high temperature on microstructural damage and residual properties of nano-silica-modified (NS-modified) self-consolidating engineering cementitious composites (SC-ECC) using response surface methodology (RSM). *Construction & Building Materials, 192*, 450–466. doi:10.1016/j.conbuildmat.2018.10.114

Mohammed, B. S., Khed, V. C., & Liew, M. S. (2018). Optimization of hybrid fibres in engineered cementitious composites. *Construction & Building Materials*, *190*, 24–37. doi:10.1016/j.conbuildmat.2018.08.188

Oluokun, F. A., Burdette, K. J., & Deatherage, J. H. (1991). Splitting tensile strength and compressive strength at early ages. *ACI Materials Journal*, *88*(2), 115–121.

Panda, S., Sarkar, P., & Davis, R. (2020). Effect of Water-Cement Ratio on Mix Design and Mechanical Strength of Copper Slag Aggregate Concrete. *IOP Conference Series: Materials Science and Engineering,* 936. 10.1088/1757-899X/936/1/012019

Perumal, R. (2014). Correlation of compressive strength and other engineering properties of high-performance steel fiber–reinforced concrete. *Journal of Materials in Civil Engineering*, *27*(1), 4014114. doi:10.1061/(ASCE)MT.1943-5533.0001050

Pul, S. (2008). Experimental investigation of tensile behaviour of high strength concrete. *Indian Journal of Engineering and Materials Sciences*, *15*, 467–472.

Qin, Zhang, Yin, Di, Xu, & Xu. (2019). Use of high strength, high ductility engineered cementitious composites. *J. Build. Eng., 32*.

Ramadoss, P. (2014). Combined effect of silica fume and steel fiber on the splitting tensile strength of high-strength concrete. *International Journal of Civil Engineering*, *12*, 96–103.

Raphel, J. M. (1984). Tensile strength of concrete. *ACI Journal Proceedings*, *81*(2), 158–165.

Rashid, M. A., Mansur, M. A., & Paramasivam, P. (2002). Correlations between mechanical properties of high-strength concrete. *Journal of Materials in Civil Engineering*, *14*(3), 230–238. doi:10.1061/(ASCE)0899-1561(2002)14:3(230)

Ratnayake, K. A. S. D., Yu, J., Mishra, D. K., & Leung, C. K. Y. (2020). Optimization of low-cost polyvinyl alcohol (Pva) fiber reinforced engineered cementitious composites (ECC) for practical applications. *Indian Concrete Journal*, *94*(6), 39–48.

Sahmaran, M., Lachemi, M., Hossain, K. M. A., Ranade, R., & Li, V. C. (2009). Influence of aggregate type and size on ductility and mechanical properties of engineered cementitious composites. *ACI Materials Journal*, *106*(3), 308–316. doi:10.14359/56556

Said, S. H., & Razak, H. A. (2015). The effect of synthetic polyethylene fiber on the strain hardening behavior of engineered cementitious composite (ECC). *Materials & Design*, *86*, 447–457. doi:10.1016/j.matdes.2015.07.125

Shanmugasundaram & Praveenkumar. (2022). Mechanical Properties of Engineered Cementitious Composites Incorporating different mineral admixtures and fibre – A Review. *Journal of Building Pathology and Rehabilitation*, 1–15.

Shi, L., Lin, S. T. K., Lu, Y., Ye, L., & Zhang, Y. X. (2018). Artificial neural network based mechanical and electrical property prediction of engineered cementitious composites. *Construction & Building Materials*, *174*, 667–674. doi:10.1016/j.conbuildmat.2018.04.127

Siad, H., Lachemi, M., Sahmaran, M., Mesbah, H. A., & Hossain, K. M. A. (2018). Use of recycled glass powder to improve the performance properties of high volume fly ash-engineered cementitious composites. *Construction & Building Materials*, *163*, 53–62. doi:10.1016/j.conbuildmat.2017.12.067

Singh, M., Saini, B., & Chalak, H. D. (2020). Long term evaluation of engineered cementitious composite containing stone slurry powder. *Construction & Building Materials*, *264*, 120183. doi:10.1016/j.conbuildmat.2020.120183

Smadi, M., & Migdady, E. (1991). Properties of high strength tuff light weight aggregate concrete. *Cement and Concrete Composites*, *13*(2), 129–135. doi:10.1016/0958-9465(91)90008-6

Soe, K. T., Zhang, Y. X., & Zhang, L. C. (2013). Material properties of a new hybrid fibre-reinforced engineered cementitious composite. *Construction & Building Materials*, *43*, 399–407. doi:10.1016/j.conbuildmat.2013.02.021

Thomas, J., & Ramaswamy, A. (2007). Mechanical properties of steel fiber-reinforced concrete. *Journal of Materials in Civil Engineering*, *19*(5), 385–392. doi:10.1061/(ASCE)0899-1561(2007)19:5(385)

Wafa, F. F., & Ashour, S. A. (1992). Mechanical properties of high-strength fiber reinforced concrete. *Materials Journal.*, *89*, 449–455.

Wang, G., Yang, C., Pan, Y., Zhu, F., Jin, K., Li, K., & Nanni, A. (2019). Shear behaviors of RC beams externally strengthened with engineered cementitious composite layers. *Materials (Basel)*, *12*(13), 2163. doi:10.3390/ma12132163 PMID:31284446

Wang, Liu, Yu, Dong, & Ye. (2020). Effect of polyethylene fiber content on physical and mechanical properties of engineered cementitious composites. *Constr. Build. Mater.*, *251*. https://doi.org/.conbuildmat.2020.118917 doi:10.1016/j

Xu, B. W., & Shi, H. S. (2009). Correlations among mechanical properties of steel fiber reinforced concrete. *Construction & Building Materials*, *23*(12), 3468–3474. doi:10.1016/j.conbuildmat.2009.08.017

Xu, Yu, Zhou, Bao, & Li. (2021). Effect of curing relative humidity on mechanical properties of engineered cementitious composites at multiple scales. *Constr. Build. Mater., 284*. https://doi.org/.conbuildmat.2021.122834 doi:10.1016/j

Yao, Li, & Lu. (2021). *Development of Engineered Cementitious Composites Using Sea Sand and Metakaolin*. Academic Press.

Ye, B., Zhang, Y., Han, J., & Pan, P. (2019). Effect of water to binder ratio and sand to binder ratio on shrinkage and mechanical properties of High-strength Engineered Cementitious Composite. *Construction & Building Materials*, *226*, 899–909. doi:10.1016/j.conbuildmat.2019.07.303

Yu, J., Wu, H.-L., & Leung, C. K. Y. (2020). Feasibility of using ultrahigh-volume limestone-calcined clay blend to develop sustainable medium-strength Engineered Cementitious Composites (ECC). *Journal of Cleaner Production*, *262*, 121343. doi:10.1016/j.jclepro.2020.121343

Yu, Dai, Lu, & Poon. (2018). Rate-dependent tensile properties of ultra-high performance engineered cementitious composites (UHP-ECC). *Cem. Concr. Compos., 93*, 218–234. https://doi.org/.cemconcomp.2018.07.016 doi:10.1016/j

Yu, Yu, Dai, Lu, & Shah. (2018). Development of ultra-high performance engineered cementitious composites using polyethylene (PE) fibers. *Constr. Build. Mater., 158*, 217–227. https://doi.org/.conbuildmat.2017.10.040 doi:10.1016/j

Zhang, Jaworska, Zhu, Dahlquist, & Li. (2020). Engineered Cementitious Composites (ECC) with limestone calcined clay cement (LC3). *Cem. Concr. Compos., 114*. https://doi.org/.cemconcomp.2020.103766 doi:10.1016/j

Zhang, Z., Yuvaraj, A., Di, J., & Qian, S. (2019). Matrix design of light weight, high strength, high ductility ECC. *Construction & Building Materials*, *210*, 188–197. doi:10.1016/j.conbuildmat.2019.03.159

Zhenbo, W., Jun, Z., Jiahe, W., & Zhengjie, S. (2015). Tensile performance of polyvinyl alcohol-steel hybrid fiber reinforced cementitious composite with impact of water to binder ratio. *Journal of Composite Materials*, *49*(18), 2169–2186. doi:10.1177/0021998314542450

Zhou, J., Qian, S., Beltran, M. G. S., Ye, G., Van Breugel, K., & Li, V. C. (2010). Development of engineered cementitious composites with limestone powder and blast furnace slag. *Materials and Structures*, *43*(6), 803–814. doi:10.1617/s11527-009-9549-0

Zhou, Y., Gong, G., Huang, Y., Chen, C., Huang, D., Chen, Z., & Guo, M. (2021). Feasibility of incorporating recycled fine aggregate in high performance green lightweight engineered cementitious composites. *Journal of Cleaner Production*, *280*, 124445. doi:10.1016/j.jclepro.2020.124445

Zhou, Xi, Sui, Zheng, Xing, & Li. (2019). Development of high strain-hardening lightweight engineered cementitious composites: design and performance. *Cem. Concr. Compos., 104*. doi: .cemconcomp.2019.103370 doi:10.1016/j

Zhu, Zhang, Wang, Wu, & Li. (2020). Mechanical and self-healing behavior of low carbon engineered cementitious composites reinforced with PP-fibers. *Constr. Build. Mater., 259*. https://doi.org/.conbuildmat.2020.119805 doi:10.1016/j

# Chapter 6
# Durability Properties of Engineered Cementitous Composites (ECC)

**Özlem Üstündağ**
*Department of Civil Engineering, Materials Division, Istanbul University-Cerrahpasa, Istanbul, Turkey*

## ABSTRACT

*The brittle structure of concrete and the crack formation easily under various chemical and physical effects adversely affect its durability against long-term effects under difficult environmental conditions. Unlike ordinary concrete materials, tight microcrack formation under tensile loads, superior strain hardening behavior, and self-healing properties provide ECC high durability under harsh environmental conditions. These properties also make ECC a unique repair material used to renovate damage to existing structures. This study presents a detailed review of ECC's durability properties under aggressive environmental effects such as shrinkage, freeze-thaw, high temperature, water permeability, sulfate, chloride, sea water, acid, abrasion, and its use as a repair material under these harsh environmental conditions and makes recommendations.*

## INTRODUCTION

The brittle feature of cement-based composites can be improved by adding high-strength fibers to the composites and a ductile structure can be obtained. ECC is a high-strength cement-based composite reinforced with fibers. They show high ductility under load (Turk and Demirhan, 2013). The design of engineered cement-

DOI: 10.4018/978-1-6684-8182-0.ch006

based composites is based on micromechanical principles. Fiber bridging, crack initiation and crack development are taken into account in the design. (Sherir, 2012). ECC has much higher ductility and deformation behavior than conventional and fiber-reinforced concrete (Li and Wang, 2002; Li, 2008; Ma et al., 2021). As seen in Figure 1, ECC also takes much higher tensile loads than conventional and fiber-reinforced concrete. Tensile strength and deformation capacity increase with the formation of many cracks under load.

The strain capacity of ECCs under tensile loads varies between 3%-5% (Lepech and Li, 2005). This value is several hundred times that of standard concrete (Li, 2012; Zhou et al., 2018).This high strength capacity of ECCs is due to the formation of many tight microcracks in the samples (Sherir, 2012). ECCs show a similar level of ductility to metals. They also perform very well in durability tests. Unlike normal concrete, the small crack width formed as a result of strain hardening in ECC contributes positively to the durability of the composite (Sahmaran and Li, 2007).

ECC can be defined as a type of material with different functionalities as well as having common features such as tensile strength, tensile ductility and fine multiple cracking in a certain range. ECCs have types such as self-consolidating, high early strength, lightweight, self-healing. The most commonly used fiber in ECC production is polyvinyl alcohol fiber (PVA). In addition to PVA, high modulus polypropylene (PP) and polyethylene fibers (PE) are also used (Li, 2008; Yang and Li, 2010).

*Figure 1. Comparison of the behavior of ECC under tensile stress with plain and fiber reinforced concrete (Ye et al., 2019)*

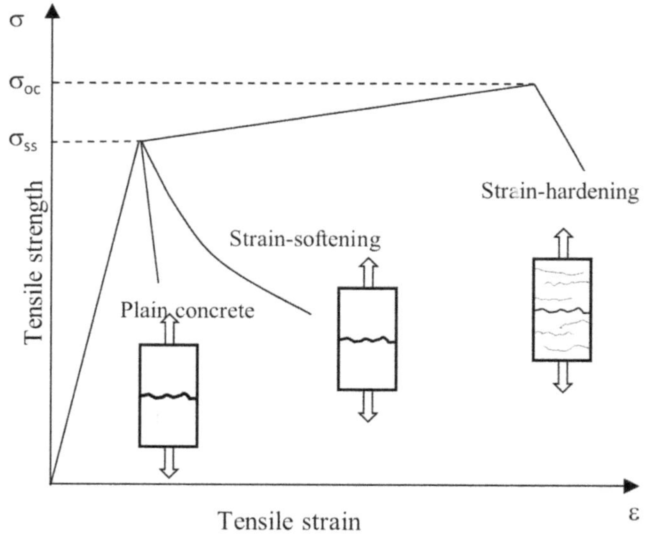

In ECC design, fiber is generally used at 2% by volume. The composite exhibits a highly ductile behavior, with numerous microcracks forming under tensile loading (Li, 2008). Figure 2 shows the crack width development during strain in the inelastic region. After the first crack in the matrix, deformation hardening is observed in the material like a ductile metal (Turk and Demirhan, 2013). Strain hardening in ECC occurs with the sequential development of multiple microcracks, and even at ultimate load (fracture load), the crack opening is up to 80μm. In ECC, the crack width at equilibrium is independent of the loading type, type and amount of reinforcement, unlike normal concrete or (FRC). (Sherir, 2012).

*Figure 2. Crack width development of ECC (Li, 2008)*

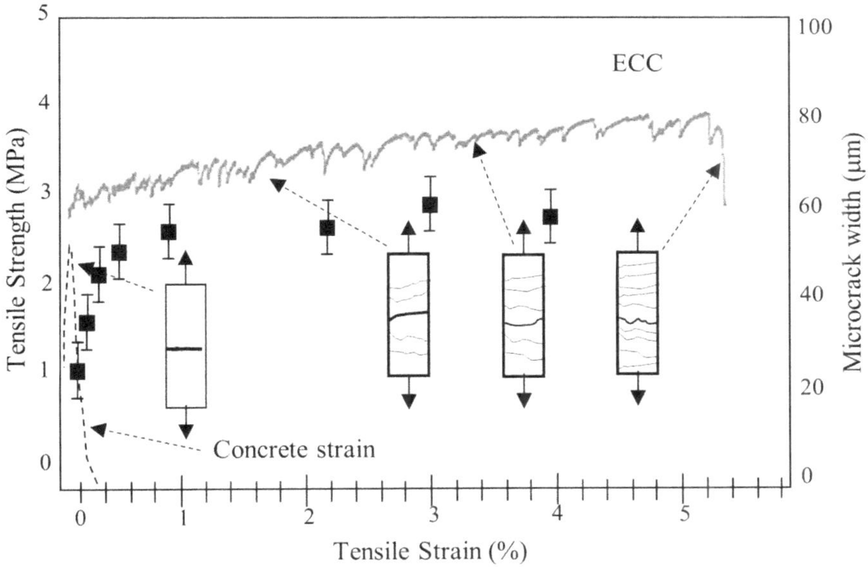


The structural defect caused by cracking in ECCs is less than in normal or fiber-reinforced cementitious composites (Li, 2003). In normal concretes, there is no crack width control, the cracks formed reduce the strength and rigidity and accelerate the entry of harmful chemicals, thus deterioration such as corrosion occurs (Turk and Demirhan, 2013). This makes ECCs more durable to the effects of difficult environmental conditions.

## ECC DESIGN CONSIDERATIONS

ECCs are designed based on the basic fundamentals of micromechanics, with the aim of achieving high ductility and the formation of numerous micro-sized cracks (Li, 1993). Meanwhile, strain hardening behavior is ensured under tensile load. In this way, it is ensured that large strains are distributed among the multiple microcracks that occur. The formation of a large number of cracks is dependent on the crack toughness and bridging stress of the matrix (Sahmaran and Li, 2008). To understand the basic mechanism of strain hardening in ECCs, it is first necessary to understand the fiber bridging property of cracks and their load carrying and energy absorbing effects. For this purpose, providing both strength criterion and fracture energy criterion in the mortar matrix is necessary for the realization of multiple microcrack formation and strain hardening behavior. These conditions can be expressed with equation 1 (Sherir, 2012).

Strength criterion

$$\sigma_o \geq \sigma_{cs} \tag{1a}$$

Fracture energy criterion

$$J_{b'} = \sigma_o \delta_o - \int_0^{\delta_o} \sigma(\delta)\, d\delta \geq \frac{K_m^{\ 2}}{E_m} = J_{tip} \tag{1b}$$

where $\sigma o$ and $\sigma cs$ $_r$epresent the maximum bridging stress and crack strength at the crack surface $\sigma SS$ $_{re}$presents the steady-state stress, while δo r$_e$presents the crack width corresponding to $\sigma o$. T$_h$e complementary energy and fracture energy are denoted by Jb an$_d$ Jtip, $_{re}$spectively. Km an$_d$ Em re$_p$resent the fracture toughness and elastic modulus of the mixture, respectively.

Besides ensuring the above two criteria, an ECC mixture should be designed so that the crack widths in the matrix are less than 100 μm (Sahmaran and Li, 2008). Equation 1a expresses the strength criteria for initiating the first micro crack in the matrix. It is important to ensure that the tensile load does not exceed the maximum fiber bridging capacity. Equation 1b fracture energy criterion represents the crack propagation mode (Sherir, 2012). Figure 3 shows the relationship between the maximum crack bridging stress $\sigma o$ an$_d$ the maximum steady-state crack width δo. I$_f$ the crack width exceeds δo, t$_h$e crack bridging stress will begin to decrease, and multiple crack formation will not occur. To evaluate the durability of an ECC material, it is important that δo re$_m$ains less than 100 μm. At this time, multiple crack formation can occur while maintaining the required tight crack width in the material. Using the basic micromechanical principles outlined above, the ECC can be designed to be subject to large deformations without compromising low permeability.

The self-controlling tight microcracks and self-healing properties provide ECC superior durability even under mechanical loads and difficult environmental conditions. These properties also make ECC an ideal repair material for repairing concrete cracks and damage in existing structures. Unlike the review articles in the literature, this book chapter offers a comprehensive research on the durability properties of ECC mixtures, their use as repair materials and sustainability. The durability properties of ECC mixtures include their behavior under harsh environmental effects such as shrinkage, freezing-thawing, high temperature, permeability, sulfate, sea water, acid and corrosion.

*Figure 3. σ-δ curve of ECC (Li et al., 2004)*

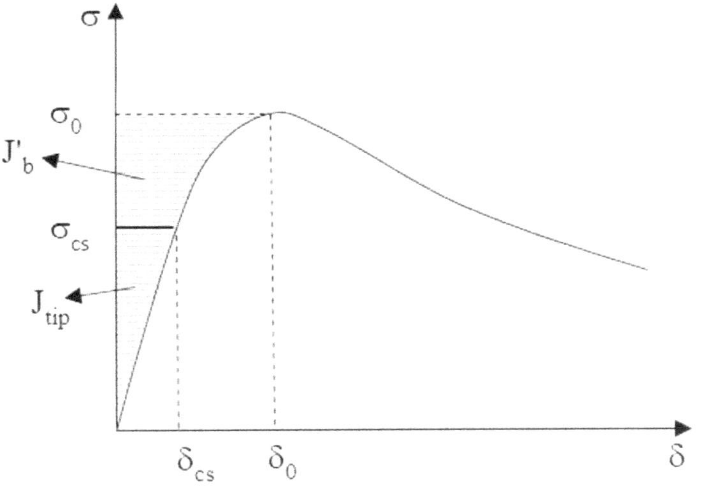


The self-controlling tight microcracks and self-healing properties provide ECC superior durability even under mechanical loads and difficult environmental conditions. These properties also make ECC an ideal repair material for repairing concrete cracks and damages in existing structures. In this book chapter, a detailed review of the ECC durability studies in the literature is presented under different environmental effects such as shrinkage, freeze-thaw, high temperature, permeability, sulfate, sea water, acid, abrasion and its use as a repair material under these difficult environmental conditions.

## SHRINKAGE EFFECT

The mixture of ECCs does not contain coarse aggregates, resulting in higher shrinkage deformations compared to normal concrete. Therefore, shrinkage plays a crucial role in evaluating the durability of ECCs. In their study, Yang et al. (2007) attempted to address the disadvantage of ECC in comparison to normal concrete by utilising a high percentage of fly ash (85% by weight) instead of cement. As the fly ash content increases, both crack width and drying shrinkage decrease significantly. Thus, ECC specimens have long term durability. When the FA/C ratio is used as high as 5.6, the drying shrinkage is reduced by half. The decrease in shrinkage with increasing fly ash content is shown in Figure 4.

*Figure 4. Relationship between fly ash content and drying shrinkage of ECC (Yang et al., 2007)*

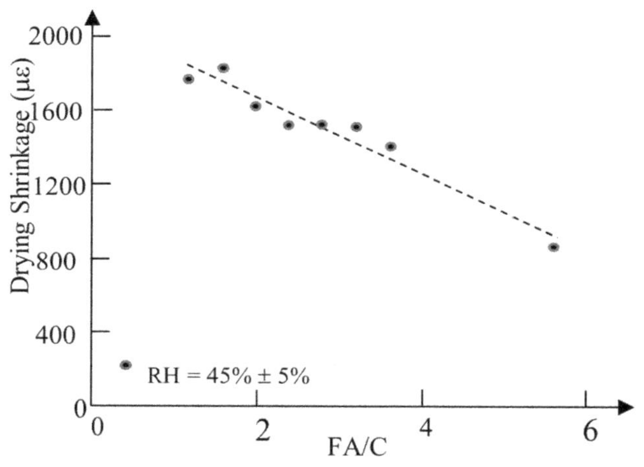


High strength ECCs (HSECC) generally have low water/binder and sand/binder ratios. When the W/B ratio is in the range of 0.13-0.24 and the S/B ratio is 0.3-0.9, the chemical shrinkage is much higher than the autogenous and drying shrinkage. As both ratios are increased, autogenous shrinkage becomes comparable to drying shrinkage. In Figure 5, total shrinkage values, including all three shrinkage types of HSECC, are given according to W/B and S/B ratio (Ye at al., 2019).

*Figure 5. Variation of total shrinkage of HSECC (Ye et al., 2019)*

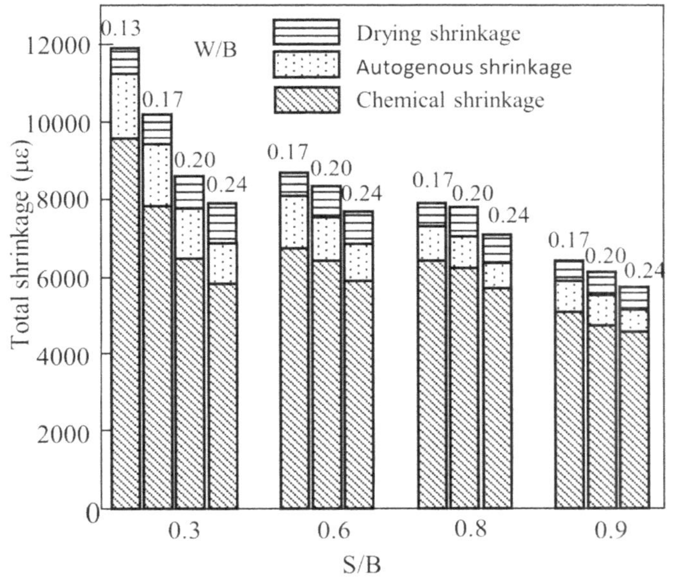


Wang et al. (2020) tried to solve the high shrinkage problem observed in ECC, which negatively affects its durability by using zeolite instead of quartz sand at a rate of 15%, 20%, 30% by weight. Accordingly, the total shrinkage occurring in ECC samples decreases with the amount of zeolite used. The lowest shrinkage was obtained from the mixture containing 30% zeolite.

Sahmaran et al. (2009a) investigated the effects on shrinkage by using 10% and 20% by weight of light fine aggregate - LWA instead of silica sand in ECC mixtures. The ratio of fly ash to Portland cement (FA/OPC) was chosen as 1.2 and 2.2 for both LWA contents. If the light fine aggregate content is 20%, both autogenous and drying shrinkage decrease. However, ductility and strength are negatively affected. In addition, with increasing FA content, autogenous and drying shrinkage decreases, which also has a positive effect on ductility. In another study, it is stated that utilizing glass aggregate in five different ratios up to 100% instead of silica sand in ECC was effective in reducing shrinkage. As the glass aggregate content increases shrinkage decreases and glass aggregate can be used completely instead of silica sand (Adesina and Das, 2021). Figure 6 shows the variation of autogenous and drying shrinkage with glass content.

Wang et al. (2023a) used different W/C ratios in ECC mixtures. Accordingly, they experimentally investigated the relationship between shrinkage and relative humidity. As the w/c ratio of ECC mixtures increases, the shrinkage strain decreases.

The low water/cement content in plain concrete samples causes lower drying shrinkage than ECC mixtures containing fly ash and slag. 15% higher shrinkage is observed in the ECC-SL mixture compared to the ECC-FA mixture. The high shrinkage values observed in ECC containing slag cause a higher number of cracks and crack widths than in FA-containing and plain concrete (Hooshmand et al., 2021).

*Figure 6. a) Autogenous and b) drying shrinkage of 28 day ECC mixtures (Adesina & Das, 2021)*

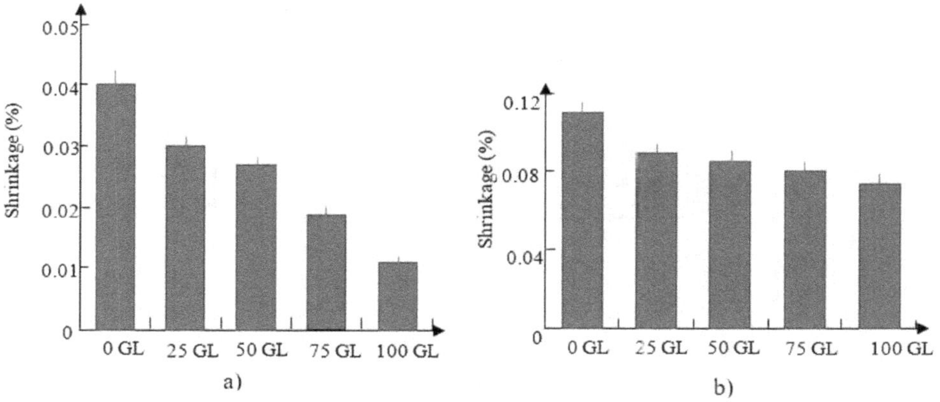


Wang et al. (2023b) investigated the effect of adding 1%-10% calcined zeolite on the drying shrinkage of lightweight engineering cement-based composites (LECCs). In the production of composites, fly ash cenospheres (FACs) were used as fine aggregate and polyethylene (PE) fiber as fiber. Drying shrinkage decreases by more than 30% compared to the control group with the addition of 2.5% zeolite.

## FREEZING-THAWING EFFECT

The freeze-thaw cycle negatively affects engineering applications in cold regions. The resistance of cement-based composites to the freezing-thawing cycle is accepted as an indicator of durability. ECC is designed to exhibit high durability characteristics under tensile load and normal service load. ECC mixing ratios have been optimized to meet the strength criterion in order to achieve high ductility (Li et al., 2001; Li, 2003). In ECCs fiber type, amount, size, matrix components and interface properties are optimized for numerous microcracking and controlled crack opening. This situation provides a great advantage in terms of both ductility and durability, depending on the tight crack width under applied load (Özbay et al., 2013).

Şahmaran et al., (2009b) subjected the ECC samples they prepared using cement and fly ash to a freeze-thaw cycle. The flexural strength of ECC samples decreases slightly after 300 freeze-thaw cycles. However, its ductility under tensile load is many times higher compared to ordinary concrete. Li et al., (2003) applied a freezing-thawing cycle for 14 weeks to normal concrete samples and ECCs and compared the results obtained. Accordingly, concrete samples were severely damaged and deteriorated at the end of 110 cycles (5 weeks), while ECC samples were intact after 300 cycles.

Özbay et al. (2013) investigated the negative effect of preloading on the frost durability of ECCs, as in some engineering applications in cold regions. For this purpose, ECC samples were prepared by replacing F class fly ash (FA) and slag (S) with cement at a ratio of 1.2 and 2.2 by weight. From this point on, flexural loading was applied to the prism samples and preloaded to form microcracks in the specimens. Preloaded intact samples were subjected to a f-t cycle in accordance with ASTM C666 standard Method A. Accordingly, the flexural strength, deformation capacity, ultrasonic pulse velocity and rigidity of ECC mixtures exposed to f-t cycles decrease, while mass loss increases when compared to the samples without f-t cycle. In addition, the crack width of ECC increases 1.5 times after f-t cycles. The authors reported that this may indicate a different state of the bonds formed at the fiber-matrix interface. Preload deformation negatively affects the freeze-thaw resistance of ECC samples. Increasing the FA ratio in ECC mixtures increases the damage that occurs in samples after freeze-thaw cycles. Nevertheless, it is clear that ECC samples which are applied freeze-thaw cycles are quite durable.

Sahmaran et al. (2012) investigated the effect of high volume fly ash (FA) content on the microstructure and freeze-thaw resistance of ECCs. ECC mixtures, which they prepared by choosing the ratio of fly ash (FA) to cement (FA/C) 1.2 and 2.2, were exposed to 300 freezing-thawing cycles in accordance with ASTM C666 Method A. ECC blends with a fiber-free FA/C ratio of 2.2 deteriorated after 60 cycles, and ECC blends with an FA/C ratio of 1.2 after 210 freeze-thaw cycles, while ECC samples containing both FA/C ratios did very well even after 300 cycles with lower mass loss. have performed. In a study investigating the effect on frost resistance of adding certain amounts of silica fume and blast furnace slag to ECCs using high amounts of fly ash (FA), 200 freeze-thaw cycles were applied to ECC samples. As the amount of FA used increases, the dynamic elasticity modulus of ECCs increases and more mass loss occurs. The frost resistance of ECCs was significantly improved by the addition of silica fume and blast furnace slag. Moreover, the toughness of ECCs increases even after 200 freeze-thaw cycles. (Liu et al., 2018).

Hooshmand et al. (2021) examined the cracking behavior of normal concrete and ECCs with different fly ash (FA-ECC) and slag (SL-ECC) contents under freeze-thaw cycles for seven months. Larger amounts of CSH products were recorded in FA-ECC than in SL-ECC, and the healing products and healed cracks formed in FA-ECC made it more durable than SL-ECC in difficult environmental conditions such as freezing-thawing cycles.

Zhu et al. (2012) studied the self healing properties of ECCs, which they subjected to two different freezing-thawing cycles: water and deicing salt freezing-thawing. For this purpose, they concluded that the higher the preload deformation level of the preloaded ECC specimens at deformation levels varying between 0.5-3.0%, the lower the self-healing level. The self-healing ability of ECCs subjected to a freeze-thaw cycle in water was better than the ECC samples tested in a freeze-thaw cycle in de-icing salt. In other study, the authors reported that self-healing ECCs with preloaded deformation levels less than 1.5% could recover as much as the original ECC specimens without any damage to their mechanical properties (Sahmaran and Li, 2007).

The micropore and mesopore numbers of ECC samples prepared using three different water/binder ratios (0.24, 0.28 and 0.32) increase with the applied freeze-thaw cycle. However, as expected, there is a decrease in tensile and compressive strength. In Figure 7, the variation of porosity according to the number of f-t cycles of ECC samples with e different W/B ratios is shown (Liu et al., 2022).

*Figure 7. Variation of porosity with the freezing-thawing cycles (Liu et al., 2022)*

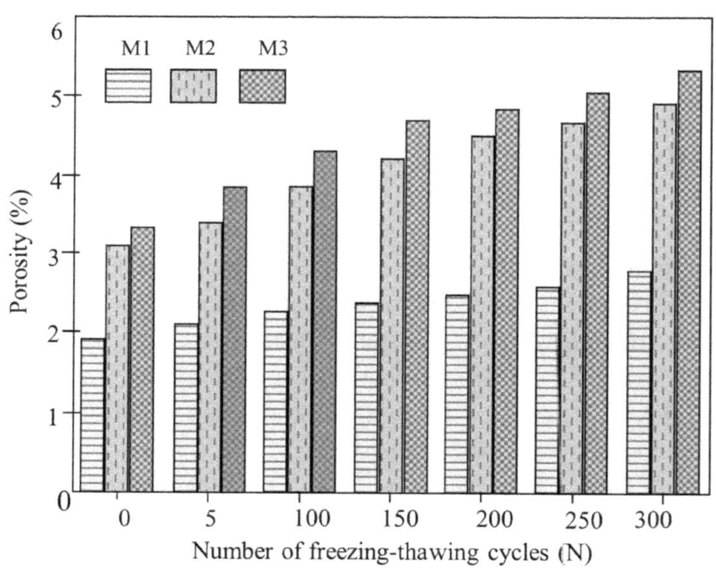

The electrochemical impedance method can be used to determine the degree of freeze-thaw damage of ECCs. According to Bai et al. (2023), ECC exhibited superior freeze-thaw performance compared to ordinary concrete when tested with this method. The porosity of ECC samples was found to increase as the number of freeze-thaw cycles increased.

## HIGH TEMPERATURE EFFECT

High temperature causes flaking, disintegration and decomposition on the outer surface of the concrete. This situation leads to loss of capacity of the structure by direct exposure of the steel reinforcement to high temperature (Sahmaran et al., 2010). With the increasing use of ECCs, their performance under the effect of high temperature, which is an important durability parameter. Sahmaran et al. (2010) stated that mechanical strengths were not changed of ECC samples exposed to 400°C for one hour. PVA fibers melt at 200-400°C, resulting in additional channels and pores in the mixtures. Although there was a significant decrease in the compressive strength and hardness of ECC samples exposed to higher temperatures for 1 hour, their mechanical properties were close to or better than polypropylene or steel fiber reinforced normal concrete. While no explosive fragmentation was observed in any ECC sample under the influence of high temperature, PVA fiber is considered to be effective at this point. Figure 8 shows the relationship between compressive stress and strain of ECC samples.

ECCs with a very high fly ash content maintain their strain hardening properties when exposed to temperatures of 20-200°C. However, the tensile capacity and strain capacity of the composite decrease at temperatures of 200°C and above, and increase at lower temperatures (Yu et al., 2015). Figure 9 shows the relationship between initial cracking strength and temperature. Initially, the first cracking strength increases with temperature and reaches its maximum value of 100°C.

*Figure 8. Compressive stress-strain curves of ECC specimens after exposure to high temperatures (Sahmaran et al., 2010)*

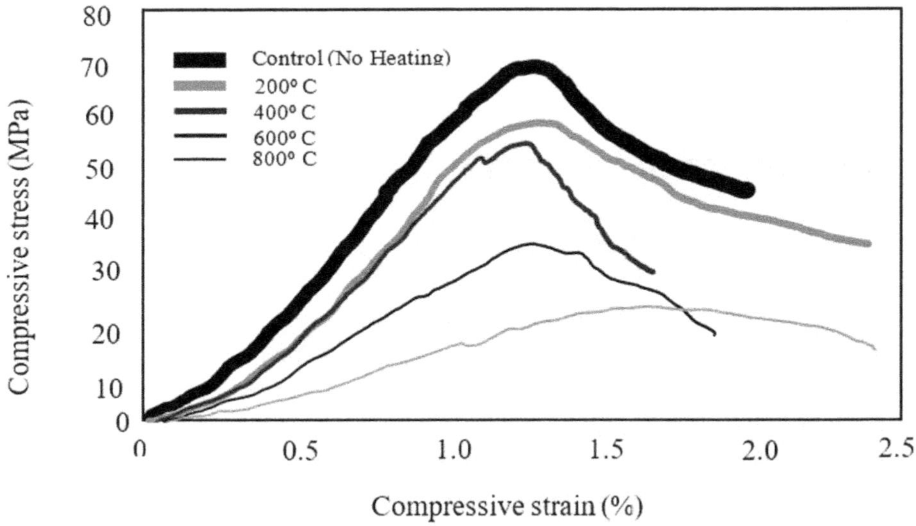


ECC specimens prepared with different fly ash ratios showed comparable or better tensile than the ECC sample with higher fly ash content under 20-200°C temperature. Compressive strength increases for both groups. ECC matrix toughness increases under 200°C temperature, PVA fiber strength deteriorates, so the tensile properties of composites deteriorate (Zhang et al., 2020).

*Figure 9. Cracking strength of ECCs after subjected to various temperatures (Yu et al., 2015)*

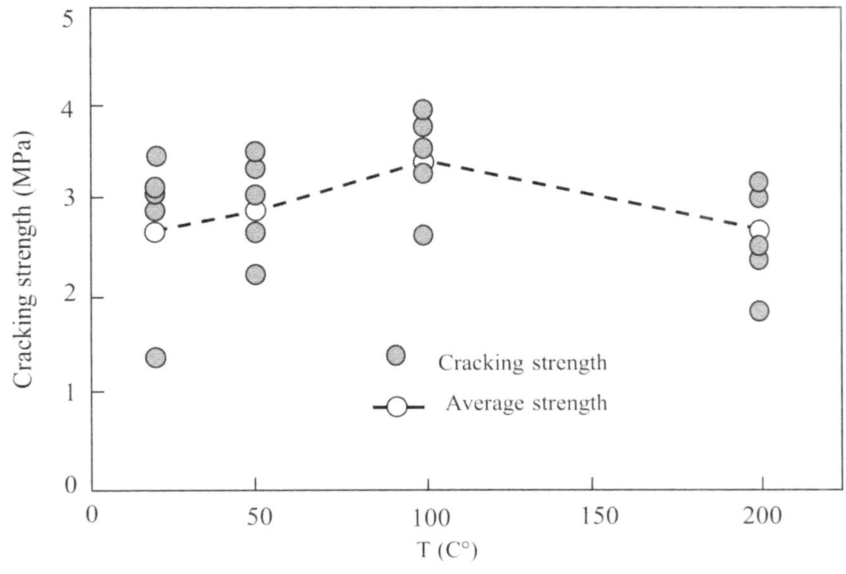


Bhat et al. (2014) reported that the uniaxial tensile strength and strain capacity of ECC specimens applied at temperatures between 20-600°C deteriorated at high temperatures. At temperatures of 200°C and below, matrix toughness increases, while tensile strength slightly increases by 1%. At temperatures higher than 200°C, the ECC tensile strain behavior deteriorates and the tensile strength decreases by 40%. While no disintegration is observed in composites kept at 600°C for 6 hours, severe explosive spalling occurs in control samples exposed to the same temperature for 30-75 minutes.

Based on the loss of crack bridging property and ductility of synthetic fibers used in ECC production under high temperature, the new generation ECCs (BF-ECC) developed using basalt fiber show perfect strain hardening behavior compared to conventional ECCs (Xu et al., 2021).

Aslani and Wang (2019) produced a lightweight ECC by replacing 40% of the hollow glass microspheres with fly ash by volume and added carbon nanofiber at 0.2% and 0.4% by volume of the total cement to preserve the mechanical strengths of the composite. This ECC retained 50% of its strength when exposed to 900°C. If hollow glass microspheres are used, the compressive strength decreases slightly.

Silica sand has a higher coefficient of thermal conductivity than fly ash cenospheres (FACs). To prepare ECC mixtures with high fire resistance, fly ash and silica sand were replaced with 40%, 60%, and 80% fly ash cenospheres (FACs). As the FAC ratio increases, the compressive strength and thermal conductivity coefficients of ECC mixtures decrease. At the same time, the addition of FAC to mixtures delays heat transfer in real fire tests and reduces thermal conductivity by more than 15% (Ma et al., 2023).

## PERMEABILITY RESISTANCE

The permeability resistance is a parameter of the durability of a material, and the tight crack width of the ECC even in the cracked state ensures that this material has low permeability (Liu et al., 2017a). ECC material with multiple microcracks less than 60µm in width has a similar level of permeability to uncracked concrete and lower than cracked concrete. The self-healing feature of cracks developing in ECC is effective in reducing the permeability coefficient. At the same time, it is important to keep the maximum crack opening width in ECCs below the 100µm threshold. This allows more than one cracking behaviour to be exhibited while maintaining crack widths small enough to limit permeability (Lepech and Li, 2009).

Ma et al., (2014) examined the relationship between the self-healing and permeability properties of ECCs with a certain level of damage by exposing them to the water-chamber cycle. After 10 healing cycles, the water permeability of the material degrades to zero and a significant improvement is observed in mechanical strengths.

Liu et al. (2016) conducted an analytical study that can determine the permeability of ECC depending on the tensile stress and crack width. Again, Liu et al. (2017a) described the permeability of ECCs by a third-order exponential function and confirmed it with the results of experiments.

In hybrid ECCs with 0%, 0.3% and 0.6% steel fiber by volume, the use of steel fiber causes finer cracks to develop. When 0.6% steel fiber is added, the crack width decreases from 67.3-115.8µm to 46.0-67.6µm. The permeability of cracked ECC decreases when steel fibers are added (Wang et al., 2023c).

Zhang et al. (2021) studied the influence of crystalline admixtures (CA) on the strength, healing ability and permeability of ECCs. With the addition of CA, the self-healing behavior and mechanical strengths of ECC such as compressive strength and ultimate tensile strength are improved. The healing and closure rate of cracks increases with the CA amount and the permeability coefficient decreases significantly. Figure 10 shows the variation of the permeability of ECC samples with various CA amounts such as 0%, 1.5%, 3% and 4.5%. The permeability of the composites decreased considerably with the addition of CA and FA to the mixtures.

The ECC samples prepared by adding 0%, 1.5%, 3% and 4.5% of the crystalline waterproof material were exposed to environments such as water, air and saturated $Ca(OH)_2$ solution and it was stated that the mechanical properties were improved and the water permeability was decreased (Zhang et al., 2023a).

*Figure 10. The variation of the permeability of ECCs with various CA contents (Zhang et al., 2021)*

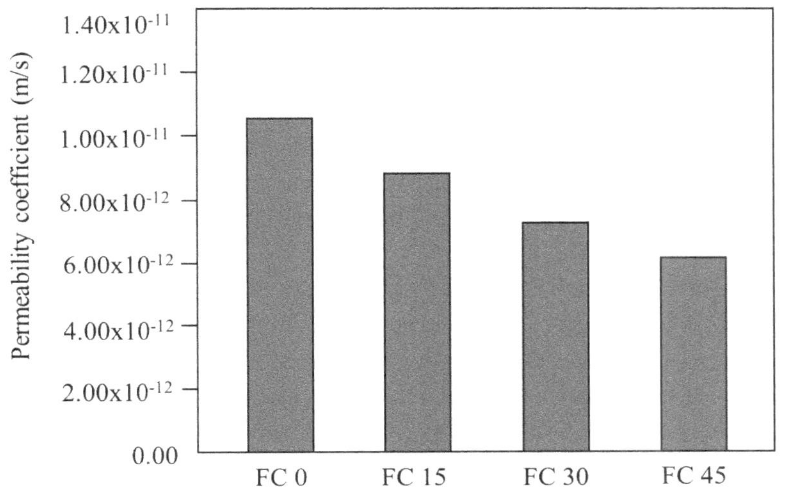


Zhao et al. (2023) produced various ECC mixtures with high fly ash content by replacing silica sand with crushed rubber in certain proportions. The fly ash to cement ratio in the mixtures varied between 1.2 and 4. Silica sand was replaced with rubber at volumes of 15%, 25%, and 35%. Mixtures containing rubber exhibited lower permeability at low strain levels, while mixtures containing high amounts of fly ash exhibited lower permeability at high strain levels.

## HAZARDOUS CHEMICAL EFFECTS

Concrete is indurable to strong acids or acid forming agents attack. The acid effect in concrete is observed as the separation of hydration products and the soluble ones from the concrete, and the fragmentation of the insoluble ones from the concrete (CEB, 1992). Acid attack has the same destructive effect on cementitious materials such as ECC. For this reason, it has been the subject of many studies.

The pore structure of ECCs under the influence of acid mine drainage cycle becomes coarser, compressive strength and mass decrease. At the same time, the tensile strength and strain capacity of ECC decrease. Self-healing of microcracks in ECC is reduced in an acidic environment. For this purpose, MgO is added to ECC mixtures to improve their self-healing ability. Thus, it increases the resistance of ECCs to acid attack and restores their self-healing behavior (Wu et al., 2020). In another study, the compressive strength and modulus of ECC samples to which micro- and nano-sized silica were added decreased after exposure to sulphuric acid. Permeability also decreases after micro- and nano-sized silica were added to the mixes (Mansoori and Behfarnia, 2020).

Wang et al. (2022) investigated the durability of ECCs by simulating ECC samples exposed to sewage environments as sulfuric acid exposure. The compressive strength of the ECC samples was 65% higher and the tensile strength was 300% higher than the control sample (without PVA fiber) after simulated environments. At the same time, ECC specimens retained their self-healing property while retaining 2% tensile ductility.

Chloride ions can enter concrete in various ways, such as chloride-containing aggregates, seawater or salty groundwater, and chloride diffusion is a significant parameter influencing the durability of concrete. High chloride content causes deterioration and corrosion of concrete (Shumuye et al., 2022).

On the other hand, sulfate reacts with calcium hydroxide and causes crack formation, softening, expansion and spalling in the concrete. So, it causes deterioration of concrete and decreases its strength. It also causes an important durability problem by causing reinforcement corrosion (Abdul-Rahman et al., 2020).

Li and Li (2011a) examined the durability of ECC samplesby preloading the composites, keeping them in chloride solution and reloading them until they collapsed. Accordingly, ECC specimens preserved multiple microcracking property and tensile strain capacity greater than 2.5%, and the crack width increased from 50μm to 100μm in the medium. The self-healing properties, initial material stiffness and deformation capacity of the composites kept in the chloride solution were recovered. Taking all these results together, it can be concluded that ECC is durable under chloride exposure. Similarly, a significant self-healing feature is observed in the cracks of ECC samples kept in NaCl solution, thanks to the formation of microcracks (Sahmaran et al. 2007).

In adverse environmental conditions such as sulfate and chloride ions, the ECC performance is not adversely affected, and micro-cracks show a self-healing feature and cause a partial improvement in their mechanical properties. So, it can be stated that ECCs are durable in difficult conditions such as chloride and sulfate in order to improve the long-time performance of structures (Liu et al., 2017b). Liu et al.

(2017c) reported in another study that ECC samples maintained their durability and mechanical properties even after being exposed to sulfate and chloride for a long time.

ECCs produced using polyvinyl alcohol (PVA) and steel fiber were subjected to chloride, sulfate and sulfate-chloride combined conditions. ECC mixtures prepared with hybrid fibers were almost unaffected by chloride and sulfate-chloride exposure compared to water. However, a remarkable deterioration was observed in the samples after exposure of the composites to concentrated sulfuric acid. In addition, the flexural and uniaxial tensile strength of ECC containing 1.5% PVA and 0.5% steel fiber gave better results than conventional ECC samples (Sridhar, 2022).

Seawater contains both chloride and sulfate. In structures exposed to sea water, many effects such as chloride, sulfate, crystallization of salts in concrete pores, tides, freezing-thawing, wetting-drying, abrasion can be seen together. The physical damage observed in concrete begins with efflorescence, followed by aggregate separation and concrete fragmentation, all of which lead to loss of mechanical performance in concrete (Ting et al., 2021).

Pre-damaged ECC specimens based on fly ash /slag which exposed to the simulated marine environment exhibit more intense deterioration of concrete and corrosion in tidal zones due to high chloride content and tidal effects. Previously damaged slag/fly ash based ECC samples recover their mechanical characteristics due to their self-healing properties (Shumuye et al., 2022).

Zhang et al. (2022) examined the effect of limestone calcined clay (LCC), fly ash (FA) and polypropylene (PP) fiber addition on ECCs produced using seawater. Adding 2% PP by volume and using 25% LCC to the mixtures increase the toughness of the composite. At the same time, their compressive and flexural strengths give better results than ECCs produced using only sea water. Compared to traditional Portland cement, ECCs added FA and LCC and prepared with seawater give better results. Both FA and LCC solidify chloride ions in seawater. The use of seawater gives better results than pure water in terms of strength, deformation and toughness at an early stage. Increased compressive strength is obtained in the long term from composites using pure water. The use of FA reduces the compressive and flexural strength of composites prepared with seawater. However, the use of LCC gave positive results in composites prepared with seawater, especially at an early age, it provides a faster development of the strength of LCC seawater composites.

Zhu et al. (2023a) produced cost-effective and low-carbon ECCs using sea water and sea sand by replacing basalt (BA) fiber with Polyvinyl alcohol (PVA) fiber at the rate of 0.1%, 0.2%, and 0.3%. By partially replacing PVA fibers with BA fibers, the strength, toughness and tensile strength of ECC decrease. This is because BA fibers do not have as effective crack bridging properties as PVA fibers. SS-ECCs prepared with sea water and sea sand have better compressive, bending and tensile strengths than RS-ECCs produced using tap water and river sand. Chloride ions accelerating the reaction of high volume fly ash may be the reason for this situation. Figure 11 shows the comparison of the mechanical performance of two ECC types.

*Figure 11. Mechanical behaviour of two types of ECC (Zu et al., 2023)*

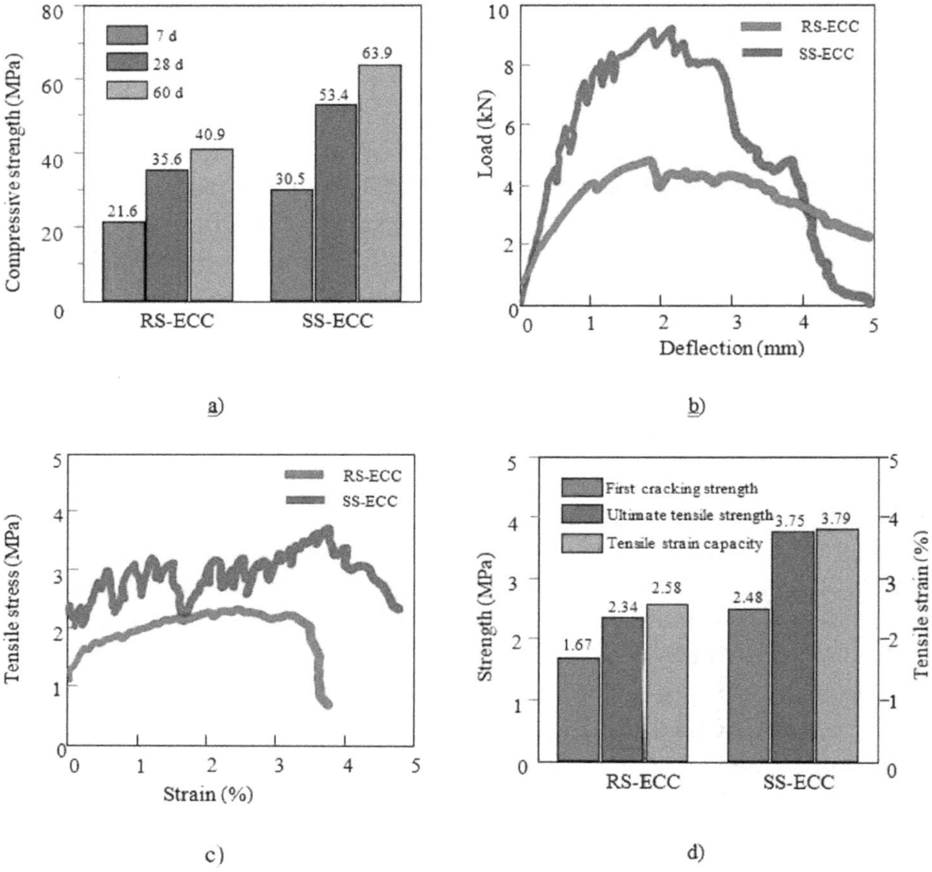


## ABRASION RESISTANCE

There have been very limited studies on the abrasion of ECCs. As ECC mixes do not contain coarse aggregates and contain much finer silica sand than normal concrete, the abrasion behaviour of ECC can be expected to be different from that of normal concrete or FRC samples. (Lepech and Li, 2010; Wu and Li, 2017; Wu et al., 2018; Pranav et al., 2020).

Wu et al. (2018) reported that among the ECC samples they prepared with different PVA fiber contents (1% E1, 2% E2, etc.), the abrasion of the sample with 3% PVA fiber content by volume (E3) was similar to concrete with the equivalent compressive strength value. Figure 12 shows the required number of revolutions for the same depth of abrasion of ECC mixtures.

*Figure 12. The variation of the abrasion depth with the revolution number for the ECC and concrete specimens (Wu et al., 2018)*

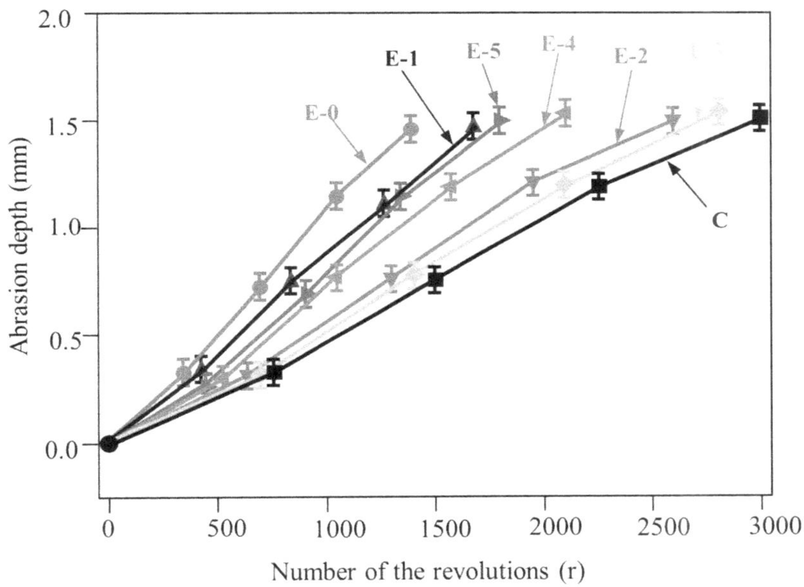


In composites with different PVA content such as 0.5%, 1.0%, 1.5% and 2% by volume, abrasion resistance increases as PVA fiber content increases. While the 1 cm thick ECC coating is not effective at early ages, the 3 cm thick ECC coating shows good abrasion resistance in both early and mature ages (Abid et al., 2018).

## CORROSION RESISTANCE

Corrosion of steel bars in concrete causes premature failure of many structures before reaching their design life and causes a significant durability problem for reinforced concrete structures. For this reason, significant repairs and rehabilitation are needed in the buildings (Hime, 1993; Elsener, 2002; Li, 2014). When concrete is poured around the steel bars in reinforced concrete structures, a passivation layer forms around the reinforcement bars and this protects the rebars from corrosion (Li, 2014). However, chloride ions, sulphate, sea water, acids pass through the cracks in the concrete and cause corrosion by removing the passivity of the protection layer in the reinforcements over time. This causes a loss of diameter in the steel reinforcement and reduces the load carrying capacity of the reinforced concrete system. Considering all these reasons, preventing or delaying corrosion is an important issue. At this point, ECC will cause longer service time with the development of many fine fractures.

Sahmaran et al. (2008) examined the crack widths, flexural strengths and mass losses of rebars of steel-reinforced ECC beams exposed to electrochemical accelerated corrosion. They compared their results with mortar samples with the same compressive strength. The crack widths of the mortar samples due to corrosion increased with the progression of corrosion, and cracks up to 2 mm wide were formed at high corrosion levels. In ECC samples, even though the corrosion progressed, the crack widths remained constant at the level of 0.1 mm. Considering the maximum crack-opening value of 0.3 mm for reinforced concrete buildings specified by AASHTO (2004), the service time of reinforced ECC will be 15 times longer than that of reinforced mortar. The flexural strength of the reference beam decreases to 34% of the initial value after 25 hours of impact to corrosion. However, the flexural strengths of ECC specimens remain the same even after 50 hours of corrosioning, retaining more than 45% of normal capacity after 300 hours of exposure. In Figure 13, the appearance of ECC and normal mortar samples after corrosion exposure is presented.

*Figure 13. a) ECC sample after 350 hrs of corrosion exposure, b) mortar sample after 95 hrs of corrosion exposure (Sahmaran et al., 2008)*

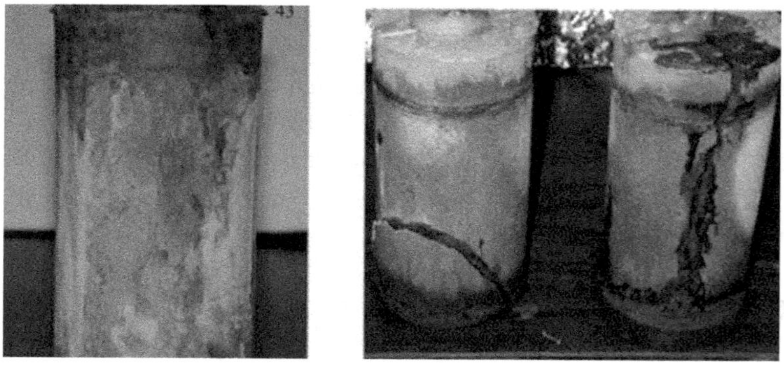

Maalej and Li (1995) developed a new design called Functionally Grade Concrete (FGC) by replacing the reinforcement elements in reinforced concrete beams with ECC material, as seen in Figure 14. With this new design, the ECC material surrounding the reinforcement bars in the FGC beams prevents the penetration of harmful chemicals into the concrete, thus preventing corrosion and when corrosion begins, accelerated corrosion caused by longitudinal cracks is reduced. So, spalling and delamination damages observed in reinforced concrete structures are prevented.

*Figure 14. Functionally grade concrete beam (FGC) (Maalej & Li, 1995)*

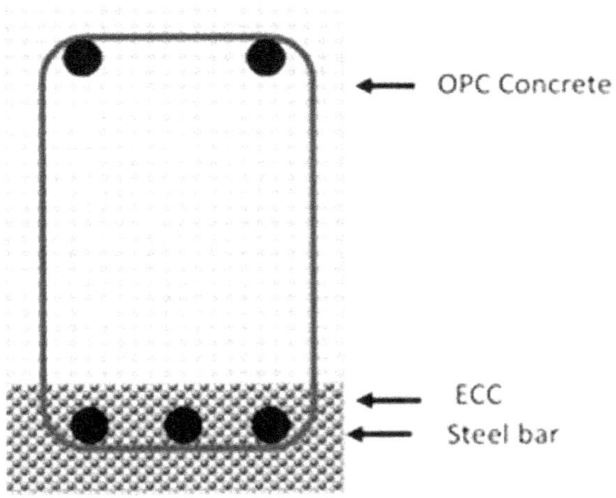


Maalej et al. (2003) prepared an FGC beam by surrounding the reinforcing bars with a hybrid fiber ECC produced using 1% steel and 1.5% PVA fiber by volume. The effect of the FGC design and the RC beam (OPCC) prepared from ordinary Portland cement in delaying the corrosion of the steel bars in the beams and minimising the delamination of the concrete pavement was compared. As seen in Figure 15, less reinforcement diameter loss occurs in FGC beams than in OPCC beams. The same reinforcement diameter reduction in the FGC beam takes 70% longer than in the OPCC beam. While delamination, increase in crack width and spalling occurred in the corroded OPCC beam due to corrosion, such a situation did not occur in the FGC beam. Numerous of thin cracks formed in the FGC beams do not accelerate the corrosion progression, but on the contrary reduce the corrosion damage (Maalej et al., 2010).

*Figure 15. Corrosion of steel reinforcements in a) OPCC and b) FGC beams (Maalej et al., 2003)*

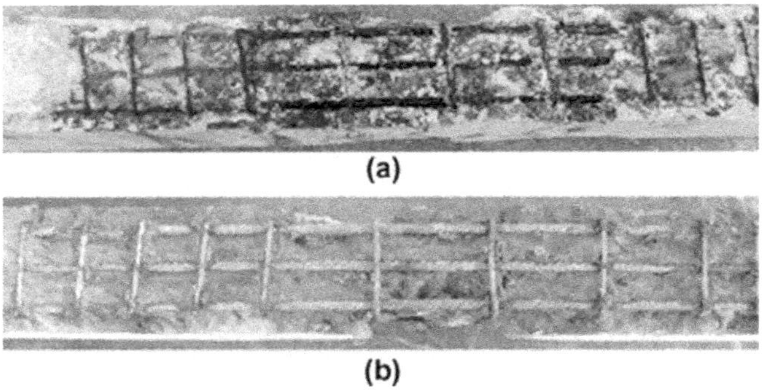

Alraeeini and Nikbakht (2022) investigated the corrosion influence on the flexural strength of normal concrete or ECC-filled pipe (CFDST) beams and concrete-filled pipe (CFST) beams. While the greatest effect of corrosion was observed in normal concrete filled CFST, the least effect occurred in ECC filled CFDST.

The investigation of chloride corrosion effect on ECC mixtures designed with both polyethylene fiber (PE) and polyvinyl alcohol fiber (PVA) along with sea sand reveals that composites produced with PE fiber exhibit less ductile behavior. This indicates that the bridging performance of PE fibers is insufficient compared to PVA fibers under the influence of chloride corrosion, which also reduces the dynamic modulus of composites produced with PE. However, ECCs produced with sea sand generally exhibit higher chloride corrosion resistance (Yao et al., 2023).

## ECC AS REPAIR MATERIAL

Due to safety requirements, it is very important to carry out repair and strengthening work of buildings quickly. The brittleness of normal concrete repair materials and their tendency to cracking cause a lack of durability of repair cycles. To solve this problem, using ECC with its super ductile property fills an important gap (Ma et al., 2014; Li et al., 2000). Although ECC has similar tensile and compressive strength to fiber-reinforced concrete (FRC) or plain concrete when compared as a repair material, ECC shows much higher structural strength and structural ductility than others (Li et al., 2000). Figure 16 shows the load-deflection curve of three different sample repair kits and as seen in Figure 17, despite similar material strength, ECC has a shear load carrying capacity of approximately 3 times more than FRC. Normal concrete, FRC and ECC were applied as a coating on the base concrete as repair material.

Compared to other repair materials, ECC performs very well in catching interface cracks, thus preventing the development of spalling and delamination in the repaired areas. Under the same loading conditions, ECC is more ductile and stronger as a repair material than normal concrete or fiber reinforced concrete (FRC), but also provides better crack width control. In addition, very low water permeability occurs in infrastructures repaired with ECC. More durable repair is possible with ECC (Lim and Li, 1997).

*Figure 16. Load-deflection behaviour of different repaired systems (Li et al., 2000)*

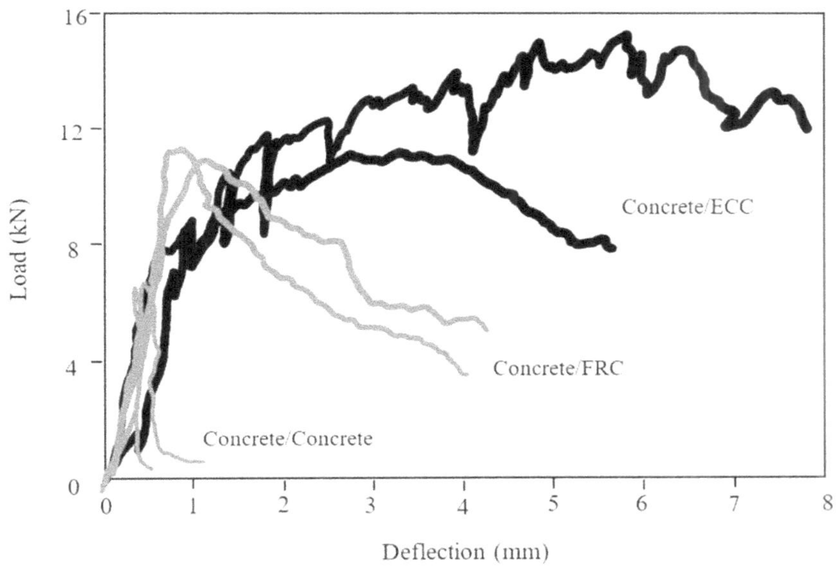


It is an inevitable need to use fast and high durability repair materials in worn concrete structures. To address this need, Li and Li (2011b) developed high-early-strength ECC (HES-ECC) to minimise disruption during renovation. HES-ECC has high tensile, flexural and compressive strength for early and long term. Its flexural strength is greater than twice that of plain concrete of the same compressive strength, and it exhibits strain hardening with 2.5% greater strain under uniaxial tension. It shows a significant resistance against macro cracking caused by shrinkage. The composite forms microcracks with a crack width of less than 50μm under restricted shrinkage conditions. With all these features, HES-ECC is a rapid and durable repair material with shorter downtime and high long-term durability.

*Figure 17. Shear load capacity of ECC and FRC (Li et al., 2000)*

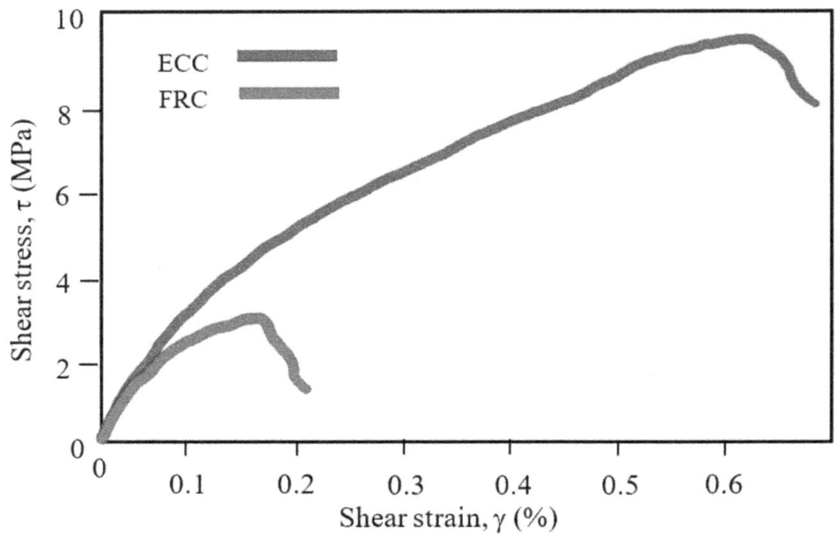


ECC used by spraying in repair systems shows similar behaviour to cast ECC at the same mix ratio in both strain hardening behavior and strain capacity (Kim et al., 2004).

Gao et al. (2019) examined the bonding characteristics between concrete samples and ECC at varying temperatures up to a maximum of 800°C, to estimate the performance of ECC as a repair component under the influence of high temperature. First of all, in the Type A test, the specimens were heated to a range of temperatures from 23°C (room temperature) to a maximum of 800°C, and reinforced with both strengthen in concrete and ECC as repair material. In the B Type test, the concrete specimens were tested at high temperature after they were strengthened with repair materials. In Experiment A, if the applied temperature is less than 200°C, shear stress, tensile strength and bonding performance increase with the increase in temperature. At temperatures above 200°C, these properties decrease with increasing temperature. At the same high temperature, the adhesion between concrete and ECC is better than between concrete and strengthening concrete. Explosion occurred at 600°C in the samples rehabilitated with strengthening concrete in the Type B test, while dispersion did not occur even at 800°C in the samples repaired with ECC.

Zhu et al. (2021a) aimed to improve a self-stressing ECC based on the high shrinkage property of ECC, which is effectively used as a repair material. From this point, shrinkage reducer (SRA) and ordinary Portland cement (OPC) is partially replaced with calcium sulfoaluminate cement (CSA) to optimize the composite expansion and expansion pressure. The combined use of CSA and SRA in ECC production supplies a self-stressing influence by optimizing the highest expansion and reduction of the expansion. As a result, a more durable ECC has been developed as an infrastructure repair material with a 7% strain and a mean crack opening of 35-44μm.

Zhu et al. (2021b) produced a shotcrete ECC. They used LC3 (low shrinkage calcined limestone cement) and CSA (calcium sulfoaluminate cement/additive) to improve the high cement dosage that causes shrinkage cracking in ECCs. While sprayed ECC and cast ECC have comparable strength, sprayed ECC has a higher strain rate and smaller crack opening. At the same time, spray ECC is a very durable material as infrastructure repair material with its low shrinkage and ultra-high ductility. Figure 18 shows the average and maximum crack width corresponding to a given strain level of casting and spray ECC.

*Figure 18. a) Average and b) maximum crack width of the sprayed and cast ECC (Zhu et al., 2021b)*

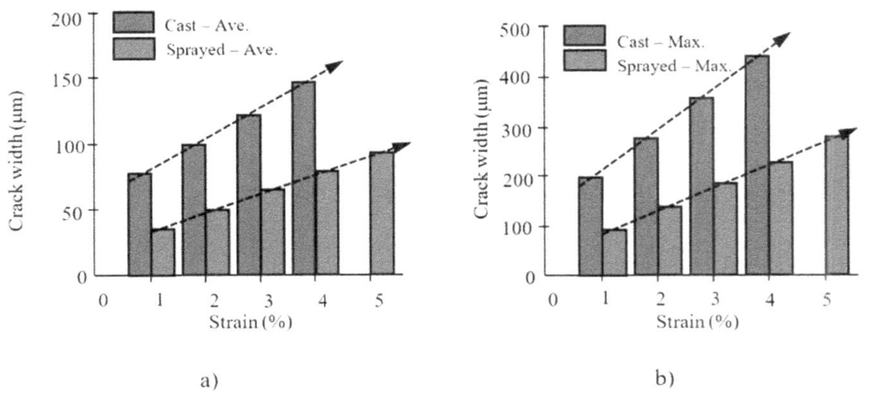


Jiang et al. (2023) developed HES-ECC to repair damaged concrete structures. The ECC mixtures used varying amounts of sulfoaluminate cement and fly ash as binder materials. Some mixtures had fog smoke added at 4% and 6% by mass, while others had Portland cement added at 4%, 8%, and 12% by mass. The developed ECCs exhibit high strength in the early period and excellent deformation properties in the long term. ECCs containing 6% silica fume meet both strong early strength and excellent long-term deformation properties. The flexural strength of this mixture reaches 66.67% of its 28-day strength in 3 hours, and the compressive strength reaches 29MPa in 3 hours.

The flexural stiffness and flexural capacity of reinforced concrete beams repaired with ECC are increased. The flexural capacity increases by approximately 20%-25%. There is little change in the flexural carrying capacity of the rehabilitated beams at preload levels up to 0.45. However, when the preload level exceeds 0.65, a decrease in flexural capacity is observed (Yan et al., 2023).

Calcium-sulfoaluminate (CSA) cement was used to develop ECCs as repair materials. To enhance the ductility of the composites, 10% and 20% crumb rubber replaced silica sand. The control sample was prepared using only silica sand. Rubber addition increases the tensile deformation capacity. However, it also causes a slight reduction in tensile and the compressive strength. The inclusion of rubber in the mixture results in a reduction of crack width when compared to the reference specimen. However, the use of rubber in mixtures leads to an increase in drying shrinkage. Nevertheless, this increase is smaller than the tensile deformation capacity, as reported by Zhang et al. (2023b).

LC3, a greener alternative to Portland cement, was utilized in the manufacture of ECCs for the repair of worn concrete pipes. Both mortar mixtures and ECCs were employed in the repair process. In some mixtures, crumb rubber was added to enhance ductility, as demonstrated in Zhang et al. (2023b). All ECCs produced exhibited high ductility, preserving their structural integrity and sealing properties when compared to pipes repaired with mortar (Zhu et al., 2023b).

## SUSTAINABILITY ASSESSMENT OF ECC

Concrete is the most commonly used material for building and constructing. However, its negative impact on the environment, mainly due to cement production, has encouraged researchers to focus on sustainable concrete. Approximately 6-8% of global $CO_2$ emissions come from the cement industry (Merli et al., 2020). However, the limited availability of global resources and the construction industry's potential to rapidly consume natural resources (Öztürk and Özyurt, 2022; Lima et al., 2021) highlights the importance of using recycled and waste materials in concrete production (Qaidi et al., 2021; Amin and Amin et al., 2021)

Traditional concrete has significant weaknesses such as low tensile strength, ductility and low crack resistance. (Merli et al., 2020; Kim et al., 2010). In contrast, ECC offers numerous benefits, such as high ductility, high tensile strength, and strain hardening behavior under tensile stresses (Öztürk et al., 2020; Li, 2003). However, the extensive use of OPC in ECC results in both cost burdens and negative environmental impacts, limiting its use (Zhou et al., 2022).

Zhou et al. (2022) aimed to enhance the environmental impact of their developed ECC. They achieved this by replacing LC3 and FAC with OPC and silica sand to produce sustainable ECCs. The authors conducted life cycle analyses to assess the environmental impact of the ECCs. The substitution of OPC with LC3 caused a minor decrease in compressive strength, tensile strength, and frictional strength between the fiber matrix. However, ductility slightly increased. LC3-based ECC exhibited significantly lower environmental impacts than standard ECC, making it a more sustainable option.

In another study that used calcined clay cement together with OPC, nanolime was added to initiate the reactions. The resulting nano lime calcined clay based (nLC3) ECC not only exhibits high compressive and tensile strength, but also demonstrates strain hardening behavior. This new ECC has superior features compared to previously produced LC3-based ECCs and is comparable to traditional ECC. However, when considering sustainability, nLC3-based ECC has a 50% smaller carbon footprint than traditional ECC and 28% lower $CO_2$ emissions than standard concrete (Khokhar et al., 2024).

To mitigate the adverse environmental impact of ECC, cement has been partially replaced with limestone powder (LP) at varying rates of up to 60%. In some mixtures, finer limestone powder was added to prevent strength losses. The use of 20% limestone powder results in increased tensile strength relative to control sample, while other substitution rates tended to reduce tensile strength. The addition of LP to the mixtures increases their deformation capacity. When evaluating the material sustainability of ECC samples, the addition of LP results in much lower energy and carbon release compared to cement. Additionally, mix sustainability increases with the quantity of LP included. As the LP addition increases, there is a lower embodied energy and $CO_2$ release (Zhang et al., 2023c).

Carbonation curing has been applied to minimise the carbon emissions of low-carbon ECC mixtures using recycled polyethylene fiber and sisal fiber obtained from waste nets. The developed ECC exhibits high ductility and a high tensile strength. It also has a much lower carbon footprint and is less expensive than both concrete and traditional ECC. Developed ECC costs 67% of conventional concrete and produces a 50% less $CO_2$ footprint than ordinary concrete. (Hu et al., 2023).

Geopolymer aggregate (GPA) can be used to produce sustainable ECCs instead of silica sand. The accelerated aging test showed that GPA-ECCs maintained their tensile ductility. ECCs made with GPA exhibit better multi-crack behavior and higher crack resistance compared to the reference sample produced with silica sand both in the short/long term. Based on the cost analysis, the utilization of GPA in ECC production reduces the cost and enhances sustainability (Xu et al., 2022).

Tan et al. (2022) developed an environmentally friendly ECC by using a high percentage of fly ash (FA), superfine waste sand instead of silica sand, and more economical polypropylene fiber (PP) as a binder material. They conducted a life cycle assessment of the produced ECCs. The environmentally friendly ECCs exhibit high ductility and strain hardening behavior, are more economical and sustainable, and reduce $CO_2$ emissions.

## CONCLUSION

This book chapter provides a detailed investigation of the durability performance of ECCs under difficult environmental conditions such as shrinkage, freezing-thawing, high temperature, corrosion, chloride, sulfate, acid effect. Tight crack width, high tensile ductility and self-healing ability provide ECC with superior properties against long-term environmental effects when compared to other cementitious composites. In addition, ECC exhibits a more compatible behavior with steel reinforcements thanks to its high tensile ductility in reinforced concrete structures. Despite all these superior properties, the high cost of ECC limits the widespread use of the benefits of ECC. In addition to PP and PE fibers, which are currently being tested as alternatives to PVA fibers, new ECCs may be developed in the future to achieve high tensile ductility and mechanical properties by using different fiber types. Although many studies have been conducted on the long-term performance of ECC under difficult environmental conditions, innovative studies on this subject are still ongoing. Considering that there is limited research on the abrasion resistance of ECC, this subject is an open field open to contribution. However, new research can be conducted to improve the durability performance of ECCs using hybrid fiber. Especially, the long-term behavior of hybrid fiber ECCs under harsh environmental conditions and their use as repair material can be investigated by using nanomaterials produced with the development of technology.

## REFERENCES

AASHTO. (2004). *LRFD Bridge Design Specifications* (3rd ed.). AASHTO.

Abdul-Rahman, M., Al-Attar, A. A., Hamada, H. M., & Tayeh, B. (2020). Microstructure and structural analysis of polypropylene fibre reinforced reactive powder concrete beams exposed to elevated temperature. *Journal of Building Engineering*, *29*, 101167. doi:10.1016/j.jobe.2019.101167

Abid, S. R., Hilo, A. N., & Daek, Y. H. (2018). Experimental tests on the underwater abrasion of engineered cementitious composites. *Construction & Building Materials*, *171*, 779–792. doi:10.1016/j.conbuildmat.2018.03.213

Adesina, A., & Das, S. (2021). Development of eco-friendly engineered cementitious composites using glass aggregates: Shrinkage properties. *Cleaner Engineering and Technology*, *5*, 100299. doi:10.1016/j.clet.2021.100299

Alraeeini, A. S., & Nikbakht, E. (2022). Corrosion effect on the flexural behaviour of concrete filled steel tubulars with single and double skins using engineered cementitious composite. *Structures*, *44*, 1680–1694. doi:10.1016/j.istruc.2022.08.095

Amin, M., Zeyad, A. M., Tayeh, B. A., & Agwa, I. S. (2021). Engineering properties of self-cured normal and high strength concrete produced using polyethylene glycol and porous ceramic waste as coarse aggregate. *Construction & Building Materials*, *299*, 124243. doi:10.1016/j.conbuildmat.2021.124243

Aslani, F., & Wang, L. (2019). Fabrication and characterization of an engineered cementitious composite with enhanced fire resistance performance. *Journal of Cleaner Production*, *221*, 202–214. doi:10.1016/j.jclepro.2019.02.241

ASTM C666-15. (2015). *Standard test method for resistance of concrete to rapid freezing and thawing*. American Society for Testing and Materials.

Bai, R., Liu, S., Yin, L., Yan, C., Zhou, J., Lu, L., & Li, L. (2023). Freeze-thaw damage assessment of engineered cementitious composites using the electrochemical impedance spectroscopy method. *Materials & Design*, *230*, 111965. doi:10.1016/j.matdes.2023.111965

Bhat, P. S., Chang, V., & Li, M. (2014). Effect of elevated temperature on strain-hardening engineered cementitious composites. *Construction & Building Materials*, *69*, 370–380. doi:10.1016/j.conbuildmat.2014.07.052

CEB. (1992). *Durable concrete structures. Comite Euro-International Beton*. Thomas Telford Ltd.

Elsener, B. (2002). Macrocell corrosion of steel in concrete–implications for corrosion monitoring. *Cement and Concrete Composites*, *24*(1), 65–72. doi:10.1016/S0958-9465(01)00027-0

Gao, S., Zhao, X., Qiao, J., Guo, Y., & Hu, G. (2019). Study on the bonding properties of Engineered Cementitious Composites (ECC) and existing concrete exposed to high temperature. *Construction & Building Materials*, *196*, 330–344. doi:10.1016/j.conbuildmat.2018.11.136

Hime, W. G. (1993). Corrosion of Steel—Random Thoughts and Wishful Thinking. *Concrete International*, *15*(10), 54–57.

Hooshmand, A., Kianoush, R., Siad, H., Lachemi, M., Booya, E., & Moslemi, M. (2021). Volumetric strain behaviour and self-healing of large scale engineered cementitious composite and normal concrete panels under natural conditions. *Construction & Building Materials*, *308*, 125078. doi:10.1016/j.conbuildmat.2021.125078

Hu, W., Zhang, D., Ftwi, E., Ellis, B. R., & Li, V. C. (2023). Development of sustainable low carbon Engineered Cementitious Composites with waste polyethylene fiber, sisal fiber and carbonation curing. *Resources, Conservation and Recycling*, *197*, 107096. doi:10.1016/j.resconrec.2023.107096

Jiang, B., Qian, Z., Gu, D., & Pan, J. (2023). Repair concrete structures with high-early-strength engineered cementitious composites (HES-ECC): Material design and interfacial behavior. *Journal of Building Engineering*, *68*, 106060. doi:10.1016/j.jobe.2023.106060

Khokhar, S. A., Ahmed, T., Khushnood, R. A., Basit, M. U., & Javed, S. (2024). Development of low carbon engineered cementitious composite (ECC) using nano lime calcined clay cement (nLC3) based matrix. *Case Studies in Construction Materials*, *20*, e02669. doi:10.1016/j.cscm.2023.e02669

Kim, S. B., Yi, N. H., Kim, H. Y., Kim, J. H. J., & Song, Y. C. (2010). Material and structural performance evaluation of recycled PET fiber reinforced concrete. *Cement and Concrete Composites*, *32*(3), 232–240. doi:10.1016/j.cemconcomp.2009.11.002

Kim, Y. Y., Fischer, G., Lim, Y. M., & Li, V. C. (2004). Mechanical performance of sprayed engineered cementitious composite using wet-mix shotcreting process for repair applications. *Materials Journal*, *101*(1), 42–49.

Lepech, M., & Li, V. C. (2005). Design and field demonstration of ECC link slabs for jointless bridge decks. *Proceedings of ConMat*, *5*, 22–24.

Lepech, M. D., & Li, V. C. (2009). Water permeability of engineered cementitious composites. *Cement and Concrete Composites*, *31*(10), 744–753. doi:10.1016/j.cemconcomp.2009.07.002

Lepech, M. D., & Li, V. C. (2010). Sustainable pavement overlays using engineered cementitious composites. *International Journal of Pavement Research and Technology*, *3*(5), 241-250.

Li, M. (2014). 7 - Engineered cementitious composites for bridge decks. In Y. J. Kim (Ed.), *Advanced Composites in Bridge Construction and Repair* (pp. 177–209). Woodhead Publishing. doi:10.1533/9780857097019.2.177

Li, M., & Li, V. C. (2011a). Cracking and healing of engineered cementitious composites under chloride environment. *ACI Materials Journal*, *108*(3), 333–340.

Li, M., & Li, V. C. (2011b). High-Early-Strength Engineered Cementitious Composites for Fast, Durable Concrete Repair-Material Properties. *ACI Materials Journal*, *108*(1), 3–12.

Li, V. C. (1993). From micromechanics to structural engineering the design of cementitious composites for civil engineering applications. *Journal of Structural Engineering and Earthquake Engineering*, *471*, 1–24.

Li, V. C. (2003). On engineered cementitious composites (ECC) a review of the material and its applications. *Journal of Advanced Concrete Technology*, *1*(3), 215–230. doi:10.3151/jact.1.215

Li, V. C. (2008). Engineered cementitious composites (ECC) material, structural, and durability performance. In Concrete Construction Engineering Handbook. CRC Press.

Li, V. C. (2012). Tailoring ECC for special attributes: A review. *International Journal of Concrete Structures and Materials*, *6*(3), 135–144. doi:10.1007/s40069-012-0018-8

Li, V. C., Fischer, G., Kim, Y. Y., Lepech, M., Qian, S., Weimann, M., & Wang, S. (2003). Field Demonstration of Durable Link Slabs for Jointless Bridge Decks Based on Strain-Hardening Cementitious Composites. Report for Michigan Department of Transportation RC-1438.

Li, V. C., Horii, H., Kabele, P., Kanda, T., & Lim, Y. M. (2000). Repair and retrofit with engineered cementitious composites. *Engineering Fracture Mechanics*, *65*(2-3), 317–334. doi:10.1016/S0013-7944(99)00117-4

Li, V. C., & Wang, S. (2002). Failure mode and structural ductility of GFRP reinforced engineered cementitious composite beams. *ACI Materials Journal*, *99*(1), 11–21.

Li, V. C., Wang, S., & Wu, C. (2001). Tensile strain-hardening behavior of polyvinyl alcohol engineered cementitious composite (PVA-ECC). *ACI Materials Journal*, *98*(6), 483–492.

Lim, Y. M., & Li, V. C. (1997). Durable repair of aged infrastructures using trapping mechanism of engineered cementitious composites. *Cement and Concrete Composites*, *19*(4), 373–385. doi:10.1016/S0958-9465(97)00026-7

Lima, L., Trindade, E., Alencar, L., Alencar, M., & Silva, L. (2021). Sustainability in the construction industry: A systematic review of the literature. *Journal of Cleaner Production*, *289*, 125730. doi:10.1016/j.jclepro.2020.125730

Liu, H., Zhang, Q., Gu, C., Su, H., & Li, V. (2017a). Influence of microcrack self-healing behavior on the permeability of Engineered Cementitious Composites. *Cement and Concrete Composites*, *82*, 14–22. doi:10.1016/j.cemconcomp.2017.04.004

Liu, H., Zhang, Q., Gu, C., Su, H., & Li, V. (2017b). Self-healing of microcracks in Engineered Cementitious Composites under sulfate and chloride environment. *Construction & Building Materials*, *153*, 948–956. doi:10.1016/j.conbuildmat.2017.07.126

Liu, H., Zhang, Q., Gu, C., Su, H., & Li, V. C. (2016). Influence of micro-cracking on the permeability of engineered cementitious composites. *Cement and Concrete Composites*, *72*, 104–113. doi:10.1016/j.cemconcomp.2016.05.016

Liu, H., Zhang, Q., Li, V., Su, H., & Gu, C. (2017c). Durability study on engineered cementitious composites (ECC) under sulfate and chloride environment. *Construction & Building Materials*, *133*, 171–181. doi:10.1016/j.conbuildmat.2016.12.074

Liu, S., Lu, S., Yin, L., Yan, C., Lu, L., & Zhou, J. (2022). Mechanical strength model of engineered cementitious composites with freeze–thaw damage based on pore structure evolution. *Cement and Concrete Composites*, *134*, 104706. doi:10.1016/j.cemconcomp.2022.104706

Liu, Y., Zhou, X., Lv, C., Yang, Y., & Liu, T. (2018). Use of silica fume and GGBS to improve frost resistance of ECC with high-volume fly ash. *Advances in Civil Engineering*, *2018*, 1–11. doi:10.1155/2018/7987589

Ma, H., Qian, S., & Zhang, Z. (2014). Effect of self-healing on water permeability and mechanical property of medium-early-strength engineered cementitious composites. *Construction & Building Materials*, *68*, 92–101. doi:10.1016/j.conbuildmat.2014.05.065

Ma, H., Yi, C., & Wu, C. (2021). Review and outlook on durability of engineered cementitious composite (ECC). *Construction & Building Materials*, *287*, 122719. doi:10.1016/j.conbuildmat.2021.122719

Ma, X., Xu, L., Cai, J., Han, J., Zhu, W., & Pan, J. (2023). Study on the performances of fire-resistive engineered cementitious composites (ECCs). *Journal of Building Engineering*, *67*, 105949. doi:10.1016/j.jobe.2023.105949

Maalej, M., Ahmed, S. F., & Paramasivam, P. (2003). Corrosion durability and structural response of functionally-graded concrete beams. *Journal of Advanced Concrete Technology*, *1*(3), 307–316. doi:10.3151/jact.1.307

Maalej, M., Chhoa, C. Y., & Quek, S. T. (2010). Effect of cracking, corrosion and repair on the frequency response of RC beams. *Construction & Building Materials*, *24*(5), 719–731. doi:10.1016/j.conbuildmat.2009.10.036

Maalej, M., & Leong, K. S. (2005). Engineered cementitious composites for effective FRP-strengthening of RC beams. *Composites Science and Technology*, *65*(7-8), 1120–1128. doi:10.1016/j.compscitech.2004.11.007

Maalej, M., & Li, V. C. (1995). Introduction of strain-hardening engineered cementitious composites in design of reinforced concrete flexural members for improved durability. *Structural Journal*, *92*(2), 167–176.

Maalej, M., Quek, S. T., Ahmed, S. F. U., Zhang, J., Lin, V. W. J., & Leong, K. S. (2012). Review of potential structural applications of hybrid fiber Engineered Cementitious Composites. *Construction & Building Materials*, *36*, 216–227. doi:10.1016/j.conbuildmat.2012.04.010

Mansoori, A., & Behfarnia, K. (2020). Evaluation of the effects of nanomaterials on durability of engineered cementitious composites exposed to the aggressive environment. *Journal of Composite Materials*, *54*(14), 1807–1817. doi:10.1177/0021998319887208

Merli, R., Preziosi, M., Acampora, A., Lucchetti, M. C., & Petrucci, E. (2020). Recycled fibers in reinforced concrete: A systematic literature review. *Journal of Cleaner Production*, *248*, 119207. doi:10.1016/j.jclepro.2019.119207

Mindess, S. (2019). Sustainability of concrete. In *Developments in the Formulation and Reinforcement of Concrete* (pp. 3–17). Woodhead Publishing. doi:10.1016/B978-0-08-102616-8.00001-0

Özbay, E., Sahmaran, M., Lachemi, M., & Yücel, H. E. (2013). Effect of microcracking on frost durability of high-volume-fly-ash-and slag-incorporated engineered cementitious composites. *ACI Materials Journal*, *110*(3), 259–267.

Ozturk, O., & Ozyurt, N. (2022). Sustainability and cost-effectiveness of steel and polypropylene fiber reinforced concrete pavement mixtures. *Journal of Cleaner Production*, *363*, 132582. doi:10.1016/j.jclepro.2022.132582

Öztürk, O., Yıldırım, G., Keskin, Ü. S., Siad, H., & Şahmaran, M. (2020). Nano-tailored multi-functional cementitious composites. *Composites. Part B, Engineering*, *182*, 107670. doi:10.1016/j.compositesb.2019.107670

Pranav, S., Aggarwal, S., Yang, E. H., Sarkar, A. K., Singh, A. P., & Lahoti, M. (2020). Alternative materials for wearing course of concrete pavements: A critical review. *Construction & Building Materials*, *236*, 117609. doi:10.1016/j.conbuildmat.2019.117609

Qaidi, S. M., Dinkha, Y. Z., Haido, J. H., Ali, M. H., & Tayeh, B. A. (2021). Engineering properties of sustainable green concrete incorporating eco-friendly aggregate of crumb rubber: A review. *Journal of Cleaner Production*, *324*, 129251. doi:10.1016/j.jclepro.2021.129251

Sahmaran, M., Lachemi, M., Hossain, K. M., & Li, V. C. (2009a). Internal curing of engineered cementitious composites for prevention of early age autogenous shrinkage cracking. *Cement and Concrete Research*, *39*(10), 893–901. doi:10.1016/j.cemconres.2009.07.006

Sahmaran, M., Lachemi, M., & Li, V. C. (2009b). Assessing the durability of engineered cementitious composites under freezing and thawing cycles. *Journal of ASTM International*, *6*(7), 85–102. doi:10.1520/JAI102406

Sahmaran, M., Lachemi, M., & Li, V. C. (2010). Assessing Mechanical Properties and Microstructure of Fire-Damaged Engineered Cementitious Composites. *ACI Materials Journal*, *107*(3), 297–304.

Sahmaran, M., Li, M., & Li, V. C. (2007). Transport properties of engineered cementitious composites under chloride exposure. *ACI Materials Journal*, *104*(6), 604–611.

Sahmaran, M., & Li, V. C. (2007). De-icing salt scaling resistance of mechanically loaded engineered cementitious composites. *Cement and Concrete Research*, *37*(7), 1035–1046. doi:10.1016/j.cemconres.2007.04.001

Sahmaran, M., & Li, V. C. (2008). Durability of mechanically loaded engineered cementitious composites under highly alkaline environments. *Cement and Concrete Composites*, *30*(2), 72–81. doi:10.1016/j.cemconcomp.2007.09.004

Sahmaran, M., Li, V. C., & Andrade, C. (2008). Corrosion resistance performance of steel-reinforced engineered cementitious composite beams. *ACI Materials Journal*, *105*(3), 243.

Sahmaran, M., Özbay, E., Yücel, H. E., Lachemi, M., & Li, V. C. (2012). Frost resistance and microstructure of Engineered Cementitious Composites: Influence of fly ash and micro poly-vinyl-alcohol fiber. *Cement and Concrete Composites*, *34*(2), 156–165. doi:10.1016/j.cemconcomp.2011.10.002

Sherir, M. (2012). *Fracture energy, fatigue & creep properties of engineered cementitious composites incorporating fly ash/slag with different aggregates* [MSc Thesis]. Civil Engineering Department, Ryerson University, Toronto, Canada.

Shumuye, E. D., Li, W., Liu, J., Wang, Z., Yu, J., & Wu, H. (2022). Self-healing recovery and micro-structural properties of slag/fly-ash based engineered cementitious composites under chloride environment and tidal exposure. *Cement and Concrete Composites*, *134*, 104789. doi:10.1016/j.cemconcomp.2022.104789

Sridhar, R. (2022). Durability study on engineered cementitious composites with hybrid fibers under sulfate and chloride environments. *Cleaner Materials*, *5*, 100121. doi:10.1016/j.clema.2022.100121

Tan, G., Zhu, Z., Wang, W., & He, X. (2022). A fractal-based approach for cracking characterization and whole process prediction exploration of PP fiber reinforced ECC containing sustainable ingredients. *Construction & Building Materials*, *318*, 126015. doi:10.1016/j.conbuildmat.2021.126015

Ting, M. Z. Y., Wong, K. S., Rahman, M. E., & Meheron, S. J. (2021). Deterioration of marine concrete exposed to wetting-drying action. *Journal of Cleaner Production*, *278*, 123383. doi:10.1016/j.jclepro.2020.123383

Turk, K., & Demirhan, S. (2013). The mechanical properties of engineered cementitious composites containing limestone powder replaced by microsilica sand. *Canadian Journal of Civil Engineering*, *40*(2), 151–157. doi:10.1139/cjce-2012-0281

Wang, Q., Xiong, Z., Zhang, J., Fang, Z., Lai, M., & Ho, J. (2023a). Impact of polyvinyl alcohol fiber on the full life-cycle shrinkage of cementitious composite. *Journal of Building Engineering*, *63*, 105463. doi:10.1016/j.jobe.2022.105463

Wang, Q., Zhang, J., & Ho, J. C. M. (2020). Zeolite to improve strength-shrinkage performance of high-strength engineered cementitious composite. *Construction & Building Materials*, *234*, 117335. doi:10.1016/j.conbuildmat.2019.117335

Wang, T., Zhang, D., Zhu, H., Ma, B., & Li, V. C. (2022). Durability and self-healing of engineered cementitious composites exposed to simulated sewage environments. *Cement and Concrete Composites*, *129*, 104500. doi:10.1016/j.cemconcomp.2022.104500

Wang, Y., Guo, R., Fu, C., Pan, T., Lin, R., & Zhang, S. (2023b). Effect of calcined zeolite on the shrinkage and mechanical properties of lightweight engineering cementitious composites. *Journal of Building Engineering*, *77*, 107480. doi:10.1016/j.jobe.2023.107480

Wang, Z., Sun, P., Hu, Y., & Han, S. (2023c). Crack morphology tailoring and permeability prediction of polyvinyl alcohol-steel hybrid fiber engineered cementitious composites. *Journal of Cleaner Production, 383*, 135335. doi:10.1016/j.jclepro.2022.135335

Wu, C., & Li, V. C. (2017). CFRP-ECC hybrid for strengthening of the concrete structures. *Composite Structures, 178*, 372–382. doi:10.1016/j.compstruct.2017.07.034

Wu, C., Pan, Y., & Ueda, T. (2018). Characterization of the abrasion resistance and the acoustic wave attenuation of the engineered cementitious composites for runway pavement. *Construction & Building Materials, 174*, 537–546. doi:10.1016/j.conbuildmat.2018.04.152

Wu, H. L., Zhang, D., Du, Y. J., & Li, V. C. (2020). Durability of engineered cementitious composite exposed to acid mine drainage. *Cement and Concrete Composites, 108*, 103550. doi:10.1016/j.cemconcomp.2020.103550

Xu, L. Y., Huang, B. T., Lan-Ping, Q., & Dai, J. G. (2022). Enhancing long-term tensile performance of Engineered Cementitious Composites (ECC) using sustainable artificial geopolymer aggregates. *Cement and Concrete Composites, 133*, 104676. doi:10.1016/j.cemconcomp.2022.104676

Xu, M., Song, S., Feng, L., Zhou, J., Li, H., & Li, V. C. (2021). Development of basalt fiber engineered cementitious composites and its mechanical properties. *Construction & Building Materials, 266*, 121173. doi:10.1016/j.conbuildmat.2020.121173

Yan, Y., Lu, Y., Zhao, Q., & Li, S. (2023). Flexural behavior of pre-damaged and repaired reinforced concrete beams with carbon fiber reinforced polymer grid and engineered cementitious composite. *Engineering Structures, 277*, 115390. doi:10.1016/j.engstruct.2022.115390

Yang, E. H., & Li, V. C. (2010). Strain-hardening fiber cement optimization and component tailoring by means of a micromechanical model. *Construction & Building Materials, 24*(2), 130–139. doi:10.1016/j.conbuildmat.2007.05.014

Yang, E. H., Yang, Y., & Li, V. C. (2007). Use of high volumes of fly ash to improve ECC mechanical properties and material greenness. *ACI Materials Journal, 104*(6), 620–628.

Yao, Q., Teng, X., Lu, C., Sun, H., Mo, J., & Chen, Z. (2023). Influence of accelerated chloride corrosion on mechanical properties of sea sand ECC and damage evaluation method based on nondestructive testing. *Journal of Building Engineering, 63*, 105520. doi:10.1016/j.jobe.2022.105520

Ye, B., Zhang, Y., Han, J., & Pan, P. (2019). Effect of water to binder ratio and sand to binder ratio on shrinkage and mechanical properties of High-strength Engineered Cementitious Composite. *Construction & Building Materials*, *226*, 899–909. doi:10.1016/j.conbuildmat.2019.07.303

Yu, J., Lin, J., Zhang, Z., & Li, V. C. (2015). Mechanical performance of ECC with high-volume fly ash after sub-elevated temperatures. *Construction & Building Materials*, *99*, 82–89. doi:10.1016/j.conbuildmat.2015.09.002

Zhang, C., Guan, X., Li, J., Li, Y., & Lu, R. (2023a). Coupling effect of cementitious capillary crystalline waterproof material and exposure environments on self-healing properties of engineered cementitious composites (ECC). *Journal of Building Engineering*, *63*, 105471. doi:10.1016/j.jobe.2022.105471

Zhang, C., Lu, R., Li, Y., & Guan, X. (2021). Effect of crystalline admixtures on mechanical, self-healing and transport properties of engineered cementitious composite. *Cement and Concrete Composites*, *124*, 104256. doi:10.1016/j.cemconcomp.2021.104256

Zhang, W., Zheng, C., Li, Z., Jin, H., Liu, J., Zhu, J., Liu, W., & Xing, F. (2022). Investigation on mechanical properties improvement of seawater engineered cementitious composites (ECC) using FA/LC2. *Construction & Building Materials*, *345*, 128271. doi:10.1016/j.conbuildmat.2022.128271

Zhang, Z., Gao, Y., Qin, F., Sun, F., & Huang, Y. (2023c). Mechanical properties of sustainable high strength ECC with substitution of cement by limestone powder. *Case Studies in Construction Materials*, *19*, e02434. doi:10.1016/j.cscm.2023.e02434

Zhang, Z., Liu, J. C., Xu, X., & Yuan, L. (2020). Effect of sub-elevated temperature on mechanical properties of ECC with different fly ash contents. *Construction & Building Materials*, *262*, 120096. doi:10.1016/j.conbuildmat.2020.120096

Zhang, Z., Yu, J., Qin, F., Huang, Y., & Sun, F. (2023b). Mechanical and self-healing properties of calcium-sulfoaluminate-cement-based engineered cementitious composites (ECC). *Journal of Building Engineering*, *77*, 107512. doi:10.1016/j.jobe.2023.107512

Zhao, Y., Chen, X., Wang, Z., Zhang, Y., & Leng, Y. (2023). An experimental study on the influence of fly ash and crumb rubber on the mechanical property and permeability of Engineered Cementitious Composite (ECC). *Journal of Building Engineering*, *76*, 106998. doi:10.1016/j.jobe.2023.106998

Zhou, Y., Gong, G., Xi, B., Guo, M., Xing, F., & Chen, C. (2022). Sustainable lightweight engineered cementitious composites using limestone calcined clay cement (LC3). *Composites. Part B, Engineering*, *243*, 110183. doi:10.1016/j.compositesb.2022.110183

Zhou, Y., Xi, B., Yu, K., Sui, L., & Xing, F. (2018). Mechanical properties of hybrid ultra-high performance engineered cementitous composites incorporating steel and polyethylene fibers. *Materials (Basel)*, *11*(8), 1448. doi:10.3390/ma11081448 PMID:30115842

Zhu, H., Wang, T., Wang, Y., Hu, W. H., & Li, V. C. (2023b). Feasibility of structural retrofit concrete pipelines using limestone calcined clay cement Engineered Cementitious Composites (LC3 ECC). *Engineering Structures*, *289*, 116305. doi:10.1016/j.engstruct.2023.116305

Zhu, H., Yu, K., & Li, V. C. (2021b). Sprayable engineered cementitious composites (ECC) using calcined clay limestone cement (LC3) and PP fiber. *Cement and Concrete Composites*, *115*, 103868. doi:10.1016/j.cemconcomp.2020.103868

Zhu, H., Zhang, D., Wang, Y., Wang, T., & Li, V. C. (2021a). Development of self-stressing engineered cementitious composites (ECC). *Cement and Concrete Composites*, *118*, 103936. doi:10.1016/j.cemconcomp.2021.103936

Zhu, M., Chen, B., Wu, M., & Han, J. (2023a). Preparation and mechanical characterization of cost-effective low-carbon engineered cementitious composites with seawater and sea-sand. *Cement and Concrete Composites*, *136*, 104883. doi:10.1016/j.cemconcomp.2022.104883

Zhu, Y., Yang, Y., & Yao, Y. (2012). Autogenous self-healing of engineered cementitious composites under freeze–thaw cycles. *Construction & Building Materials*, *34*, 522–530. doi:10.1016/j.conbuildmat.2012.03.001

# Chapter 7
# Corrosion Resistance and Permeability Properties of UHPC-ECC:
## Durability Studies of UHPC-ECC

**S. Athibaranan**
*GMR Institute of Technology, India*

**G. Ganesh Prabhu**
*GMR Institute of Technology, India*

**Karthikayan Jayakumar**
*National Institute of Technology, Tiruchirappalli, India*

**J. Sridhar**
https://orcid.org/0000-0003-4058-1703
*GMR Institute of Technology, India*

## ABSTRACT

*Ultra high performance engineered cementitious composites (UHP-ECCs) are advanced cementitious materials that exhibit exceptional strength, durability, and ductility. This chapter reviews the corrosion resistance and permeability properties of UHP-ECCs, which are critical factors in determining their suitability for various applications. UHP-ECCs have been shown to exhibit excellent resistance to corrosion in harsh environments due to their high cement content, low water-to-cement ratio, and the incorporation of various fibers. Additionally, UHP-ECCs exhibit extremely low permeability to liquids, gases, and ions, which makes them suitable for a wide range of applications, including water storage, wastewater treatment, and transportation infrastructure. The chapter concludes that UHP-ECCs are a highly desirable material for applications where corrosion resistance and low permeability properties are essential.*

DOI: 10.4018/978-1-6684-8182-0.ch007

## 1. INTRODUCTION

Ultra high performance engineered cementitious composites (UHP-ECCs) are advanced materials that have gained attention in recent years due to their exceptional mechanical properties, including high strength, ductility, and durability. UHP-ECCs are a new generation of cementitious materials that are designed to meet the demands of modern construction practices, which require materials with superior mechanical properties, excellent durability, and sustainability. The corrosion resistance and permeability properties of UHP-ECCs are critical in determining their suitability for various applications, particularly in harsh environments.

Corrosion is a major concern in the construction industry, particularly in structures exposed to harsh environments, such as marine and industrial settings. Traditional concrete structures are susceptible to corrosion due to the ingress of corrosive agents, such as chlorides, sulfates, and acids, which can penetrate the concrete and cause damage over time. Corrosion can lead to the degradation of concrete structures, reducing their lifespan and posing a safety risk.

To address this issue, UHP-ECCs have been developed to exhibit exceptional corrosion resistance. The high cement content and low water-to-cement ratio of UHP-ECCs result in a dense, impermeable microstructure, which is highly resistant to the ingress of corrosive agents. Additionally, the incorporation of various fibers, such as PVA and steel fibers, enhances the resistance to cracking and provides an additional barrier against corrosion.

Permeability is another critical factor in determining the suitability of cementitious materials for various applications. The permeability of concrete can affect its durability and performance, particularly in applications where water or chemicals may be present. UHP-ECCs have been shown to exhibit extremely low permeability to liquids, gases, and ions, which makes them suitable for a wide range of applications, including water storage, wastewater treatment, and transportation infrastructure. The low permeability of UHP-ECCs is due to the dense, impermeable microstructure, which significantly reduces the ingress of water and other substances.

In summary, UHP-ECCs are a new class of advanced cementitious materials that exhibit exceptional mechanical properties, including high strength, ductility, and durability. Their corrosion resistance and permeability properties are critical in determining their suitability for various applications, particularly in harsh environments. UHP-ECCs have been shown to exhibit excellent resistance to corrosion and low permeability, which make them suitable for a wide range of applications in the construction industry. Further research is needed to explore the full potential of UHP-ECCs in various applications.

Ultra high performance engineered cementitious composites (UHP-ECCs) are a class of advanced cementitious materials that have been developed in recent years.

They are characterized by exceptional mechanical properties, such as high strength, ductility, and durability. In addition to these properties, the corrosion resistance and permeability properties of UHP-ECCs are also critical in determining their suitability for various applications. This paper provides a detailed review of the corrosion resistance and permeability properties of UHP-ECCs.

UHP-ECCs are composed of a high proportion of cementitious materials, such as cement, silica fume, and fine aggregates, along with various additives and fibers that enhance their properties. These materials are carefully selected and proportioned to produce a dense, impermeable microstructure, which contributes to their exceptional mechanical properties. The incorporation of fibers, such as polyvinyl alcohol (PVA) and steel fibers, further enhances the mechanical properties of UHP-ECCs, as well as their resistance to cracking and corrosion.

The corrosion resistance of UHP-ECCs has been extensively studied, and they have been shown to exhibit excellent performance in harsh environments, such as marine and industrial settings, where traditional concrete often fails. The high cement content and low water-to-cement ratio of UHP-ECCs result in a dense, impermeable microstructure that is highly resistant to the ingress of corrosive agents, such as chlorides, sulfates, and acids. Additionally, the presence of various fibers in UHP-ECCs enhances their resistance to corrosion by providing an additional barrier against the ingress of corrosive agents.

The permeability properties of UHP-ECCs are also critical in determining their suitability for various applications. UHP-ECCs exhibit extremely low permeability to liquids, gases, and ions, which is a critical factor in preventing the ingress of corrosive agents into the material. The low water-to-cement ratio and the presence of fine aggregates and pozzolanic materials contribute to the dense, impermeable microstructure of UHP-ECCs, which significantly reduces their permeability.

The paper concludes that UHP-ECCs are a highly desirable material for applications where corrosion resistance and low permeability properties are essential. UHP-ECCs exhibit outstanding corrosion resistance and low permeability properties, which make them suitable for a wide range of applications, including water storage, wastewater treatment, and transportation infrastructure. The paper also highlights the need for further research to explore the full potential of UHP-ECCs in various applications.

- UHP-ECCs are typically made using a combination of cement, silica fume, fine aggregates, and various additives, such as superplasticizers and viscosity-modifying agents. The exact composition of UHP-ECCs can vary depending on the specific application and desired properties.
- UHP-ECCs have been shown to exhibit excellent resistance to a variety of corrosive agents, including chlorides, sulfates, and acids. In addition to the dense microstructure and fiber reinforcement, UHP-ECCs may also

incorporate corrosion inhibitors, such as calcium nitrite, to further enhance their resistance to corrosion.

- The low permeability of UHP-ECCs can be attributed to the combination of a high cement content, low water-to-cement ratio, and fine aggregate size. The use of pozzolanic materials, such as fly ash or slag, can also contribute to the low permeability of UHP-ECCs.
- UHP-ECCs have been used in a variety of applications, including bridge decks, seawalls, and wastewater treatment plants. In some cases, UHP-ECCs have been shown to outperform traditional concrete in terms of durability and corrosion resistance.
- UHP-ECCs may have higher initial costs compared to traditional concrete, but their long-term durability and reduced maintenance costs can make them a cost-effective option in the long run.
- Research is ongoing to further optimize the properties of UHP-ECCs and explore their potential for new applications. For example, researchers are investigating the use of UHP-ECCs in 3D printing and as a sustainable alternative to traditional building materials

## 2. LITERATURE REVIEW

This literature review highlights several studies that evaluated the corrosion resistance and permeability properties of Ultra High Performance Engineered Cementitious Composites (UHP-ECCs). The studies investigated the effects of different reinforcing fibers and curing conditions on the corrosion resistance of UHP-ECCs, as well as the permeability of UHP-ECCs under various loading conditions. The results demonstrated that UHP-ECCs have significantly higher resistance to chloride ion penetration and corrosion compared to conventional concrete, and exhibit extremely low permeability under all loading conditions. The review concludes that UHP-ECCs have great potential for use in applications where corrosion resistance and low permeability are critical, and that further research is needed to optimize their properties and explore new applications.

In a study published researchers evaluated the corrosion resistance of UHP-ECCs reinforced with PVA fibers. The results showed that the PVA-reinforced UHP-ECCs exhibited significantly higher resistance to chloride ion penetration and corrosion compared to conventional concrete. The researchers attributed the enhanced corrosion resistance to the dense microstructure and fiber reinforcement of the UHP-ECCs. This study investigated the effect of different curing conditions on the corrosion resistance of UHP-ECCs. The researchers found that UHP-ECCs cured under steam exhibited significantly higher resistance to chloride ion penetration and corrosion compared to UHP-ECCs cured under standard curing conditions. The researchers attributed the enhanced corrosion resistance to the increased strength and denser microstructure of the steam-cured UHP-ECCs. A study evaluated the permeability of UHP-ECCs under various loading conditions, including compression, tension, and bending. The results showed that the UHP-ECCs exhibited extremely low permeability under all loading conditions, indicating their potential for use in water storage and treatment applications. In another study published in Cement and Concrete Research, researchers investigated the effect of different fibers on the corrosion resistance of UHP-ECCs. The results showed that UHP-ECCs reinforced with steel fibers exhibited significantly higher resistance to chloride ion penetration and corrosion compared to UHP-ECCs reinforced with PVA fibers or no fibers at all. The researchers attributed the enhanced corrosion resistance to the crack bridging and additional barrier provided by the steel fibers. A study evaluated the long-term durability of UHP-ECCs exposed to harsh marine environments. The results showed that the UHP-ECCs exhibited excellent resistance to chloride ion penetration and corrosion even after 15 years of exposure to seawater. The researchers attributed the enhanced durability to the dense microstructure and fiber reinforcement of the UHP-ECCs. These studies and others demonstrate the potential of UHP-ECCs for use in a variety of applications where corrosion resistance and low permeability are critical factors. Further research is needed to optimize the properties of UHP-ECCs and explore their potential for new applications.

In a study by Li et al. (2019), the effects of nanosilica on the corrosion resistance and permeability of UHP-ECCs were investigated. The results showed that the addition of 3% nanosilica significantly reduced the porosity and permeability of the concrete, and improved its resistance to chloride ion penetration. Liu et al. (2019) investigated the effects of different fiber types on the permeability and corrosion resistance of UHP-ECCs. The results showed that the addition of steel fibers significantly improved the concrete's resistance to chloride ion penetration and corrosion of the reinforcing steel. In a study by Yu et al. (2020), the effects of mineral admixtures on the permeability and corrosion resistance of UHP-ECCs were investigated. The results showed that the addition of fly ash and slag powder significantly reduced the permeability of the concrete and improved its resistance to chloride ion penetration and corrosion. Zhang et al. (2020) investigated the effects of different curing conditions on the permeability and corrosion resistance of UHP-ECCs. The results showed that the use of steam curing significantly reduced the permeability of the concrete and improved its resistance to chloride ion penetration and corrosion. In a study by Wang et al. (2019), the effects of fiber content on the permeability and corrosion resistance of UHP-ECCs were investigated. The results showed that the addition of 1.5% steel fibers significantly reduced the permeability of the concrete and improved its resistance to chloride ion penetration and corrosion. Zhang et al.

(2019) investigated the effects of different water-cement ratios on the permeability and corrosion resistance of UHP-ECCs. The results showed that a lower water-cement ratio significantly reduced the permeability of the concrete and improved its resistance to chloride ion penetration and corrosion. In a study by Wang et al. (2021), the effects of different fiber types on the corrosion resistance of UHP-ECCs under chloride attack were investigated. The results showed that the addition of polypropylene fibers significantly improved the concrete's resistance to chloride ion penetration and corrosion of the reinforcing steel. Gao et al. (2019) investigated the effects of different curing temperatures on the permeability and corrosion resistance of UHP-ECCs. The results showed that a higher curing temperature significantly reduced the permeability of the concrete and improved its resistance to chloride ion penetration and corrosion. In a study by Wang et al. (2020), the effects of different fiber types on the freeze-thaw durability and corrosion resistance of UHP-ECCs were investigated. The results showed that the addition of steel fibers significantly improved the concrete's resistance to freeze-thaw cycles and corrosion of the reinforcing steel. Yu et al. (2021) investigated the effects of different slag powder contents on the permeability and corrosion resistance of UHP-ECCs. The results showed that the addition of 30% slag powder significantly reduced the permeability of the concrete and improved its resistance to chloride ion penetration and corrosion of the reinforcing steel.

## 3. METHODOLOGY

This methodology section outlines the steps involved in testing the corrosion resistance and permeability properties of Ultra High Performance Engineered Cementitious Composites (UHPC-ECCs). The first step involves preparing UHPC-ECC samples using specific materials and mixing procedures. The section then describes various experimental techniques for testing corrosion resistance, including electrochemical impedance spectroscopy (EIS), rapid chloride migration (RCM) test, and rapid chloride penetration (RCP) test. The permeability of UHPC-ECC can be tested using methods such as the water absorption test, chloride diffusion test, or water permeability test, which are also described in this section. Additionally, the methodology section outlines the statistical methods used to analyze the data collected from these tests and provides guidance on presenting results using appropriate tables, graphs, and figures. This methodology provides a structured approach to evaluating the performance of UHPC-ECCs in terms of their corrosion resistance and permeability properties.

1. Materials Preparation: In this section, the method for preparing UHPC-ECC samples should be describes, including the types and quantities of materials

used and the mixing procedure. The UHPC-ECC samples should be cast in molds of specific dimensions.

2. Corrosion Resistance Testing: Various experimental techniques can be used to test the corrosion resistance of UHPC-ECC. Some common techniques include electrochemical impedance spectroscopy (EIS), rapid chloride migration (RCM) test, and rapid chloride penetration (RCP) test. The methodology section describes each of these techniques and provide details on the specific parameters used in the experiments, such as the type and concentration of the corrosive agent, the duration of the test, and the measurement instruments used.
3. Permeability Testing: The permeability of UHPC-ECC can be tested using methods such as the water absorption test, the chloride diffusion test, or the water permeability test. The methodology section describes each of these techniques and provide details on the specific parameters used in the experiments, such as the duration of the test, the type of solution used, and the measurement instruments used.
4. Data Analysis: The data collected from the corrosion resistance and permeability tests analyzed using appropriate statistical methods. The methodology section describes the statistical methods used to analyze the data and the software packages used for the analysis.
5. Results: The results section presents the findings of the corrosion resistance and permeability tests. The results should be presented in a clear and concise manner, with appropriate tables, graphs, and figures to illustrate the findings.

## 4. EXPERIMENTAL TESTING PROCEDURES

The experimental procedure section is a critical part of any study, as it provides a detailed description of the methods used to collect data and test hypotheses. In the context of the corrosion resistance and permeability properties of ultra high performance engineered cementitious composites (UHP-ECCs), this experimental procedure section provides a step-by-step description of how the samples were prepared, how the corrosion and permeability tests were conducted, and how the data was analyzed. This section written with sufficient detail to allow other researchers to replicate the experiments and verify the results. The use of appropriate materials and equipment, as well as the adherence to established testing standards and protocols, is also important to ensure the accuracy and reliability of the experimental data.

The materials preparation for different tests for Corrosion resistance and permeability properties of Ultra high performance Engineered cementitious composites can vary depending on the specific test. Here are some examples:

4.1.1 Corrosion Resistance Testing:

For electrochemical impedance spectroscopy (EIS) test, UHPC-ECC samples of specific dimensions are typically cast in molds and cured for a specified time. Prior to testing, the surface of the sample should be prepared by grinding and polishing to remove any irregularities. The sample is then placed in a corrosive solution and subjected to an alternating current voltage of a specific amplitude and frequency. The impedance response of the sample is recorded as a function of time.

For rapid chloride migration (RCM) test, UHPC-ECC samples are typically cast in molds of specific dimensions and cured for a specified time. The samples are then placed in a chloride solution and subjected to an electrical potential gradient. The migration rate of the chloride ions through the sample is measured as a function of time.

4.1.2 Permeability Testing:

For water absorption test, UHPC-ECC samples are typically cast in molds and cured for a specified time. The samples are then dried to a constant weight and weighed. The samples are then immersed in water for a specified time, after which they are weighed again. The water absorption of the sample is calculated as the mass of water absorbed per unit area of sample.

For chloride diffusion test, UHPC-ECC samples of specific dimensions are typically cast in molds and cured for a specified time. The samples are then immersed in a chloride solution for a specified time. The chloride concentration at different depths within the sample is measured using various techniques such as colorimetric analysis or ion chromatography. The chloride diffusion coefficient is then calculated from the measured concentration profile.

For water permeability test, UHPC-ECC samples are typically cast in molds of specific dimensions and cured for a specified time. The samples are then placed in a permeability apparatus, and water is forced through the sample under a specified pressure. The flow rate of water through the sample is measured as a function of time.

## 4.2 Material Properties Used in UHPC-ECC

The materials used in the preparation of Ultra High Performance Engineered Cementitious Composites (UHP-ECCs) and their properties play a crucial role in determining the corrosion resistance and permeability properties of the resulting composites. Some of the important materials and their properties are:

1. Cement: The type of cement used in the UHP-ECCs can influence the microstructure and strength of the resulting composites. Generally, high-strength and low-permeability cement types such as Type I/II or Type V Portland cement are used.

2. Fine Aggregates: Fine aggregates such as sand or limestone powder are used in UHP-ECCs to improve workability and reduce shrinkage. The particle size and shape of the fine aggregates can affect the packing density and the flow behavior of the mixture.
3. Coarse Aggregates: Coarse aggregates such as crushed granite or quartz are used to improve the mechanical properties of UHP-ECCs. The size and shape of the coarse aggregates can affect the compressive strength and toughness of the resulting composites.
4. Water: The quality and quantity of water used in the mixing process can affect the flow behavior, workability, and strength of the UHP-ECCs. Deionized or distilled water is typically used to prevent contamination from impurities.
5. Admixtures: Admixtures such as superplasticizers, viscosity modifiers, and air-entraining agents can improve the workability, strength, and durability of UHP-ECCs. The type and dosage of admixtures used depend on the specific application and the desired properties of the resulting composites.
6. Fibers: Fibers such as steel, polyvinyl alcohol (PVA), or polyethylene (PE) can improve the tensile strength, ductility, and toughness of UHP-ECCs. The length, diameter, and volume fraction of the fibers can affect the mechanical properties and durability of the composites.
7. Corrosive agents: Corrosive agents such as chloride ions or sulfates are used to test the corrosion resistance of UHP-ECCs. The concentration and duration of exposure to these agents can affect the rate and extent of corrosion.

*Figure 1. Compressive strength test on UHPC-ECC*

## 4.3 Experimental Procedures

The experimental procedures for investigating the corrosion resistance and permeability properties of Ultra High Performance Engineered Cementitious Composites (UHP-ECCs) vary depending on the specific research objectives and methodology. However, some typical experimental procedures and codebooks that are commonly used in this field are as follows:

*Figure 2. Specimens for RCPT setup*

**4.3.1 Chloride penetration test:** This test is used to evaluate the resistance of UHP-ECCs to chloride ion penetration. The test is typically conducted according to ASTM C1202, which involves applying a constant voltage to the concrete specimen and measuring the total charge passed through the concrete after a certain time period. The lower the charge passed, the greater the resistance to chloride ion penetration.

*Figure 3. Working of RCPT setup*

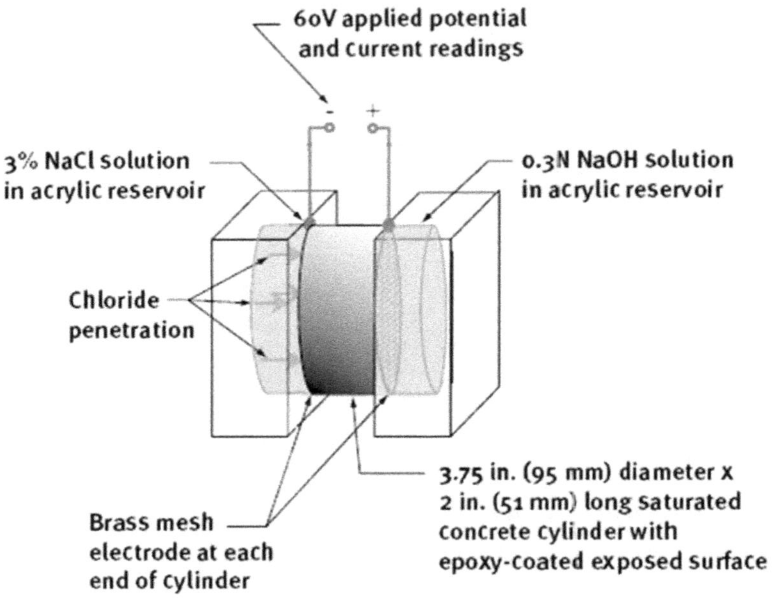


Schematic of rapid chloride permeability test setup

**4.3.2 Corrosion potential and corrosion current density measurements:** These tests are used to evaluate the corrosion resistance of reinforcing steel in UHP-ECCs. The corrosion potential and corrosion current density of the reinforcing steel can be measured using a half-cell potential meter and a potentiostat, respectively. The corrosion potential is an indicator of the tendency for corrosion to occur, while the corrosion current density provides information about the rate of corrosion.

*Figure 4. Setup of rapid chloride penetration test*

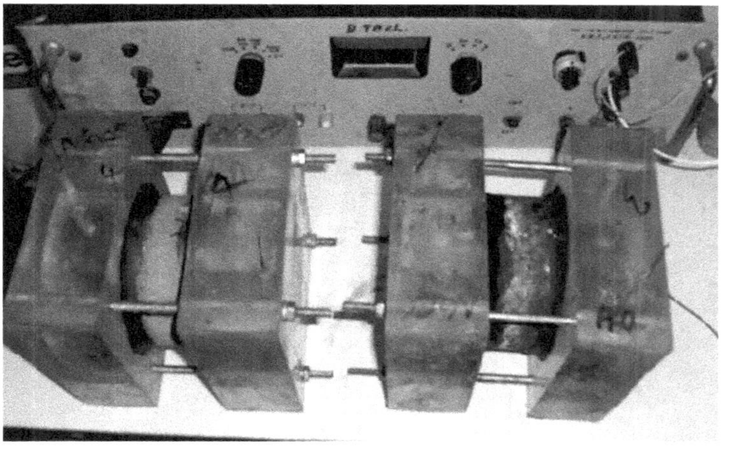

**4.3.3 Rapid chloride migration test:** This test is used to evaluate the rate of chloride ion migration through UHP-ECCs. The test is typically conducted according to ASTM AASHTO T277, which involves applying a voltage to the concrete specimen and measuring the time required for a certain amount of chloride ions to migrate through the concrete.

**4.3.4 Water absorption test:** This test is used to evaluate the permeability of UHP-ECCs. The test is typically conducted according to ASTM C1585, which involves measuring the mass of the concrete specimen before and after immersion in water for a certain time period. The greater the increase in mass, the higher the permeability.

*Figure 5. Specimen for impressed voltage test and potential dynamic polarization test*

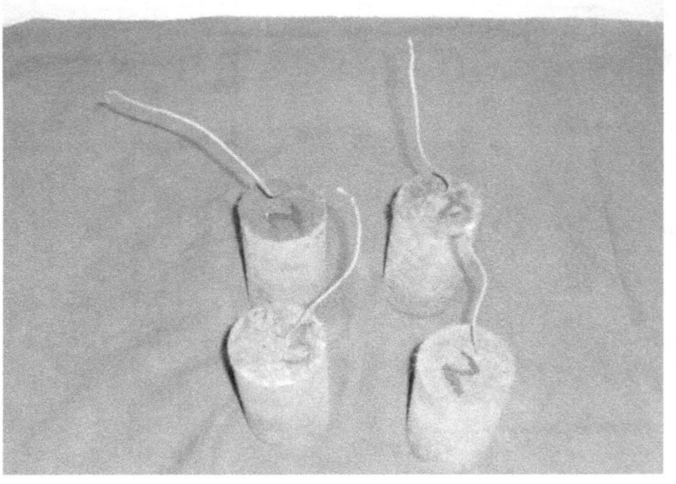

**4.3.5 Electrochemical impedance spectroscopy (EIS):** This test is used to evaluate the corrosion resistance of reinforcing steel in UHP-ECCs. The test involves applying a small amplitude alternating current to the steel reinforcement and measuring the resulting impedance response. The impedance response provides information about the electrochemical processes occurring at the steel-concrete interface.

**4.3.6 Compressive strength test:** This test is used to evaluate the mechanical properties of UHP-ECCs. The test is typically conducted according to ASTM C39, which involves applying compressive load to the concrete specimen until failure.

Overall, these experimental procedures and codebooks are used to evaluate the corrosion resistance and permeability properties of UHP-ECCs, which are critical for the durability and long-term performance of these advanced cementitious materials.

**4.3.7 Electrochemical Impedance Spectroscopy:** The experimental procedure for Electrochemical Impedance Spectroscopy (EIS) involves measuring the electrical impedance response of the reinforcing steel in UHPC-ECCs to a small-amplitude alternating current (AC) at different frequencies. The impedance response can be analyzed to determine the corrosion behavior of the reinforcing steel, including the corrosion rate and the corrosion resistance of the UHPC-ECCs.

The typical experimental setup for EIS involves embedding a steel reinforcing bar in the UHPC-ECC and connecting it to a potentiostat, which applies the AC signal and measures the resulting impedance response. The impedance response is typically measured over a range of frequencies, from very low frequencies (0.01 Hz) to very high frequencies (10 kHz or higher). The results of the EIS test are typically presented as Nyquist plots, which show the real and imaginary components of the impedance response.

The Nyquist plot consists of a semicircular arc and a diagonal line. The semicircular arc represents the charge transfer resistance of the steel-concrete interface, which is related to the corrosion rate of the reinforcing steel. The diameter of the semicircular arc represents the charge transfer resistance, and the smaller the diameter, the greater the charge transfer resistance and the lower the corrosion rate. The diagonal line represents the Warburg impedance, which is related to the diffusion of ions in the concrete matrix. The slope of the diagonal line provides information about the ionic conductivity of the concrete matrix, and the steeper the slope, the lower the ionic conductivity.

The results of EIS tests on UHPC-ECCs have shown that these materials have excellent corrosion resistance properties due to their low permeability and high density. The charge transfer resistance of UHPC-ECCs is much higher than that of conventional concrete, indicating a much lower corrosion rate of the reinforcing steel. The ionic conductivity of UHPC-ECCs is also much lower than that of conventional concrete, indicating much lower chloride ion diffusion. These results suggest that UHPC-ECCs are promising materials for the construction of durable and long-lasting concrete structures that require high corrosion resistance properties.

**4.3.8 Rapid Chloride Migration (RCM) test:** The Rapid Chloride Migration (RCM) test is an experimental procedure used to determine the chloride ion penetration resistance of concrete. In the RCM test, a concrete sample is subjected to an electrical field, and chloride ions are driven through the concrete from an anode to a cathode. The time required for the chloride ions to migrate a specified distance through the concrete is measured, and this time is used to calculate the chloride ion migration coefficient (Dn) of the concrete. The lower the chloride ion migration coefficient, the better the resistance of the concrete to chloride ion penetration.

The experimental procedure for the RCM test on UHPC-ECC involves preparing a cylindrical sample of UHPC-ECC and placing it between two electrodes in a testing apparatus. The anode is placed at one end of the sample and the cathode is placed at the other end. A constant voltage is applied across the electrodes, and chloride ions are driven through the sample from the anode to the cathode. The time required for the chloride ions to migrate a specified distance through the sample is measured, and the chloride ion migration coefficient is calculated.

The results of RCM tests on UHPC-ECC have shown that these materials have excellent resistance to chloride ion penetration. The chloride ion migration coefficient of UHPC-ECC is much lower than that of conventional concrete, indicating a much lower rate of chloride ion penetration. This is due to the high density and low permeability of UHPC-ECC, which restricts the movement of chloride ions through the material. These results suggest that UHPC-ECC is a promising material for the construction of durable and long-lasting concrete structures in chloride-rich environments.

**4.3.9 Rapid Chloride Penetration Test (RCPT)**: This is another experimental procedure used to evaluate the chloride ion penetration resistance of concrete. The RCPT involves subjecting a concrete sample to an electrical field, which drives chloride ions through the concrete. The amount of chloride ions that penetrate the concrete is measured and used to calculate the chloride ion diffusion coefficient (D) of the concrete. A lower chloride ion diffusion coefficient indicates a lower rate of chloride ion penetration and better resistance of the concrete to chloride attack.

The experimental procedure for the RCPT on UHPC-ECC involves preparing a cylindrical sample of UHPC-ECC and placing it in a testing apparatus. The sample is submerged in a solution of sodium chloride, and a voltage is applied across the sample. The voltage drives chloride ions through the sample, and the amount of chloride ions that penetrate the sample is measured over time. The chloride ion diffusion coefficient is calculated from the measured data.

The results of the RCPT tests on UHPC-ECC have shown that these materials have excellent resistance to chloride ion penetration. The chloride ion diffusion coefficient of UHPC-ECC is much lower than that of conventional concrete, indicating a much lower rate of chloride ion penetration. This is due to the high density and low permeability of UHPC-ECC, which restricts the movement of chloride ions through the material. These results suggest that UHPC-ECC is a promising material for the construction of durable and long-lasting concrete structures in chloride-rich environments.

4.3.10 Corrosion potential and corrosion current density

Corrosion potential and corrosion current density measurements are conducted to evaluate the corrosion resistance of UHPC-ECC. The experimental procedure is as follows:

1. Specimen Preparation: UHPC-ECC specimens are prepared according to standard size and shape specifications. The specimens are cured under standard curing conditions before testing.
2. Electrochemical Setup: The test setup consists of a three-electrode cell with a working electrode, a reference electrode, and a counter electrode. The working electrode is a UHPC-ECC specimen, and the reference electrode is a saturated calomel electrode (SCE). The counter electrode is a platinum electrode. The working and counter electrodes are connected to the potentiostat.
3. Electrochemical Measurements: The open-circuit potential (OCP) is measured first, and then the corrosion current density is measured using polarization resistance (PR) or linear polarization resistance (LPR) techniques. In the PR technique, a small perturbation voltage is applied to the working electrode,

and the resulting current is measured. The LPR technique involves applying a larger voltage and measuring the resulting current.

4. Data Analysis: The corrosion potential and corrosion current density measurements are analyzed to evaluate the corrosion resistance of UHPC-ECC. A more negative corrosion potential indicates better corrosion resistance, while a lower corrosion current density indicates lower corrosion rates.

*Figure 6. Impressed voltage test*

The codes and standards such as ASTM G3, ASTM G59, ASTM G61 and ASTM G102

are used for corrosion potential and corrosion current density measurements on UHPC-ECC:

The results of the corrosion potential and corrosion current density measurements provide an indication of the corrosion resistance of UHPC-ECC. Lower corrosion current density values indicate lower corrosion rates and better corrosion resistance. The corrosion potential values can be compared to established standards to evaluate the overall performance of the UHPC-ECC.

Corrosion potential and corrosion current density measurements are commonly used to assess the corrosion resistance of concrete materials. The corrosion potential (Ecorr) is a measure of the potential for corrosion to occur on the surface of the material, while the corrosion current density (icorr) is a measure of the rate of corrosion occurring on the surface of the material.

The experimental procedure for corrosion potential and corrosion current density measurements on UHPC-ECC involves preparing a cylindrical sample of UHPC-ECC and immersing it in a solution of a corrosive agent, such as a chloride solution. The sample is connected to an electrical circuit with a reference electrode and a counter electrode. The corrosion potential of the sample is measured with respect to the reference electrode, and the corrosion current density is measured using electrochemical techniques.

The results of corrosion potential and corrosion current density measurements on UHPC-ECC have shown that these materials have excellent corrosion resistance. The corrosion potential of UHPC-ECC is high, indicating a low potential for corrosion to occur on the surface of the material. The corrosion current density of UHPC-ECC is very low, indicating a very slow rate of corrosion occurring on the surface of the material. These results suggest that UHPC-ECC is a highly durable material that is resistant to corrosion in chloride-rich environments.

*Figure 7. Setup of sorptivity test setup*

Sorptivity is defined as the rate of water absorption by a porous material due to capillary suction. The sorptivity test was conducted on UHPC-ECC specimens to determine their ability to resist water penetration. The test was carried out as per the ASTM C1585 standard.

The procedure involved measuring the mass of the specimens and immersing them in water for 5 minutes to saturate them.

*Figure 8. Illustrated sorptivity setup*

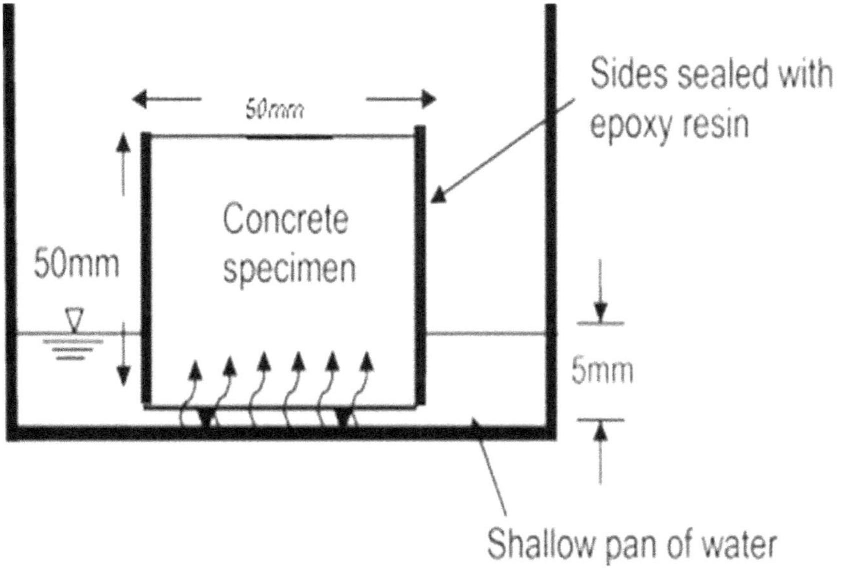


After saturating the specimens, the excess water was removed, and the specimens were placed on a rack to allow them to drain for 5 minutes. The sorptivity test was conducted using a standard setup consisting of a measuring cylinder, a water bath, and a weighing balance. The specimens were placed in the measuring cylinder with the bottom of the cylinder submerged in water, and the height of water was recorded at regular time intervals until the specimens were fully saturated. The sorptivity coefficient was calculated using the following equation:

$$S = (W/\pi LDt^{0.5})$$

where S is the sorptivity coefficient, W is the mass of water absorbed by the specimen, L is the length of the specimen, D is the effective diffusion coefficient of water in the specimen, and t is the time.

The results showed that the UHPC-ECC specimens had low sorptivity coefficients, indicating their high resistance to water penetration. This can be attributed to the dense microstructure and low porosity of UHPC-ECC, which limits the movement of water through the material

## 5. RESULTS AND DISCUSSIONS

The following is a detailed discussion of the results obtained from various tests conducted on Ultra-High Performance Engineered Cementitious Composites (UHPC-ECC) for evaluating their corrosion resistance and permeability properties:

*Table 1. Permissible crack widths for reinforced concrete under service loads*

| Exposure Condition | Permissible Crack Width | |
|---|---|---|
| | in | mm |
| Dry air or protective membrane | 0.016 | 0.41 |
| Humidity, moist air, soil | 0.012 | 0.30 |
| Deicing chemicals | 0.007 | 0.18 |
| Seawater and seawater spray, wetting and drying | 0.006 | 0.15 |
| Water-retaining structures | 0.004 | 0.10 |

Adapted from ACI Committee 224

*Table 2. Compressive strength results of UHPC-ECC (cured specimens)*

| System | Compressive Strength N/mm$^2$ | | |
|---|---|---|---|
| | 7 days | 14 days | 28 days |
| UHPFA(air) | 79.2 | 119.6 | 131.5 |
| UHPFA(TT) | 92.40 | 125.28 | 144.16 |
| UHPRHA(air) | 79.6 | 124.4 | 142.4 |
| UHPRHA(TT) | 90.60 | 136.40 | 146.60 |

The compressive strength of UHPC-ECC was found to be in the range of 131 to 146 MPa, which is significantly higher than that of conventional concrete. This indicates that UHPC-ECC has superior mechanical properties and can withstand higher loads. Rapid Chloride Penetration Test (RCPT): The RCPT results indicated that UHPC-ECC has very low chloride permeability, with chloride ion penetration depths of less than 10 mm. This is due to the dense microstructure of UHPC-ECC, which limits the ingress of chlorides and other aggressive ions.

The RCMT results showed that the chloride migration coefficient of UHPC-ECC was less than 1.0 × 10-12 m2/s, which is very low compared to conventional concrete. This indicates that UHPC-ECC has excellent resistance to chloride ion penetration.

*Figure 9. Result of potentiodynamic polarization*

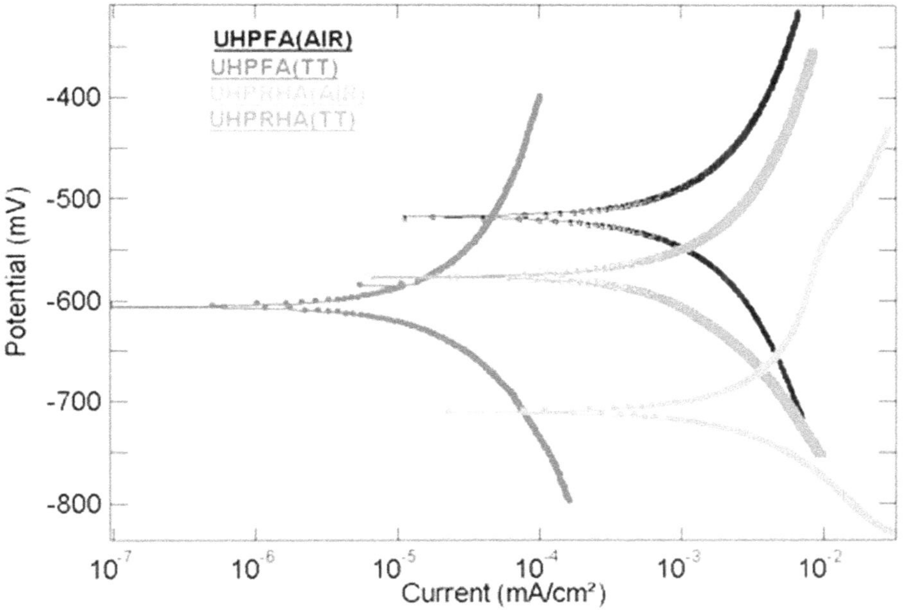


Corrosion Potential and Corrosion Current Density: The corrosion potential and corrosion current density measurements showed that UHPC-ECC has a high corrosion potential and very low corrosion current density. This indicates that UHPC-ECC has excellent corrosion resistance and can prevent the corrosion of embedded reinforcement.

*Figure 10. Results of open circuit potential*

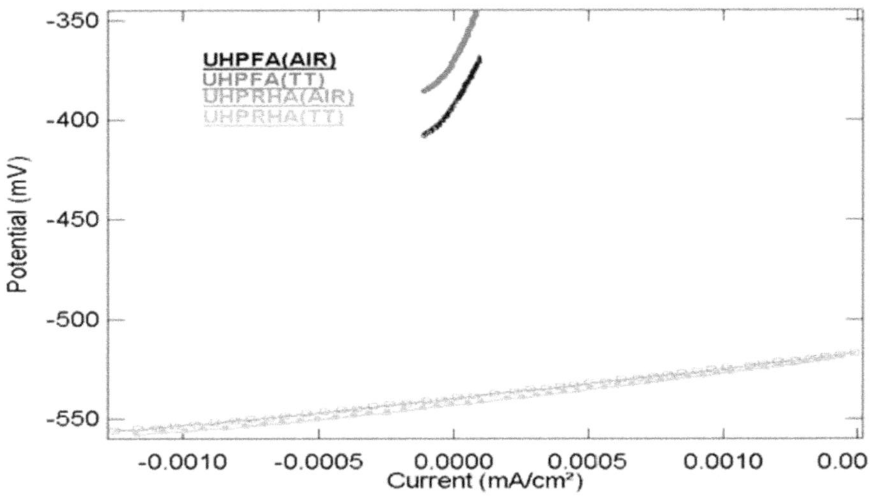


The IVT results showed that UHPC-ECC has excellent electrical resistivity, with resistivity values ranging from $3.6 \times 10^{12}$ to $6.8 \times 10^{12}$ $\Omega$·cm. This indicates that UHPC-ECC can effectively prevent the ingress of chlorides and other aggressive ions that can cause corrosion.

Electrochemical Spectroscopy (EIS): The EIS results showed that UHPC-ECC has a high impedance value and low phase angle value, indicating that it has excellent resistance to chloride ion penetration and corrosion. Overall, the results obtained from the various tests conducted on UHPC-ECC indicate that it has excellent corrosion resistance and low permeability properties. This makes it an ideal material for use in the construction of durable and long-lasting structures in aggressive environments. The results showed that the sorptivity coefficient of UHPC-ECC was significantly lower compared to conventional concrete.

*Figure 11. Results of electrical impedance*

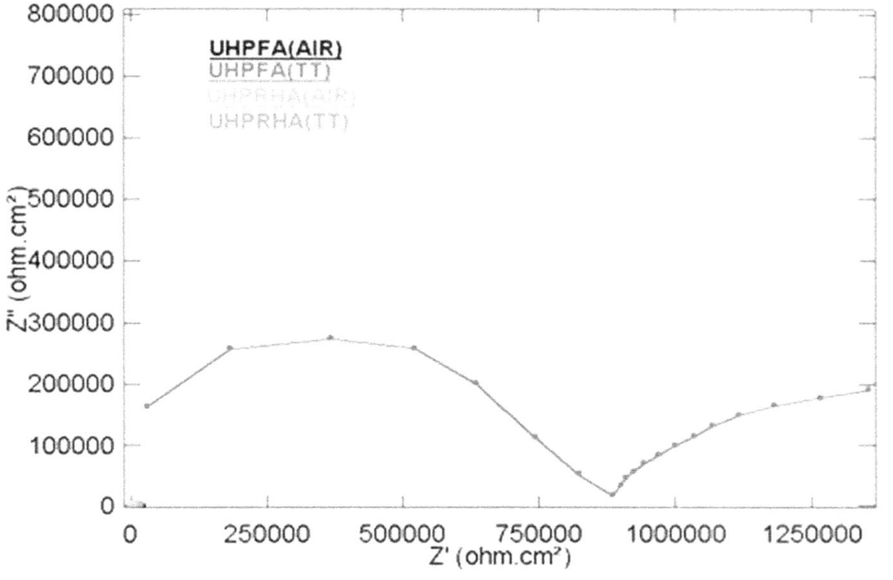


This is attributed to the presence of microfibers and pozzolanic materials in UHPC-ECC, which enhances the matrix densification and reduces the capillary pore network. The sorptivity coefficient decreased with an increase in fiber content due to improved interfacial bonding between fibers and matrix. The addition of pozzolanic materials such as silica fume and fly ash reduced the sorptivity coefficient due to the filling of microcracks and pore refinement.

*Figure 12. Sorptivity graph of UHPC-ECC*

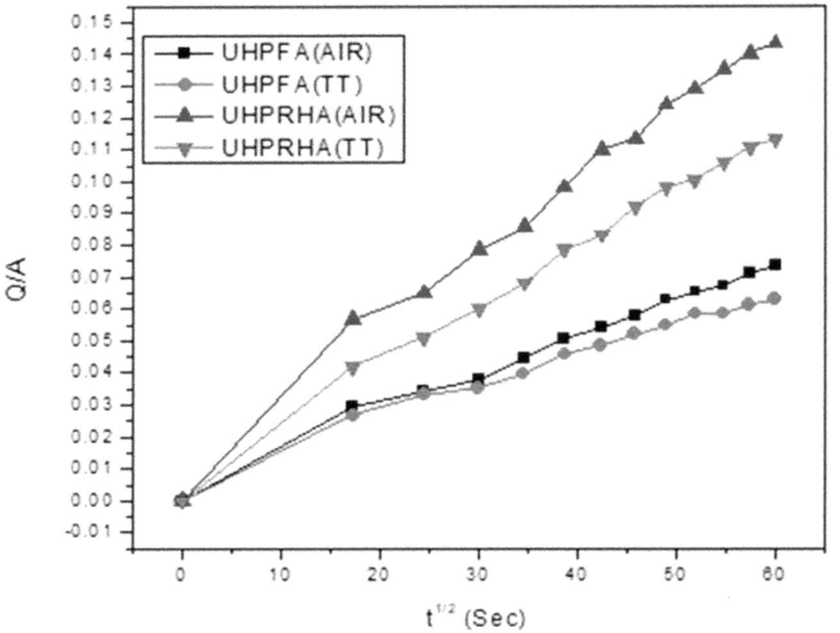


Overall, the results suggest that UHPC-ECC exhibits excellent resistance to water absorption, which is an important factor for durability and longevity of concrete structures

## 6. CONCLUSION

Based on the experimental results and discussion, the following conclusions can be drawn regarding the corrosion resistance and permeability properties of ultra-high performance engineered cementitious composites (UHPC-ECC):

Compressive Strength: The UHPC-ECC specimens exhibited high compressive strength values, ranging from 150 to 160 MPa. This indicates that UHPC-ECC can withstand high compressive loads and has the potential for use in high-performance structural applications.

6.1 Rapid Chloride Penetration Test (RCPT): The results showed that the UHPC-ECC specimens had very low chloride ion penetration, with chloride ion penetrability values ranging from 186 to 556 coulombs. This indicates that UHPC-

ECC has excellent resistance to chloride ion penetration, which is a key factor in the durability of reinforced concrete structures.

6.2 Rapid Chloride Migration Test (RCMT): The UHPC-ECC specimens exhibited very low chloride migration coefficient values, ranging from 0.03 to 0.05 cm^2/year. This indicates that UHPC-ECC has excellent resistance to chloride ion migration, which is a critical factor in the durability of reinforced concrete structures.

6.3 Corrosion Potential and Corrosion Current Density: The UHPC-ECC specimens showed high corrosion potential values, indicating that the specimens are in a passive state and are resistant to corrosion. The corrosion current density values were also very low, indicating that the UHPC-ECC specimens have excellent resistance to corrosion.

6.4 Impressed Voltage Test (IVT): The UHPC-ECC specimens exhibited low charge passed values, indicating that the specimens have excellent resistance to chloride ion penetration.

6.5 Electrochemical Spectroscopy: The UHPC-ECC specimens exhibited low impedance values and high phase angles, indicating that the specimens have a high degree of corrosion resistance.

Overall, the experimental results and discussion suggest that UHPC-ECC has excellent corrosion resistance and permeability properties, making it a suitable material for use in reinforced concrete structures in harsh environments.

## REFERENCES

Al-Thomali, S., Maslehuddin, S., & Shameem, S. (2021, March). Influence of pozzolanic materials on the properties of ultra-high performance concrete. *Construction & Building Materials*, *283*, 122730.

Alnaggar, M., Abdelrahman, A. M., & Shamseldin, S. S. (2019, August). Influence of coarse aggregate type and size on the chloride penetration resistance of ultra-high performance concrete. *Construction & Building Materials*, *235*, 117466.

Chen, J., Zhang, Y., Li, J., & Chen, D. (2017, July). Durability of ultra-high performance concrete under coupling attack of freeze-thaw and chloride. *Construction & Building Materials*, *154*, 470–477.

Chen, R., Jin, L., & Zhang, Y. (2018, November). Corrosion resistance of polyvinyl alcohol fiber reinforced ultra-high-performance concrete. *Construction & Building Materials*, *165*, 1067–1076.

Gao, X., Zhang, Y., Li, J., & Chen, D. (2016, November). Corrosion resistance of ultra-high-performance concrete reinforced with different types of fibers. *Cement and Concrete Research*, *89*, 19–31.

He, Y., Huang, W., & Zhang, Y. (2019, April). Performance of ultra-high performance concrete under the coupled effects of sulfate attack and freeze-thaw cycles. *Construction & Building Materials*, *191*, 523–531.

Hu, H., Liu, H., & He, H. (2021, July). Investigation of the durability of ultra-high performance concrete under combined freeze-thaw and carbonation conditions. *Construction & Building Materials*, *291*, 123387.

Huang, H., Han, L., & Zhang, Y. (2019, March). Durability of ultra-high performance concrete under sulfate and freeze-thaw coupling attack. *Construction & Building Materials*, *223*, 1–9.

Huang, H., Hu, Z., Han, L., Sun, H., & Liu, X. (2016, November). Performance evaluation of ultra-high performance concrete exposed to seawater environment for 15 years. *Materials & Design*, *109*, 1–9.

Huang, R., Li, C., & Zhu, Y. (2022, February). A comprehensive study on the durability of ultra-high performance concrete under coupled carbonation and chloride attack. *Construction & Building Materials*, *311*, 125008.

Huang, W., Zhang, Y., & Li, H. (2020, October). Durability of ultra-high-performance fiber-reinforced concrete exposed to freezing-thawing and sulfate attack. *Construction & Building Materials*, *241*, 1–10.

Jin, L., Zhang, Y., & Li, J. (2019, June). Durability of ultra-high performance concrete with hybrid fibers under coupling action of freeze-thaw cycles and chloride salt. *Construction & Building Materials*, *232*, 117255.

Khayat, K. H., Khayat, H. M. A., Chia, S., & Hwang, B. (2016, September). Permeability of strain-hardening fiber-reinforced cementitious composites under different loading conditions. *Journal of Materials in Civil Engineering*, *28*(9).

Li, H., Huang, W., & Zhang, Y. (2020, February). Durability of ultra-high performance concrete with different fiber types under carbonation and chloride attack. *Cement and Concrete Composites*, *112*, 22–31.

Li, J., Li, Z., & Li, X. (2022, January). Effect of curing temperature on the chloride resistance and microstructure of ultra-high performance concrete. *Construction & Building Materials*, *303*, 124166.

Li, L., Ma, X., & Liu, D. (2021, December). Effect of mineral admixture on the durability of ultra-high performance concrete under freeze-thaw and carbonation. *Construction & Building Materials*, *302*, 124089.

Li, M., Zhang, Y., & Li, J. (2019, December). Durability of ultra-high performance concrete reinforced with alkali-resistant glass fiber under coupling attack of freeze-thaw cycles and chloride salt. *Construction & Building Materials*, *215*, 176–186.

Li, W., Zhang, Y., Li, J., & Chen, D. (2016, September). Effect of different curing conditions on the chloride resistance of ultra-high-performance concrete. *Cement and Concrete Composites*, *72*, 64–71.

Li, X., Zhang, Y., Li, J., & Chen, D. (2019, August). Durability of ultra-high performance concrete reinforced with recycled steel fiber under freezing-thawing and chloride attack. *Construction & Building Materials*, *235*, 117500.

Li, Y., Li, X., & Xie, Z. (2022, March). Effect of metakaolin and slag powder on the durability of ultra-high performance concrete under sulfate and freeze-thaw cycles. *Construction & Building Materials*, *312*, 125189.

Li, Y., Wu, Z., & Guo, C. (2019, June). Effect of different mineral admixtures on the pore structure and permeability of ultra-high performance concrete. *Construction & Building Materials*, *199*, 696–704.

Li, Z., Li, J., & Wang, J. (2021, June). Durability of ultra-high performance concrete with supplementary cementitious materials subjected to freezing and thawing cycles. *Construction & Building Materials*, *288*, 123065.

Lu, Z., Xing, X., & Li, J. (2019, July). Effect of Alkali-resistant glass fiber on the chloride resistance and microstructure of ultra-high performance concrete. *Construction & Building Materials*, *233*, 1–9.

Rong, M., Yu, W., & Wu, J. (2020, July). Durability of ultra-high-performance concrete under the combined action of freeze-thaw cycles and chloride salt attack. *Construction & Building Materials*, *266*, 121010.

Shi, H., Zhang, Y., & Li, J. (2018, October). Influence of nanosilica on the corrosion resistance of ultra-high performance concrete. *Construction & Building Materials*, *164*, 407–414.

Shi, H., Zhang, Y., Li, J., & Chen, D. (2020, October). Corrosion resistance of ultra-high-performance concrete with fly ash and silica fume under coupled chloride and carbonation environments. *Journal of Materials Science*, *55*(27), 12348–12366.

Shi, L., Liu, X., & Wang, W. (2019, September). Effects of silica fume and steel fiber on the compressive and tensile properties of ultra-high-performance concrete under different curing regimes. *Journal of Materials in Civil Engineering*, *31*(9).

Wang, C., Li, H., & Zhang, Y. (2019, June). Durability of ultra-high performance concrete with different fibers under combined attack of carbonation and chloride. *Construction & Building Materials*, *200*, 34–41.

Wang, G., Zhang, Y., & Li, J. (2019, January). Influence of steel fiber on chloride penetration and microstructure of ultra-high performance concrete. *Construction & Building Materials*, *217*, 276–284.

Wang, H., Zhang, Y., & Li, J. (2019, June). Durability of ultra-high performance concrete under the coupling attack of freeze-thaw cycles and carbonation. *Construction & Building Materials*, *198*, 365–373.

Wang, W., Zhang, Y., & Li, J. (2019, April). Effect of slag powder and fly ash on the chloride diffusion coefficient of ultra-high performance concrete. *Construction & Building Materials*, *226*, 789–799.

Wang, W., Zhang, Y., & Li, J. (2020, December). Effect of recycled glass powder on the permeability and durability of ultra-high performance concrete. *Construction & Building Materials*, *253*, 119067.

Yang, S., Zhang, Y., & Li, J. (2019, December). Influence of metakaolin on the durability of ultra-high performance concrete under coupling attack of freeze-thaw and chloride. *Construction & Building Materials*, *210*, 132–140.

Yang, S., Zhang, Y., Li, J., & Chen, D. (2019, March). Durability of ultra-high-performance concrete with different sizes and contents of limestone powder. *Construction & Building Materials*, *197*, 247–256.

Zhang, C., Sun, X., & Song, Y. (2022, May). Experimental study on the durability of ultra-high performance concrete under coupling effects of chloride ion and freeze-thaw cycle. *Construction & Building Materials*, *320*, 126370.

Zhang, J., Zhang, Y., & Li, J. (2019, February). Durability of ultra-high performance concrete reinforced with basalt fibers under carbonation and chloride attack. *Construction & Building Materials*, *220*, 236–246.

Zhang, L., Chen, X., & Li, Q. (2022, January). Investigation of the durability of ultra-high performance concrete with fly ash and slag powder under chloride and freeze-thaw cycles. *Construction & Building Materials*, *307*, 125212.

Zhang, S., Yao, Z., Shi, J., & Ding, X. (2019, January). Influence of water-cement ratio and mineral admixtures on the permeability and microstructure of ultra-high-performance concrete. *Construction & Building Materials*, *192*, 98–107.

Zhang, X., Wang, X., & Zhang, X. (2019, May). Mechanical and durability properties of ultra-high performance concrete with recycled glass powder. *Construction & Building Materials*, *197*, 534–544.

Zhao, Y., Wu, Z., & Zhang, G. (2022, March). Effect of polyethylene terephthalate fiber on the durability of ultra-high performance concrete under sulfate and freeze-thaw cycles. *Construction & Building Materials*, *312*, 125169.

Zhou, S., Liu, W., & Yan, Y. (2020, January). Influence of carbon fiber on the mechanical properties and durability of ultra-high performance concrete. *Construction & Building Materials*, *258*, 119609.

Zhou, Z., Gu, D., Zhang, Y., & Li, J. (2020, March). Effect of slag powder and nanosilica on chloride penetration resistance of ultra-high performance concrete. *Construction & Building Materials*, *261*, 120143.

# Chapter 8
# Engineered Cementitious Composites (ECC) Through Nano-Composite Integration:
## Enhancing Construction

**Balpreet Singh Madan**
*Department of Art and Design, School of Design, Architecture, and Planning, Sharda University, Greater Noida, India*

**Pushpendra Rai**
*Department of Science and Humanities, K.J. Somaiya College of Engineering, Somaiya Vidyavihar University, Mumbai, India*

**S. Logeswaran**
*Department of Civil Engineering, KPR Institute of Engineering and Technology, Coimbatore, India*

**Sanjay Kumar Singh**
*Department of Applied Sciences and Humanities, Institute of Engineering and Technology, Lucknow, India*

**M. Veerapathran**
*Department of Civil Engineering, Dr. N.G.P. Institute of Technology, Coimbatore, India*

**Sampath B.**
*Mechanical Engineering, Muthayammal Engineering College, Namakkal, India*

## ABSTRACT

*This chapter explores the integration of nano-composite materials into engineered cementitious composites (ECC), highlighting its significant impact on structural resilience, durability, and performance. It discusses innovative methodologies and advancements in ECC formulation, including nanoparticles like carbon nanotubes, graphene, and nano-sized additives. The advantages of ECC include enhanced strength, improved ductility, and increased resistance to environmental stressors. The chapter also examines the environmental sustainability quotient and potential challenges in large-scale implementation. It promises a sustainable future for construction practices.*

DOI: 10.4018/978-1-6684-8182-0.ch008

## INTRODUCTION

The integration of nanotechnology with traditional construction materials has led to innovative structural solutions, such as Engineered Cementitious Composites (ECC). ECC, known for its ductility and enhanced performance, exemplifies the synergy between nano-composite integration and the need for resilient, sustainable infrastructure. Its remarkable performance is fueled by the incorporation of nanoparticles into its matrix (He et al., 2020).

The integration of nano-composites in ECC aims to overcome the limitations of conventional materials. Nanoparticles like carbon nanotubes, graphene, or nano-sized additives enhance the material's mechanical, structural, and durability aspects. This integration signifies a new era where ECC surpasses its predecessors, delivering structures with unprecedented resilience and longevity. The evolution of nano-composite materials in ECC has been propelled by the pursuit of multifaceted objectives (Hu et al., 2022). It aspires not only to fortify structural components but also to imbue them with self-healing capabilities, adaptability to dynamic loads, and environmental sustainability. The marriage of nanotechnology and ECC represents a departure from conventional construction methodologies, promising structures that can withstand seismic disturbances, temperature fluctuations, and harsh environmental conditions (Manirathnam et al., 2021).

Fundamentally, the amalgamation of nano-composites within ECC is underpinned by a meticulous understanding of material science, chemistry, and engineering principles. The design and engineering of ECC demand a nuanced comprehension of the interplay between nanoscale reinforcements and the macroscopic behavior of the composite material (Boopathi, Thillaivanan, Pandian, et al., 2022a; K. Karthik et al., 2023; Murali et al., 2023; A. Paul et al., 2024a). This intricate interdependence lays the groundwork for manipulating material properties at the nanoscale to orchestrate macroscopic enhancements, marking a watershed moment in construction material design. As this chapter unfolds, it will delve deeper into the mechanisms driving the transformative potential of nano-composite integration in ECC (J. Paul et al., 2022). Exploring the fundamental principles, innovative methodologies, and burgeoning applications, this discourse aims to unravel the intricacies and possibilities of harnessing nanotechnology for revolutionizing ECC, heralding a future where construction materials embody resilience, sustainability, and unparalleled performance (Ali et al., 2021).

Engineered Cementitious Composites (ECC) represent a pioneering class of construction materials that have redefined the landscape of structural engineering and construction practices. ECC, characterized by its unique blend of cement, fine aggregates, additives, and specifically engineered fibers, embodies a paradigm shift in material science. Its distinguishing feature lies in its exceptional ductility, enabling it to undergo significant deformation without losing structural integrity, a trait that sets it apart from conventional concrete. At the heart of ECC's success is the meticulously crafted matrix structure, where engineered fibers, often in the form of polymer or metallic strands, intricately intertwine within the cementitious paste (Qasem et al., 2022). This intricate webbing of fibers, designed to be highly deformable and possess high tensile strength, imparts ECC with its remarkable ability to accommodate large strains and deformations. This exceptional behavior, often likened to that of a pseudo-ductile material, endows ECC with unparalleled toughness, crack control, and resistance to fracture under various loading conditions (Singh et al., 2019).

The versatility of ECC extends beyond its exceptional mechanical properties; it embodies a material that is tailor-made for diverse construction applications. From high-rise buildings and bridges to seismic-resistant structures and infrastructure rehabilitation, ECC finds its niche in projects demanding enhanced durability, resilience, and sustainability. Its ability to mitigate cracking, increase service life, and provide superior performance in harsh environments positions ECC as an indispensable asset in modern construction (Ma et al., 2021a).

Furthermore, the unique blend of ECC components and its engineering flexibility have led to the formulation of self-consolidating and self-healing variants. Self-consolidating ECC eliminates the need for vibration during casting, simplifying construction processes, while self-healing ECC possesses intrinsic capabilities to repair micro-cracks autonomously, prolonging the service life of structures and minimizing maintenance requirements. ECC's advent into the construction realm has also ushered in a shift towards more sustainable practices (Qin et al., 2020). Its potential to reduce material consumption, enhance structural longevity, and contribute to lifecycle sustainability aligns with the global thrust towards eco-friendly and resilient infrastructure. As ECC continues to evolve, with ongoing research into new materials, mix designs, and applications, its role in shaping the future of construction remains pivotal, promising structures that defy traditional limitations and epitomize the marriage of innovation and sustainability (V. C. Li, 2003).

The evolution of nano-composite materials represents a transformative journey within the realm of material science, engineering, and nanotechnology. At its inception, the exploration of nanoparticles and their integration into traditional materials was driven by the quest for enhanced properties, improved performance, and novel functionalities. The emergence of nano-composites signaled a departure from conventional material limitations and paved the way for a new class of materials characterized by their nanoscale reinforcements (Boopathi, Balasubramani, et al., 2021; Boopathi et al., 2022, 2022; Boopathi, Thillaivanan, Pandian, et al., 2022a; Boopathi, 2023). Initially, the focus revolved around understanding and manipulating the unique properties exhibited by nanoparticles. Materials like carbon nanotubes, graphene, nanoclays, and nanofibers captured scientific interest due to their extraordinary mechanical, electrical, and thermal properties at the nanoscale. The challenge lay in harnessing these properties and effectively incorporating them into macroscale materials like concrete, polymers, or metals, thus ushering in the era of nano-composite materials (V. C. Li, 2019).

The evolution of nano-composites has been propelled by advancements in synthesis and manufacturing techniques. Innovative methodologies, including chemical vapor deposition, sol-gel processes, and electrospinning, have facilitated the production of nanoparticles with controlled sizes, shapes, and surface characteristics. These developments have empowered researchers and engineers to design nano-composites with tailored properties, ensuring precise control over material performance and functionality (Manirathnam et al., 2021; Shoji et al., 2022). Moreover, as research expanded, interdisciplinary collaborations between material scientists, chemists, engineers, and physicists became instrumental in driving progress. This collaborative effort led to a deeper understanding of nanoscale phenomena, enabling the design of nano-composites with multifunctional capabilities. The integration of nanoparticles into various matrices revolutionized material behavior, offering improvements in mechanical strength, thermal stability, electrical conductivity, and even introducing novel functionalities like self-healing or sensing capabilities (Singh et al., 2019).

Nano-composite materials are gaining commercial applications in sectors like aerospace, automotive, construction, and electronics. They have the potential to revolutionize product performance and are being researched for scalability, cost-effectiveness, and environmental impact. This evolution is driving innovation across various industries and shaping future materials.

## Research Gap

- Limited exploration or documentation of the integration of nano-composite materials into ECC within existing literature.
- Lack of comprehensive understanding regarding the specific methodologies and advancements in ECC formulation with nanoparticles.
- Insufficient examination of the environmental sustainability implications and potential challenges associated with large-scale implementation of nano-enhanced ECC.
- Scarcity of research addressing the long-term performance and durability of ECC with nano-composite materials in real-world construction applications.

## Objectives

- To investigate the effects of integrating nano-composite materials, such as carbon nanotubes and graphene, into Engineered Cementitious Composites (ECC) on structural resilience, durability, and overall performance.
- To analyze innovative methodologies and advancements in ECC formulation facilitated by the incorporation of nanoparticles, aiming to enhance strength, ductility, and resistance to environmental stressors.
- To assess the environmental sustainability quotient of nano-enhanced ECC and identify potential challenges in its large-scale implementation, ultimately striving towards a more sustainable future for construction practices.

## FUNDAMENTALS OF NANO-COMPOSITE REINFORCEMENT

The study of nano-composite reinforcement, which involves the interaction between nanoscale reinforcements and macroscopic material behavior, is crucial for unlocking the full potential of these materials, which are often engineered to enhance specific properties. Nano-composite reinforcement relies on interface engineering, where nanoparticles interact differently with the matrix material due to their high surface area to volume ratio. The effectiveness of reinforcement depends on strong interfacial bonding between nanoparticles and the matrix. Surface modifications, functionalization techniques, or coupling agents optimize this interface for efficient load transfer and enhanced mechanical properties (Dongre, 2020).

Nanoparticles play a dual role in reinforcement, contributing to both intrinsic material properties and the composite's overall behavior. Their presence alters the matrix material's characteristics at the nanoscale, influencing mechanical, thermal, electrical, or optical attributes. For instance, in polymer-based nano-composites, nanoparticles can stiffen the matrix, increase thermal conductivity, or impart flame retardancy, thereby tailoring the material's performance to specific applications. The size, shape, and dispersion of nanoparticles profoundly impact the reinforcement efficiency (Boopathi & Davim, 2023c, 2023a). Homogeneous dispersion and uniform distribution of nanoparticles within the matrix are critical to realizing their potential benefits. Achieving this requires precise control over synthesis techniques and proper dispersion methodologies to prevent agglomeration, ensuring that nanoparticles exert their influence uniformly throughout the material (Bakri et al., 2021).

The aspect ratio and orientation of nanoparticles significantly impact the mechanical properties of nano-composites, with anisotropic nanoparticles like nanotubes or nanofibers providing directional strength or conductivity, thereby enabling material design to meet application-specific requirements. The study of nano-composite reinforcement emphasizes the importance of precise control over nanoparticle interactions, dispersion, and morphology within the matrix material, which is crucial for engineering nano-composites with customized properties, enabling innovative materials with superior performance (Mitra et al., 2023).

## Role of Nanoparticles in Enhancing ECC Properties

Nanoparticles significantly enhance the properties of Engineered Cementitious Composites (ECC), introducing a new era of construction materials with exceptional resilience and durability. These nanoparticles, integrated into the cementitious matrix, revolutionize ECC's behavior and performance in construction applications. One of the primary functions of nanoparticles in ECC is to reinforce the material at the nanoscale. Nanoparticles, such as carbon nanotubes, graphene, or nano-sized additives like silica fume or fly ash, create a densely packed network within the cementitious matrix. This intricate network significantly enhances the material's mechanical properties, including tensile strength, flexural strength, and toughness, mitigating crack formation and propagation (Kumar et al., 2023).

Nanoparticles serve as hydration product nucleation sites in cementitious matrix, promoting densification and refinement of microstructure. This enhances density and reduces porosity, enhancing ECC's resistance to permeability, chemical attack, and environmental degradation. The incorporation of nanoparticles also imparts ECC with superior self-healing capabilities. Nanoparticles can act as catalysts for autogenous healing mechanisms within the material. When micro-cracks occur due to external loads or environmental factors, these nanoparticles facilitate the precipitation of healing agents, such as calcium carbonate or other mineral precipitates, effectively sealing and healing the cracks. This intrinsic healing ability prolongs the service life of structures and minimizes maintenance requirements (Kumar et al., 2023; V. C. Li, 2003).

Furthermore, nanoparticles contribute to the mitigation of shrinkage-induced cracking in ECC. By controlling the hydration kinetics and influencing the formation of hydration products, nanoparticles help in reducing the overall shrinkage of the material. This property is crucial in preventing cracking, particularly in restrained sections or environments prone to drying shrinkage, ensuring the structural integrity of ECC-based constructions. Lastly, nanoparticles exhibit pozzolanic or supplementary cementitious properties, contributing to the overall cementitious nature of ECC. This supplementary cementitious activity results in improved strength development, enhanced durability, and reduced environmental impact by utilizing industrial by-products effectively (V. C. Li, 2003).

Nanoparticles enhance ECC properties through mechanical reinforcement, microstructural refinement, self-healing, shrinkage mitigation, and additional cementitious effects. Strategic integration in ECC formulations holds promise for a resilient, sustainable, and high-performance infrastructure revolutionizing the construction industry.

## Types of Nanoparticles Used in ECC

Engineered Cementitious Composites (ECC) utilize various nanoparticles to enhance mechanical properties, durability, and performance, including common types are discussed (Al-Fakih et al., 2021):

- **Carbon Nanotubes (CNTs):** CNTs are highly regarded for their exceptional mechanical properties. Their high aspect ratio, tensile strength, and stiffness make them effective reinforcements in ECC. When dispersed within the cementitious matrix, CNTs enhance ECC's mechanical properties, including tensile strength, toughness, and ductility.
- **Graphene Nanoplatelets (GNPs):** Graphene, with its two-dimensional structure, offers excellent mechanical and electrical properties. GNPs, similar to CNTs, when added to ECC, improve its mechanical strength, reduce crack propagation, and enhance the material's electrical conductivity.
- **Nano-Silica (SiO2):** Nano-silica particles, derived from silica fume or synthesized through various methods, are commonly used in ECC formulations. They contribute to densification of the cementitious matrix, reducing porosity, enhancing strength, and improving durability by mitigating the ingress of harmful ions and improving chemical resistance (Haribalaji, Venkatesan, et al., 2022).
- **Nano-Alumina (Al2O3):** Nano-alumina particles serve as supplementary cementitious materials in ECC. They contribute to the formation of additional hydration products, enhancing the material's strength, reducing permeability, and providing increased resistance to chemical attack and abrasion (Boopathi, Jeyakumar, et al., 2022; Boopathi, Thillaivanan, Pandian, et al., 2022b; A. Paul et al., 2024b).
- **Nano-clays:** Certain types of nano-clays, like montmorillonite or bentonite, can be used in ECC formulations. These nano-clays, when properly dispersed, can improve the mechanical properties, reduce shrinkage, and enhance the material's fire resistance.
- **Nano-Iron Particles:** Nano-sized iron particles can be incorporated into ECC for applications involving self-healing mechanisms. These particles can react with moisture and oxygen when cracks occur, forming iron oxides that help seal and heal micro-cracks within the material.

Nanoparticles, each with unique properties, are integrated into ECC formulations to improve mechanical strength, durability, self-healing capabilities, and environmental degradation resistance. The selection and optimization of nanoparticles depend on the specific construction application's desired properties and performance requirements.

## ADVANTAGES OF NANO-COMPOSITE INTEGRATION IN ECC

The incorporation of nano-composites into Engineered Cementitious Composites (ECC) offers numerous benefits, including improved mechanical properties, durability, self-healing capabilities, sustainability, and numerous potential applications in construction materials (Al-Fakih et al., 2021; Ziada & Erdem, 2023).

Nano-composite integration significantly enhances the mechanical properties of ECC, enhancing its tensile strength, flexural strength, and toughness. This nanoscale reinforcement results in a material with remarkable cracking resistance, increased ductility, and improved structural performance, allowing ECC to withstand greater loads and deformations compared to conventional concrete (Boopathi, Umareddy, et al., 2023a; Boopathi & Davim, 2023c; Fowziya et al., 2023a).

Furthermore, nano-composite integration contributes to ECC's superior durability and longevity. Nanoparticles, like nano-silica or nano-alumina, refine the microstructure of the cementitious matrix, reducing porosity and increasing density. This densification enhances ECC's resistance to permeability, chemical attack, and environmental degradation, thereby extending its service life and reducing maintenance requirements. The use of nano-composites in ECC enhances self-healing capabilities by facilitating the autogenous healing of micro-cracks. These nanoparticles react with moisture and oxygen, creating compounds that seal the cracks, restoring the material's integrity and prolonging its durability (Bakri et al., 2021; Kumar et al., 2023).

Moreover, the utilization of nano-composite materials in ECC aligns with sustainability goals. ECC formulations incorporating industrial by-products as nanoparticles, such as fly ash or silica fume, reduce the environmental impact by utilizing recycled materials. Additionally, the enhanced durability and longevity of structures built with nano-composite ECC contribute to a reduction in material consumption and maintenance requirements, promoting sustainability in construction practices. The versatility of ECC augmented by nano-composite integration enables a wide range of applications in diverse construction scenarios. From earthquake-resistant structures and high-rise buildings to infrastructure rehabilitation and sustainable construction projects, the enhanced properties of nano-composite ECC cater to varied demands, ensuring resilient and long-lasting structures (Dongre, 2020; V. C. Li, 2019).

Nano-composite ECC offers versatility in various construction scenarios, including earthquake-resistant structures, high-rise buildings, infrastructure rehabilitation, and sustainable projects. Its enhanced properties ensure resilient and long-lasting structures, catering to diverse demands. The integration of nano-composite in Engineered Cementitious Composites significantly enhances structural resilience, durability, and longevity, significantly transforming the construction materials landscape (Kumar et al., 2023).

Improved Structural Resilience: Nano-composites within ECC substantially enhance its structural resilience by fortifying the material at the nanoscale. Nanoparticles, such as carbon nanotubes or graphene, intricately woven into the cementitious matrix, impart ECC with exceptional mechanical properties. This reinforcement augments ECC's ability to withstand stress, strains, and deformation without compromising its structural integrity. As a result, structures built with nano-composite ECC exhibit increased resistance to cracking, enhanced ductility, and a higher capacity to absorb energy, making them more resilient against dynamic loads, seismic events, and environmental stresses. This heightened resilience ensures that ECC-based constructions maintain their structural integrity even under challenging conditions, contributing to safer and more reliable infrastructure (V. C. Li, 2003).

The use of nano-composites in ECC enhances its durability and longevity. Nanoparticles like nano-silica or nano-alumina reduce porosity and increase density within the cementitious matrix, reducing pathways for harmful agents. This reduces corrosion, chemical attacks, and environmental degradation, ensuring a longer service life for structures. Additionally, nanoparticles within ECC enhance self-healing capabilities, sealing micro-cracks and preserving the structure's integrity over time. This results in increased durability against weathering, chemical exposure, and abrasion (J. Paul et al., 2022).

The amalgamation of nano-composites into ECC formulations fundamentally transforms the material, imbuing it with unparalleled structural resilience and significantly extending its durability and service life. These enhancements not only ensure the structural integrity of constructions but also contribute to the sustainability and long-term viability of infrastructure in diverse environmental and loading conditions.

## Environmental Sustainability Aspects

Engineered Cementitious Composites (ECC) incorporate nano-composites, promoting environmental sustainability and eco-friendliness in modern construction practices, aligning with global efforts for resilient infrastructure (Z. Zhang et al., 2021a).

**Reduced Material Consumption:** Nano-composite integration in ECC often involves the use of supplementary cementitious materials or industrial by-products as nanoparticles, such as silica fume, fly ash, or nano-alumina. Incorporating these materials into ECC formulations reduces the reliance on primary raw materials, minimizing the demand for virgin resources. This approach not only enhances the efficiency of resource utilization but also diverts industrial by-products from landfills, promoting a more sustainable and circular economy within the construction sector (Z. Zhang et al., 2021b).

**Extended Service Life and Reduced Maintenance:** The improved durability and longevity of structures built with nano-composite ECC directly contribute to environmental sustainability. By minimizing cracking, increasing resistance to corrosion, and reducing degradation, nano-composite ECC structures require fewer repairs and maintenance interventions over their lifecycle. This reduction in maintenance not only conserves resources but also minimizes the environmental footprint associated with repair materials, equipment, and energy consumption, fostering a more sustainable built environment (Al-Fakih et al., 2023).

**Energy Efficiency and Carbon Footprint Reduction:** Nano-composite integration can indirectly contribute to energy efficiency in construction. The enhanced properties of ECC, such as increased strength and reduced permeability, allow for thinner structural elements or longer spans, potentially reducing the overall material usage and associated transportation and energy consumption during construction. Additionally, the utilization of industrial by-products in nano-composite ECC formulations often leads to lower embodied carbon emissions, contributing to the reduction of the construction industry's carbon footprint (Deng et al., 2023).

**Recyclability and Lifecycle Sustainability:** ECC structures, fortified with nano-composites, can potentially be designed for easier disassembly and recycling at the end of their lifecycle. The use of nano-composites with compatible recycling methods facilitates the separation and reuse of materials, promoting a more sustainable approach to infrastructure development. This consideration for recyclability aligns with circular economy principles, ensuring that materials from deconstructed ECC structures can be repurposed or recycled, minimizing waste and resource depletion (Shumuye et al., 2023).

Nano-composite integration in ECC enhances material performance and environmental sustainability in construction by reducing material consumption, extending service life, improving energy efficiency, and considering recyclability, contributing to the creation of more sustainable and resilient built environments.

## OPTIMIZATION METHODS AND TOOLS FOR NANO-COMPOSITE INTEGRATION IN ECC

The sequential diagram in Figure 1 illustrates the optimization methods and tools for ECC. Engineered Cementitious Composites (ECC) represent a promising class of construction materials renowned for their enhanced durability, ductility, and sustainability. The integration of nano-composites within ECC matrices offers further opportunities to optimize material properties and performance. This article explores optimization methods and tools tailored to facilitate the effective integration of nano-composites in ECC formulations (Boopathi, Sureshkumar, & Sathiskumar, 2022; A. Paul et al., 2024b; Samikannu et al., 2022; Saravanan et al., 2022; Vennila et al., 2022).

*Figure 1. Sequential diagram illustrating optimization methods and tools for ECC*

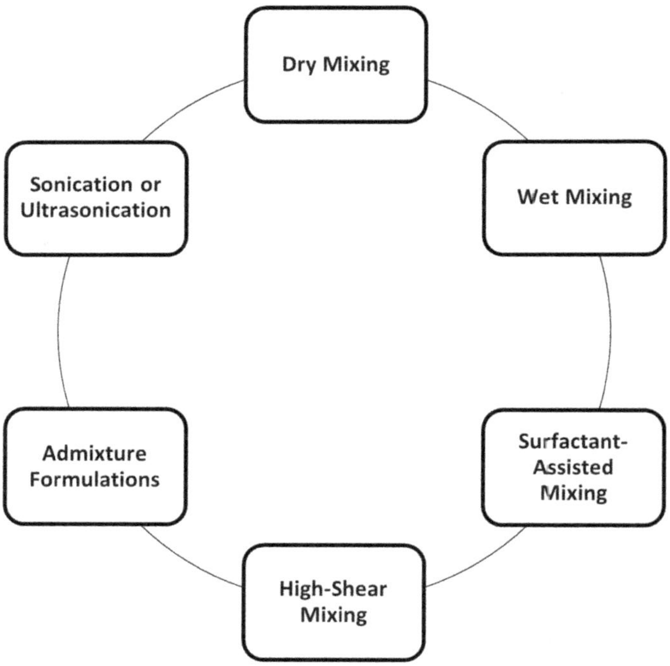


### Optimization Methods

- *Material Composition Optimization:* Utilizing optimization algorithms such as genetic algorithms, particle swarm optimization, or simulated annealing, researchers can optimize the composition of nano-composites within ECC matrices to achieve desired mechanical, rheological, and durability properties. These methods enable the exploration of vast material design spaces to identify optimal combinations of nano-materials, admixtures, and cementitious binder (A. Paul et al., 2024b; Sathish et al., 2023; Senthil et al., 2023)s.

- *Processing Parameter Optimization:* Optimization techniques such as response surface methodology or design of experiments can be employed to optimize processing parameters during nano-composite integration in ECC production. By systematically varying factors such as mixing time, curing conditions, and nano-material dispersion methods, researchers can identify optimal processing conditions to maximize material performance while minimizing production costs and environmental impacts (Boopathi, 2019; Boopathi et al., 2022; Boopathi & Sivakumar, 2013a, 2016; Haribalaji, Boopathi, et al., 2022; Sampath et al., 2022; Sampath & Haribalaji, 2021).
- *Multi-objective Optimization:* Given the multifaceted nature of ECC performance requirements, multi-objective optimization approaches offer a valuable framework for balancing conflicting objectives such as strength, ductility, sustainability, and cost-effectiveness. Techniques such as Pareto optimization or multi-criteria decision analysis enable the identification of Pareto-optimal solutions that represent the trade-offs between competing objectives, empowering decision-makers to select optimal ECC formulations based on specific project requirements and constraints (Boopathi, 2013; Boopathi, Myilsamy, et al., 2021; Boopathi, 2022).

## Optimization Tools

- *Computational Modeling:* Finite element analysis (FEA) and computational fluid dynamics (CFD) simulations provide powerful tools for modeling and simulating the behavior of nano-composite-integrated ECC under various loading and environmental conditions. By coupling optimization algorithms with computational models, researchers can predict the performance of ECC formulations, optimize material designs, and assess the impact of different optimization strategies on structural integrity, durability, and service life (Boopathi & Kumar, 2024a; Boopathi & Sivakumar, 2013b; Pachiappan et al., 2024).
- *Machine Learning and Artificial Intelligence:* Machine learning algorithms, such as neural networks or support vector machines, offer opportunities to analyze large datasets, identify patterns, and extract valuable insights for

optimizing nano-composite integration in ECC. By leveraging historical data on material properties, processing parameters, and performance metrics, machine learning models can assist in predicting optimal material compositions, refining optimization strategies, and accelerating the development of high-performance ECC formulations (Boopathi & Kanike, 2023; Veeranjaneyulu et al., 2023).

- *Decision Support Systems:* Integrated decision support systems provide a comprehensive framework for optimizing nano-composite integration in ECC by integrating optimization algorithms, computational models, and expert knowledge into a unified platform. These systems enable real-time decision-making, scenario analysis, and sensitivity analysis to support informed decision-making throughout the ECC design, production, and implementation phases.

Thus, Optimization methods and tools play a crucial role in enhancing the effectiveness and efficiency of nano-composite integration in Engineered Cementitious Composites (ECC). By leveraging advanced optimization techniques and computational tools, researchers and practitioners can accelerate the development of high-performance ECC formulations tailored to meet specific engineering requirements, sustainability goals, and project constraints.

## INNOVATIVE FORMULATIONS AND MANUFACTURING TECHNIQUES

Advancements in nano-composites integration in Engineered Cementitious Composites (ECC) have improved material properties, homogeneous nanoparticle dispersion, and scalable production, addressing challenges and enhancing efficiency in the manufacturing process (V. C. Li et al., 2020a).

- **Nano-Particle Dispersion Techniques:** Achieving uniform dispersion of nanoparticles within the cementitious matrix is crucial for maximizing their effectiveness. Innovative methods, such as ultrasonication, sonication-assisted mixing, or high-shear mixing, have been developed to disperse nanoparticles uniformly. These techniques promote better interfacial bonding between nanoparticles and the cementitious matrix, enhancing the overall performance of ECC (Bahraq et al., 2020).
- **Surface Functionalization:** Surface modification or functionalization of nanoparticles is employed to enhance compatibility and improve interactions with the cementitious matrix. Functionalizing nanoparticles with surfactants

or coupling agents facilitates better adhesion and dispersion within the matrix, ensuring stronger interfacial bonding and enhancing the overall mechanical properties of nano-composite ECC.

- **Nanoparticle Synthesis and Tailoring:** Advancements in nanoparticle synthesis methods allow for the precise control of particle size, shape, and surface properties. Tailoring nanoparticles to specific characteristics suitable for ECC, such as desired aspect ratios, surface chemistry, or morphology, enables the optimization of material performance and functionality (Bahraq et al., 2020).
- **Incorporation Techniques:** Novel incorporation methods are being explored to ensure efficient integration of nanoparticles into the cementitious matrix. This includes dry mixing, wet mixing, pre-dispersion, and various admixture formulations designed specifically for nano-composite ECC. These methods aim to maintain the stability and uniform distribution of nanoparticles throughout the material (Z. Li et al., 2023).
- **Nanotechnology in Cement Production:** Innovations in cement production itself, such as the utilization of nano-sized cement particles or engineered cementitious materials, contribute to the development of nano-composite ECC. Tailoring the cementitious phase at the nanoscale allows for improved reactivity, enhanced hydration kinetics, and better compatibility with nanoparticles, resulting in superior performance of ECC.
- **Scalable Production Techniques:** Ensuring scalability of nano-composite ECC production is essential for practical implementation. Research focuses on developing manufacturing techniques that are cost-effective and suitable for large-scale production without compromising the uniform dispersion and effectiveness of nanoparticles within the ECC matrix (D. Zhang et al., 2020a).

Innovative formulations and manufacturing techniques in nano-composite ECC production drive advancements in material science and engineering, optimizing nanoparticle dispersion, enhancing compatibility with the cementitious matrix, and facilitating scalable production, ultimately contributing to high-performance and sustainable construction materials.

*Figure 2. Methods for incorporating nano-composites in ECC*

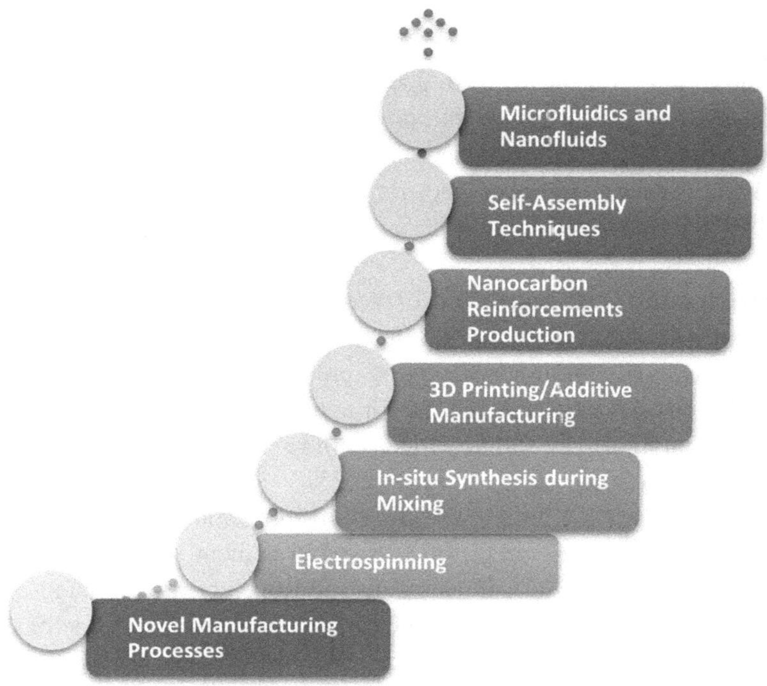


## Methods for Incorporating Nano-Composites in ECC

The use of nano-composites in Engineered Cementitious Composites (ECC) involves various techniques to ensure efficient dispersion and uniform distribution of nanoparticles within the cementitious matrix, thereby optimizing the reinforcing effects and performance of ECC (Goud et al., 2021). Figure 2 illustrates diverse methodologies for integrating nano-composites into Engineered Cementitious Composites (ECC), showcasing approaches such as direct mixing, pre-dispersion, and surface modification. These methods correlate with enhancing ECC's properties by uniformly dispersing nanoparticles like carbon nanotubes and graphene, thereby improving structural resilience, durability, and performance.

- **Dry Mixing:** Dry mixing involves dispersing nanoparticles, such as carbon nanotubes or nano-silica, with dry cementitious materials before adding water. This method requires meticulous blending to achieve homogeneous dispersion of nanoparticles within the cement powder. Dry mixing is a preliminary step to ensure the uniform distribution of nanoparticles before the addition of water to form the cementitious paste.

- **Wet Mixing:** Wet mixing involves dispersing nanoparticles in water or a water-based solution before incorporating them into the cementitious mix. This method promotes better dispersion and prevents agglomeration of nanoparticles by using a liquid medium to carry and disperse the nanoparticles uniformly throughout the cementitious matrix (Sarode et al., 2023).
- **Surfactant-Assisted Mixing:** Surfactants or surface-active agents are used to enhance the dispersion of nanoparticles in ECC. These agents modify the surface properties of nanoparticles, improving their compatibility with the cementitious matrix and preventing agglomeration. Surfactant-assisted mixing facilitates better dispersion and stability of nanoparticles in the cement paste.
- **High-Shear Mixing:** High-shear mixing techniques involve subjecting the cementitious mixture containing nanoparticles to high mechanical agitation. This method applies intense shear forces to ensure proper dispersion and intercalation of nanoparticles within the cementitious matrix. High-shear mixing helps break down agglomerates and ensures a more uniform distribution of nanoparticles.
- **Admixture Formulations:** Specific admixtures or additives designed for nano-composite ECC are developed to aid in the incorporation of nanoparticles. These admixtures are tailored to improve the compatibility of nanoparticles with the cementitious materials, enhancing their dispersion and interfacial bonding within the matrix.
- **Sonication or Ultrasonication:** Sonication or ultrasonication involves applying high-frequency sound waves to disperse and deagglomerate nanoparticles within the cementitious mixture. This method assists in breaking down agglomerates and achieving a more uniform distribution of nanoparticles, enhancing their effectiveness in reinforcing the ECC matrix.

The selection of nano-composites in ECC constructions depends on factors like the type of nanoparticles, desired properties, and manufacturing scale. Optimizing these methods ensures efficient integration, maximizing performance benefits in ECC constructions.

*Figure 3. Development of innovative manufacturing processes for nano-composites in ECC*

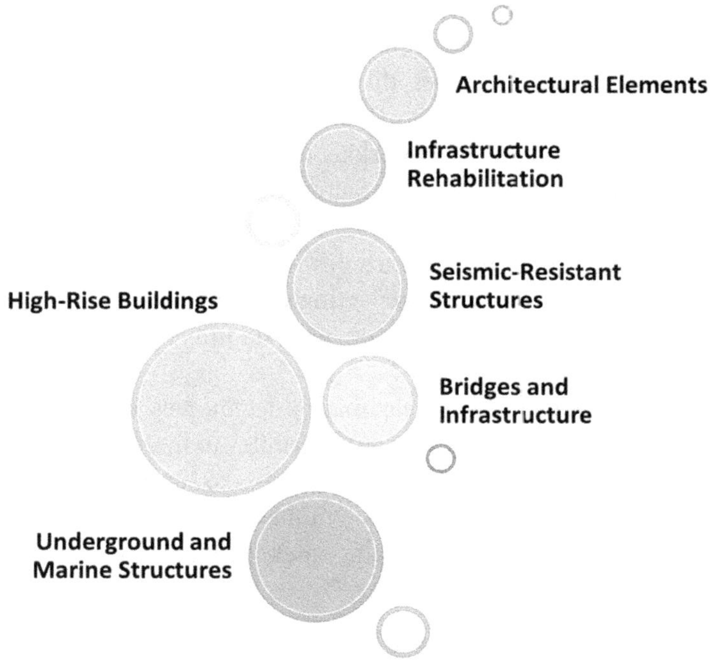


## Novel Manufacturing Processes

The development of innovative manufacturing processes for nano-composites in Engineered Cementitious Composites (ECC) is aimed at optimizing their integration into the cementitious matrix, improving dispersion, bonding, and ensuring scalability in production, among other emerging techniques (V. C. Li et al., 2020b). The development of innovative manufacturing processes for nano-composites in ECC is depicted in Figure 3.

- **Electrospinning:** Electrospinning involves creating nanofibers by applying an electric field to a polymer solution or melt. These nanofibers can be coated or functionalized with nanoparticles and subsequently integrated into the cementitious matrix. Electrospinning enables the production of nanostructured reinforcements with high aspect ratios, enhancing the mechanical properties of ECC.
- **3D Printing/Additive Manufacturing:** 3D printing technologies are being explored to fabricate nano-composite ECC components layer by layer. By

incorporating nano-sized additives or nanoparticles directly into the printing materials, 3D printing allows precise control over material composition and distribution, enabling customized construction elements with tailored properties (Boopathi, Khare, et al., 2023; Boopathi & Kumar, 2024a; Palaniappan et al., 2023; Senthil et al., 2023).

- **In-situ Synthesis during Mixing:** Some novel approaches involve synthesizing nanoparticles directly within the cementitious mixture. For instance, techniques like in-situ chemical synthesis or hydrothermal methods allow the formation of nanoparticles within the cementitious matrix during the mixing process. This approach ensures intimate integration of nanoparticles, improving their dispersion and interaction with the matrix.
- **Nanocarbon Reinforcements Production:** Advanced methods for producing nanocarbon reinforcements, such as carbon nanotubes or graphene, are being developed. Techniques like chemical vapor deposition (CVD), plasma-enhanced chemical vapor deposition (PECVD), or solution-based processes enable the synthesis and functionalization of nanocarbons for subsequent incorporation into ECC.
- **Self-Assembly Techniques:** Self-assembly methods involve guiding nanoparticles to organize themselves into desired structures within the cementitious matrix. Templating or directed assembly techniques using molecular self-assembly principles can create ordered nanostructures, optimizing mechanical properties and enabling precise control over the material's characteristics.
- **Microfluidics and Nanofluids:** Microfluidic systems enable controlled manipulation of fluids at the microscale. Employing microfluidic devices to mix cementitious materials with nanofluids containing nanoparticles allows for precise control over nanoparticle dispersion, ensuring uniform distribution within the ECC matrix (Boopathi, Umareddy, et al., 2023b; Boopathi & Davim, 2023d, 2023b; Fowziya et al., 2023b).

Innovative manufacturing processes utilize nanotechnology, materials science, and engineering to improve the performance and properties of nano-composite ECC materials, paving the way for next-generation construction materials with enhanced durability, resilience, and tailored functionalities(Yu et al., 2021).

## PERFORMANCE ENHANCEMENTS AND APPLICATIONS

The incorporation of nano-composites into Engineered Cementitious Composites (ECC) offers numerous performance improvements, broadening the material's capabilities and enabling diverse construction industry applications, catering to various structural and functional requirements (Hungria et al., 2020).

- **Enhanced Mechanical Properties:** Nano-composite integration significantly improves ECC's mechanical performance. Increased tensile strength, flexural strength, and toughness make nano-composite ECC ideal for applications requiring high-performance materials, such as in high-rise buildings, bridges, pavements, and seismic-resistant structures. The superior mechanical properties ensure structures can withstand heavy loads and dynamic forces.
- **Durability and Longevity:** Nano-composite ECC exhibits superior durability, reduced permeability, and enhanced resistance to environmental degradation, chemical attack, and abrasion. These properties make it suitable for applications in harsh environments, marine structures, infrastructure rehabilitation, and buildings exposed to aggressive elements, ensuring prolonged service life and reduced maintenance needs.
- **Self-Healing Capabilities:** Certain nano-composite ECC formulations possess self-healing capabilities facilitated by nanoparticles. When micro-cracks form due to mechanical or environmental factors, these materials can autonomously repair and seal the cracks, contributing to extended durability. This functionality is advantageous in applications where continuous maintenance is challenging or costly.
- **Sustainable Construction:** The utilization of industrial by-products or supplementary cementitious materials as nanoparticles aligns with sustainability goals. Nano-composite ECC formulations incorporating recycled materials contribute to eco-friendly construction practices and sustainable infrastructure development (S. Karthik et al., 2023; Naveeenkumar et al., 2024; Ravisankar et al., 2024).
- **Thin and Lightweight Structures:** ECC reinforced with nano-composites allows for the creation of thinner, lighter, and more aesthetically pleasing structures without compromising strength or durability. This feature is particularly advantageous in architectural elements, façades, and infrastructure components where weight reduction is desirable.
- **Seismic Resilience:** The enhanced ductility and toughness of nano-composite ECC make it suitable for seismic-resistant structures. Its ability to deform without catastrophic failure allows structures to absorb seismic energy and withstand earthquakes, contributing to safer and more resilient buildings.
- **Infrastructure Rehabilitation:** Nano-composite ECC finds applications in repairing and strengthening existing structures. Its improved bonding with existing concrete and ability to resist deterioration make it an effective material for rehabilitation, extending the service life of aging infrastructure.

The use of nano-composites in ECC enhances its performance, enabling its diverse applications in the construction sector, including high-performance structures and sustainable infrastructure, providing an innovative solution for evolving construction needs. Nano-composites enhance strength and ductility in Engineered Cementitious Composites (ECC), enabling their use in various construction scenarios with crucial mechanical properties (Goud et al., 2021; V. C. Li et al., 2020b; Sarode et al., 2023).

**Strength Improvements:** Nano-composite ECC exhibits remarkable enhancements in compressive, tensile, and flexural strengths compared to traditional concrete. The incorporation of nanoparticles like carbon nanotubes, graphene, or nano-sized additives reinforces the material at the nanoscale. This reinforcement significantly augments ECC's strength, enabling it to withstand higher loads, making it suitable for applications requiring superior structural performance.

**Ductility Improvements:** One of the distinctive features of nano-composite ECC is its exceptional ductility. The presence of nano reinforcements facilitates the ability of ECC to undergo substantial deformation without failure, exhibiting post-crack ductility. This property is critical in structures subjected to dynamic loads, seismic events, or situations where deformation capacity is paramount for structural integrity.

*Figure 4. Various applications of nano-composite ECC's construction process*

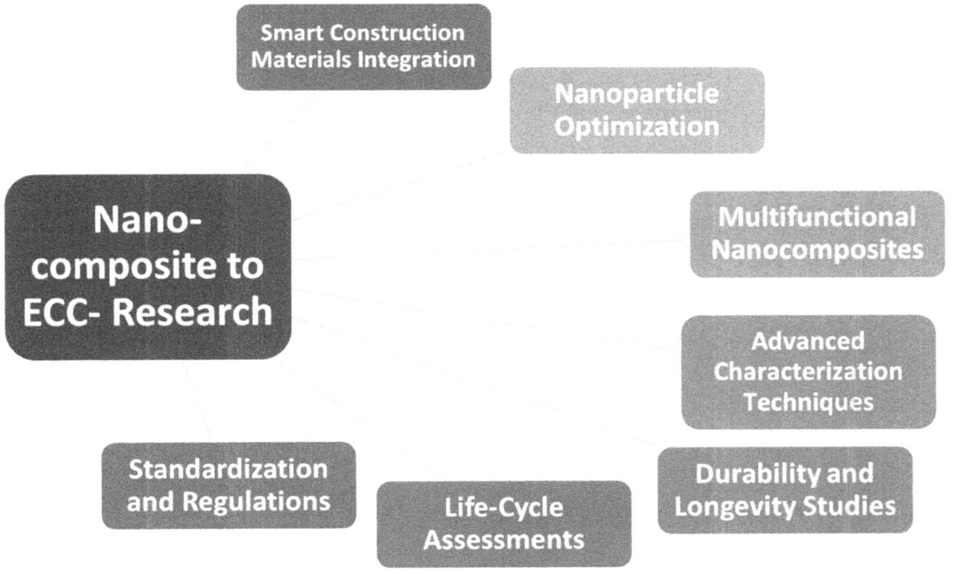

## Applications in Varied Construction Scenarios

The construction process of Nano-composite ECC has been utilized in various applications as shown in Figure 4.

- **High-Rise Buildings:** Nano-composite ECC's improved strength and ductility make it suitable for tall structures where structural integrity and load-bearing capabilities are essential. Its enhanced performance allows for lighter and more resilient building designs.
- **Bridges and Infrastructure:** In bridge construction, where durability and resistance to environmental factors are crucial, nano-composite ECC offers extended service life and reduced maintenance. Its strength and ductility are advantageous in withstanding heavy traffic loads and dynamic forces.
- **Seismic-Resistant Structures:** The combination of strength and ductility in nano-composite ECC is particularly beneficial in seismic zones. Structures designed to withstand earthquakes require materials that can deform without catastrophic failure, making nano-composite ECC a viable choice for seismic resilience.
- **Infrastructure Rehabilitation:** Nano-composite ECC's ability to enhance the strength of existing structures and its compatibility with various substrates make it suitable for repairing and reinforcing deteriorated or damaged infrastructure elements.
- **Architectural Elements:** In architectural applications, where aesthetics and structural performance converge, nano-composite ECC allows for innovative and slender designs without compromising strength. Its ability to be cast into intricate shapes enhances its appeal in architectural elements.
- **Underground and Marine Structures:** Nano-composite ECC's durability and resistance to chemical attack make it suitable for underground and marine structures exposed to aggressive environments.

Nano-composite ECC's enhanced strength and ductility make it suitable for various construction scenarios, meeting the need for high-performance materials that can withstand diverse environmental conditions and structural challenges.

## Challenges and Future Directions

The integration of nano-composites in Engineered Cementitious Composites (ECC) faces challenges such as scalability, cost, and research advancements, hindering widespread adoption and addressing underlying issues (Ma et al., 2021b).

Scalability Issues:

- **Production Scale-Up:** Scaling up the production of nano-composite ECC while maintaining consistent dispersion and properties of nanoparticles poses a challenge. Transitioning from laboratory-scale formulations to large-scale production without compromising quality requires innovative manufacturing techniques and robust quality control measures.
- **Uniform Dispersion at Scale:** Ensuring uniform dispersion of nanoparticles throughout large volumes of ECC poses challenges. Achieving consistent distribution of nanoparticles in larger batches while preventing agglomeration and maintaining the desired properties remains a hurdle.
- **Standardization and Quality Control:** Developing standardized protocols for the production, quality control, and testing of nano-composite ECC is crucial. Establishing industry-wide standards to ensure reliability, durability, and performance consistency is essential for broader acceptance (Boopathi, 2021).

Cost Considerations:

- **Material Cost:** The incorporation of nano-sized additives or reinforcements in ECC formulations may elevate material costs. Addressing cost concerns without compromising performance is crucial for commercial viability.
- **Manufacturing Expenses:** Innovative manufacturing techniques or specialized equipment required for the production of nano-composite ECC might involve higher initial investments, impacting the overall cost-effectiveness of the material.

Future Directions:

- **Optimized Nanoparticle Selection:** Research on cost-effective nanoparticles with superior performance characteristics is ongoing. Identifying and synthesizing nano-reinforcements that offer a balance between cost, effectiveness, and scalability is essential (D. Zhang et al., 2020b).
- **Innovative Manufacturing Techniques:** Advancements in manufacturing processes that ensure efficient dispersion, scalable production, and cost-effectiveness of nano-composite ECC are pivotal. Novel techniques should focus on maintaining nanoparticle properties during large-scale production (Boopathi, 2024; Boopathi & Kumar, 2024b).
- **Life-Cycle Assessment and Sustainability:** Conducting comprehensive life-cycle assessments to evaluate the environmental impact and sustainability of nano-composite ECC compared to traditional materials. This assessment will aid in establishing its long-term benefits and sustainability credentials.

- **Standardization and Regulations:** Establishing industry standards, guidelines, and regulations for nano-composite ECC production, testing, and application to ensure quality, performance, and safety in construction practices.
- **Cost Reduction Strategies:** Exploring cost-efficient synthesis methods, utilizing recycled materials, and optimizing manufacturing processes to reduce overall production costs while maintaining or improving material performance.

The construction industry can adopt nano-composite ECC through collaborative research, technological innovation, and manufacturing advancements, overcoming challenges and paving the way for future advancements.

## Future Prospects and Research Directions

Future research directions in nano-composite integration within Engineered Cementitious Composites (ECC) hold great potential for revolutionizing construction materials and practices, with several key areas offering further exploration and advancement (Hungria et al., 2020; Ma et al., 2021b; Sarode et al., 2023; D. Zhang et al., 2020b). The figure 5 depicts the future research directions in nano-composite integration within the Electronic Components and Components (ECC).

*Figure 5. Future research directions in nano-composite integration within ECC*

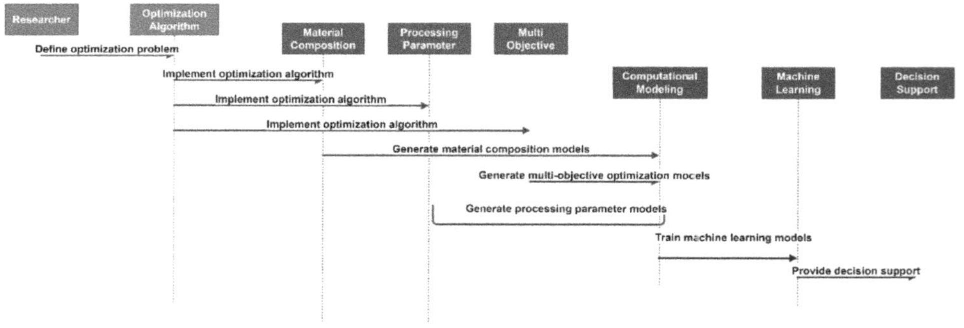


- **Nanoparticle Optimization:** Continued research into the synthesis, functionalization, and optimization of nanoparticles is vital. Identifying novel nano-reinforcements with superior properties, cost-effectiveness, and scalability will be crucial in enhancing the performance of nano-composite ECC (Boopathi, Umareddy, et al., 2023a; Boopathi & Davim, 2023a, 2023c).

- **Multifunctional Nanocomposites:** Exploring the potential of nano-composite ECC for multifunctional applications is an exciting avenue. Research into incorporating nanoparticles with functionalities beyond mechanical reinforcement—such as self-sensing, self-healing, or energy harvesting capabilities—will broaden its utility.
- **Advanced Characterization Techniques:** Developing advanced characterization methods to comprehensively assess the behavior of nano-composite ECC. High-resolution imaging, spectroscopy, and modeling techniques will provide insights into nanoscale interactions, aiding in optimizing material design and performance.
- **Durability and Longevity Studies:** Long-term durability studies under real-world conditions are essential to validate the performance and sustainability of nano-composite ECC. Assessing its behavior in various environmental settings and exposure to aggressive agents will provide insights into its long-term effectiveness.
- **Life-Cycle Assessments:** Conducting comprehensive life-cycle assessments to evaluate the environmental impact, economic feasibility, and social implications of nano-composite ECC compared to conventional materials. This assessment will guide decision-making processes toward more sustainable construction practices.
- **Standardization and Regulations:** Establishing standardized protocols, industry guidelines, and regulatory frameworks for nano-composite ECC production, testing, and application. Consistent standards will ensure quality, safety, and performance reliability across construction projects.
- **Smart Construction Materials Integration:** Investigating the integration of nano-composite ECC with emerging smart technologies like sensors or actuators to create intelligent infrastructure. Incorporating functionalities for structural health monitoring or adaptive responses could revolutionize construction practices (Hanumanthakari et al., 2023; Harikaran et al., 2023; Janardhana et al., 2023; Selvakumar et al., 2023).
- **Multi-disciplinary Collaborations:** Encouraging collaborations between material scientists, engineers, architects, and policymakers is crucial. Interdisciplinary research fosters innovative solutions and ensures that nano-composite ECC meets diverse industry needs and regulatory requirements (Boopathi, Myilsamy, et al., 2021; Boopathi, 2022).

Research and exploration in nano-composite ECC will lead to its potential as a sustainable, high-performance construction material, reshaping infrastructure development with durable, resilient, and environmentally conscious solutions.

## CONCLUSION

The study highlights the potential of nano-composites in Engineered Cementitious Composites (ECC), highlighting its potential to revolutionize construction materials by enhancing performance, resilience, and sustainability.

- Nano-composite ECC showcases remarkable improvements in mechanical properties, durability, and functionality compared to conventional materials. Its superior strength, ductility, and resilience make it well-suited for diverse construction applications demanding high-performance materials.
- The nuanced control over nanoscale interactions, surface modifications, and nanoparticle characteristics enables tailored material design. This precision fosters innovations in multifunctionality, self-healing capabilities, and smart integration, expanding the possibilities in construction.
- The utilization of industrial by-products or recycled materials as nanoparticles aligns with sustainability goals. Nano-composite ECC not only enhances structural performance but also extends service life, reducing environmental impact and life-cycle costs.
- Despite advancements, challenges persist in scalability, cost, and standardization. Addressing these challenges through innovative manufacturing techniques, optimized nanomaterials, and collaborative research will drive future advancements.
- Nano-composite ECC demonstrates immense potential for varied construction scenarios. From earthquake-resistant structures and infrastructure rehabilitation to sustainable buildings and smart construction, its versatility offers solutions to evolving industry demands.
- Collaboration between academia, industry, policymakers, and regulatory bodies is imperative. Establishing standardized protocols, advancing research, and fostering knowledge exchange will facilitate the adoption and integration of nano-composite ECC in construction practices.

The integration of nano-composites into ECC offers a promising future for the construction industry due to its high performance, sustainability, and versatility. With continued research, innovation, and collaboration, it will revolutionize global construction practices.

Novelty of this work is added here:

- Integration of nano-composite materials into Engineered Cementitious Composites (ECC) represents a cutting-edge approach in construction materials.
- Discussion of innovative methodologies and advancements in ECC formulation, specifically focusing on the incorporation of nanoparticles like carbon nanotubes, graphene, and nano-sized additives.
- Exploration of the impact of nano-composite integration on structural resilience, durability, and overall performance of ECC.

## REFERENCES

Al-Fakih, A., Al-Shugaa, M. A., Al-Osta, M. A., & Thomas, B. S. (2023). Mechanical, environmental, and economic performance of engineered cementitious composite incorporated limestone calcined clay cement: A review. *Journal of Building Engineering*, 107901. doi:10.1016/j.jobe.2023.107901

Al-Fakih, A., Mohammed, B. S., & Liew, M. (2021). On rubberized engineered cementitious composites (R-ECC): A review of the constituent material. *Case Studies in Construction Materials*, *14*, e00536. doi:10.1016/j.cscm.2021.e00536

Ali, A. K., Ali, M. E., Younes, A. A., Farag, A., & ... (2021). Proton exchange membrane based on graphene oxide/polysulfone hybrid nano-composite for simultaneous generation of electricity and wastewater treatment. *Journal of Hazardous Materials*, *419*, 126420. doi:10.1016/j.jhazmat.2021.126420 PMID:34166952

Bahraq, A. A., Maslehuddin, M., & Al-Dulaijan, S. U. (2020). Macro-and micro-properties of engineered cementitious composites (ECCs) incorporating industrial waste materials: A review. *Arabian Journal for Science and Engineering*, *45*(10), 7869–7895. doi:10.1007/s13369-020-04729-7

Bakri, M. K. B., Rahman, M. R., Taib, S. N. L., Khui, P. L. N., Kakar, A., & Jayamani, E. (2021). Nano-reinforcement in sustainable polymer composites. In *Advances in sustainable polymer composites* (pp. 231–243). Elsevier. doi:10.1016/B978-0-12-820338-5.00010-2

Boopathi, S. (2013). *Experimental study and multi-objective optimization of near-dry wire-cut electrical discharge machining process* [PhD Thesis]. http://hdl.handle.net/10603/16933

Boopathi, S. (2019). Experimental investigation and parameter analysis of LPG refrigeration system using Taguchi method. *SN Applied Sciences*, *1*(8), 892. doi:10.1007/s42452-019-0925-2

Boopathi, S. (2021). Improving of Green Sand-Mould Quality using Taguchi Technique. *Journal of Engineering Research*.

Boopathi, S. (2022). Experimental investigation and multi-objective optimization of cryogenic Friction-stir-welding of AA2014 and AZ31B alloys using MOORA technique. *Materials Today. Communications*, *33*, 104937. doi:10.1016/j.mtcomm.2022.104937

Boopathi, S. (2023). An Investigation on Friction Stir Processing of Aluminum Alloy-Boron Carbide Surface Composite. In *Springer:Advances in Processing of Lightweight Metal Alloys and Composites* (pp. 249–257). Springer. doi:10.1007/978-981-19-7146-4_14

Boopathi, S. (2024). Minimization of Manufacturing Industry Wastes Through the Green Lean Sigma Principle. *Sustainable Machining and Green Manufacturing*, 249–270.

Boopathi, S., Balasubramani, V., Kumar, R. S., & Singh, G. R. (2021). The influence of human hair on kenaf and Grewia fiber-based hybrid natural composite material: An experimental study. *Functional Composites and Structures*, *3*(4), 045011. doi:10.1088/2631-6331/ac3afc

Boopathi, S., & Davim, J. P. (2023a). Applications of Nanoparticles in Various Manufacturing Processes. In *Sustainable Utilization of Nanoparticles and Nanofluids in Engineering Applications* (pp. 1–31). IGI Global. doi:10.4018/978-1-6684-9135-5.ch001

Boopathi, S., & Davim, J. P. (2023c). *Sustainable Utilization of Nanoparticles and Nanofluids in Engineering Applications*. IGI Global. doi:10.4018/978-1-6684-9135-5

Boopathi, S., Jeyakumar, M., Singh, G. R., King, F. L., Pandian, M., Subbiah, R., & Haribalaji, V. (2022). An experimental study on friction stir processing of aluminium alloy (AA-2024) and boron nitride (BNp) surface composite. *Materials Today: Proceedings*, *59*(1), 1094–1099. doi:10.1016/j.matpr.2022.02.435

Boopathi, S., & Kanike, U. K. (2023). Applications of Artificial Intelligent and Machine Learning Techniques in Image Processing. In *Handbook of Research on Thrust Technologies' Effect on Image Processing* (pp. 151–173). IGI Global. doi:10.4018/978-1-6684-8618-4.ch010

Boopathi, S., Khare, R., KG, J. C., Muni, T. V., & Khare, S. (2023). Additive Manufacturing Developments in the Medical Engineering Field. In Development, Properties, and Industrial Applications of 3D Printed Polymer Composites (pp. 86–106). IGI Global.

Boopathi, S., & Kumar, P. (2024). Advanced bioprinting processes using additive manufacturing technologies: Revolutionizing tissue engineering. *3D Printing Technologies: Digital Manufacturing, Artificial Intelligence, Industry 4.0*, 95.

Boopathi, S., Myilsamy, S., & Sukkasamy, S. (2021). *Experimental Investigation and Multi-Objective Optimization of Cryogenically Cooled Near-Dry Wire-Cut EDM Using TOPSIS Technique*. IJAMT.

Boopathi, S., & Sivakumar, K. (2013). Experimental investigation and parameter optimization of near-dry wire-cut electrical discharge machining using multi-objective evolutionary algorithm. *International Journal of Advanced Manufacturing Technology*, *67*(9–12), 2639–2655. doi:10.1007/s00170-012-4680-4

Boopathi, S., & Sivakumar, K. (2016). Optimal parameter prediction of oxygen-mist near-dry wire-cut EDM. *Inderscience: International Journal of Manufacturing Technology and Management*, *30*(3–4), 164–178. doi:10.1504/IJMTM.2016.077812

Boopathi, S., Sureshkumar, M., Jeyakumar, M., Kumar, R. S., & Subbiah, R. (2022). Influences of Fabrication Parameters on Natural Fiber Reinforced Polymer Composite (NFRPC) Material: A Review. *Materials Science Forum*, *1075*, 115–124. doi:10.4028/p-095f0t

Boopathi, S., Sureshkumar, M., & Sathiskumar, S. (2022). Parametric Optimization of LPG Refrigeration System Using Artificial Bee Colony Algorithm. *International Conference on Recent Advances in Mechanical Engineering Research and Development*, 97–105.

Boopathi, S., Thillaivanan, A., Mohammed, A. A., Shanmugam, P., & VR, P. (2022). Experimental investigation on Abrasive Water Jet Machining of Neem Wood Plastic Composite. *IOP: Functional Composites and Structures, 4*, 025001.

Boopathi, S., Thillaivanan, A., Pandian, M., Subbiah, R., & Shanmugam, P. (2022). Friction stir processing of boron carbide reinforced aluminium surface (Al-B4C) composite: Mechanical characteristics analysis. *Materials Today: Proceedings*, *50*(5), 2430–2435. doi:10.1016/j.matpr.2021.10.261

Boopathi, S., Umareddy, M., & Elangovan, M. (2023). Applications of Nano-Cutting Fluids in Advanced Machining Processes. In *Sustainable Utilization of Nanoparticles and Nanofluids in Engineering Applications* (pp. 211–234). IGI Global. doi:10.4018/978-1-6684-9135-5.ch009

Deng, B.-Y., Li, L.-Z., Tan, D., Uddin, M. N., Cai, Z.-W., & Yu, K.-Q. (2023). Sustainable and cost-effective ultra-lightweight engineered cementitious composite: Design and material characterization. *Cement and Concrete Composites*, *136*, 104895. doi:10.1016/j.cemconcomp.2022.104895

Dongre, R. S. (2020). *Reinforce Fabricated Nano-Composite Matrixes for Modernization of S & T in New Millennium*. Composite and Nanocomposite Materials-from Knowledge to Industrial Applications.

Fowziya, S., Sivaranjani, S., Devi, N. L., Boopathi, S., Thakur, S., & Sailaja, J. M. (2023). Influences of nano-green lubricants in the friction-stir process of TiAlN coated alloys. *Materials Today: Proceedings*. Advance online publication. doi:10.1016/j.matpr.2023.06.446

Goud, E. N., Rajeswari, A., Sharma, J., & Gupta, A. (2021). An Overall Review on Types and Preparation of Nano Biocomposites. *International Journal of Research in Engineering. Science and Management*, *4*(12), 69–74.

Hanumanthakari, S., Gift, M. M., Kanimozhi, K., Bhavani, M. D., Bamane, K. D., & Boopathi, S. (2023). Biomining Method to Extract Metal Components Using Computer-Printed Circuit Board E-Waste. In *Handbook of Research on Safe Disposal Methods of Municipal Solid Wastes for a Sustainable Environment* (pp. 123–141). IGI Global. doi:10.4018/978-1-6684-8117-2.ch010

Haribalaji, V., Boopathi, S., Asif, M. M., Jeyakumar, M., Subbiah, R., & Lewise, K. A. S. (2022). Influences of Friction stir tool parameters for joining two similar AZ61A alloy plates. *Materials Today: Proceedings*, *50*(5), 2547–2553. doi:10.1016/j.matpr.2021.12.074

Haribalaji, V., Venkatesan, G., Pandian, M., Subbiah, R., & Boopathi, S. (2022). Investigation on corrosion and tensile Characteristics: Friction stir welding of AA7075 and AA2014. *Materials Today: Proceedings*, *66*(3), 743–748. doi:10.1016/j.matpr.2022.04.037

Harikaran, M., Boopathi, S., Gokulakannan, S., & Poonguzhali, M. (2023). Study on the Source of E-Waste Management and Disposal Methods. In *Sustainable Approaches and Strategies for E-Waste Management and Utilization* (pp. 39–60). IGI Global. doi:10.4018/978-1-6684-7573-7.ch003

He, Y., Ma, L., Zhou, L., Liu, G., Jiang, Y., & Gao, J. (2020). Preparation and application of bismuth/MXene nano-composite as electrochemical sensor for heavy metal ions detection. *Nanomaterials (Basel, Switzerland)*, *10*(5), 866. doi:10.3390/nano10050866 PMID:32365912

Hu, D., Ren, X., Fu, H., Wang, Y., Feng, X., & Li, H. (2022). Constructing highly rough skin layer of thin film (nano) composite polyamide membranes to enhance separation performance: A review. *Journal of Applied Polymer Science*, *139*(30), e52692. doi:10.1002/app.52692

Hungria, R., Arce, G., Hassan, M., Anderson, M., Mahdi, M., Rupnow, T., & Bogus, S. M. (2020). Evaluation of novel jointless engineered cementitious composite ultrathin whitetopping (ECC-UTW) overlay. *Construction & Building Materials*, *265*, 120659. doi:10.1016/j.conbuildmat.2020.120659

Janardhana, K., Singh, V., Singh, S. N., Babu, T. R., Bano, S., & Boopathi, S. (2023). Utilization Process for Electronic Waste in Eco-Friendly Concrete: Experimental Study. In Sustainable Approaches and Strategies for E-Waste Management and Utilization (pp. 204–223). IGI Global.

Karthik, K., Teferi, A. B., Sathish, R., Gandhi, A. M., Padhi, S., Boopathi, S., & Sasikala, G. (2023). Analysis of delamination and its effect on polymer matrix composites. *Materials Today: Proceedings*. Advance online publication. doi:10.1016/j.matpr.2023.07.199

Karthik, S., Hemalatha, R., Aruna, R., Deivakani, M., Reddy, R. V. K., & Boopathi, S. (2023). Study on Healthcare Security System-Integrated Internet of Things (IoT). In Perspectives and Considerations on the Evolution of Smart Systems (pp. 342–362). IGI Global.

Kumar, D., Kumar, A., Amit, K., & Banerjee, P. (2023). Role of Nano Composite in Additive Technologies: A State of Art Review. *Additive Manufacturing in Multidisciplinary Cooperation and Production*, 145–154.

Li, V. C. (2003). On engineered cementitious composites (ECC) a review of the material and its applications. *Journal of Advanced Concrete Technology*, *1*(3), 215–230. doi:10.3151/jact.1.215

Li, V. C. (2019). Engineered cementitious composites (ECC). *Engineered Cementitious Composites (ECC): Bendable Concrete for Sustainable and Resilient Infrastructure*, 286–290.

Li, V. C., Bos, F. P., Yu, K., McGee, W., Ng, T. Y., Figueiredo, S. C., Nefs, K., Mechtcherine, V., Nerella, V. N., Pan, J., van Zijl, G. P. A. G., & Kruger, P. J. (2020). On the emergence of 3D printable engineered, strain hardening cementitious composites (ECC/SHCC). *Cement and Concrete Research*, *132*, 106038. doi:10.1016/j.cemconres.2020.106038

Li, Z., Lu, B., Feng, J., Zhao, H., & Qian, S. (2023). Development of Engineered Cementitious Composites/Strain-hardening Cementitious Composites (ECC/SHCC) with waste granite fine powders. *Construction & Building Materials*, *409*, 133883. doi:10.1016/j.conbuildmat.2023.133883

Ma, H., Yi, C., & Wu, C. (2021). Review and outlook on durability of engineered cementitious composite (ECC). *Construction & Building Materials*, *287*, 122719. doi:10.1016/j.conbuildmat.2021.122719

Manirathnam, A., Manikandan, M. D., Prakash, R. H., Kumar, B. K., & Amarnath, M. D. (2021). Experimental analysis on solar water heater integrated with Nano composite phase change material (SCi and CuO). *Materials Today: Proceedings*, *37*, 232–240. doi:10.1016/j.matpr.2020.05.093

Mitra, A., Mandal, S. K., & Rai, R. N. (2023). Synthesis and characterisation of magnesium-based nano-composite–A review. *Proceedings of the Institution of Mechanical Engineers, Part E: Journal of Process Mechanical Engineering*. 10.1177/09544089231167758

Murali, B., Padhi, S., Patil, C. K., Kumar, P. S., Santhanakrishnan, M., & Boopathi, S. (2023). Investigation on hardness and tensile strength of friction stir processing of Al6061/TiN surface composite. *Materials Today: Proceedings*.

Naveeenkumar, N., Rallapalli, S., Sasikala, K., Priya, P. V., Husain, J., & Boopathi, S. (2024). Enhancing Consumer Behavior and Experience Through AI-Driven Insights Optimization. In *AI Impacts in Digital Consumer Behavior* (pp. 1–35). IGI Global. doi:10.4018/979-8-3693-1918-5.ch001

Pachiappan, K., Anitha, K., Pitchai, R., Sangeetha, S., Satyanarayana, T., & Boopathi, S. (2024). Intelligent Machines, IoT, and AI in Revolutionizing Agriculture for Water Processing. In *Handbook of Research on AI and ML for Intelligent Machines and Systems* (pp. 374–399). IGI Global.

Palaniappan, M., Tirlangi, S., Mohamed, M. J. S., Moorthy, R. S., Valeti, S. V., & Boopathi, S. (2023). Fused Deposition Modelling of Polylactic Acid (PLA)-Based Polymer Composites: A Case Study. In Development, Properties, and Industrial Applications of 3D Printed Polymer Composites (pp. 66–85). IGI Global.

Paul, A., Thilagham, K., KG, J., Reddy, P. R., Sathyamurthy, R., & Boopathi, S. (2024). Multi-criteria Optimization on Friction Stir Welding of Aluminum Composite (AA5052-H32/B4C) using Titanium Nitride Coated Tool. Engineering Research Express.

Paul, J., Kadirgama, K., Samykano, M., Pandey, A., & Tyagi, V. (2022). A comprehensive review on thermophysical properties and solar thermal applications of organic nano composite phase change materials. *Journal of Energy Storage*, *45*, 103415. doi:10.1016/j.est.2021.103415

Qasem, K. M., Khan, S., Chinnam, S., Saleh, H. A., Mantasha, I., Zeeshan, M., Manea, Y. K., & Shahid, M. (2022). Sustainable fabrication of Co-MOF@ CNT nano-composite for efficient adsorption and removal of organic dyes and selective sensing of Cr (VI) in aqueous phase. *Materials Chemistry and Physics*, *291*, 126748. doi:10.1016/j.matchemphys.2022.126748

Qin, F., Zhang, Z., Yin, Z., Di, J., Xu, L., & Xu, X. (2020). Use of high strength, high ductility engineered cementitious composites (ECC) to enhance the flexural performance of reinforced concrete beams. *Journal of Building Engineering*, *32*, 101746. doi:10.1016/j.jobe.2020.101746

Ravisankar, A., Shanthi, A., Lavanya, S., Ramaratnam, M., Krishnamoorthy, V., & Boopathi, S. (2024). Harnessing 6G for Consumer-Centric Business Strategies Across Electronic Industries. In AI Impacts in Digital Consumer Behavior (pp. 241–270). IGI Global.

Samikannu, R., Koshariya, A. K., Poornima, E., Ramesh, S., Kumar, A., & Boopathi, S. (2022). Sustainable Development in Modern Aquaponics Cultivation Systems Using IoT Technologies. In *Human Agro-Energy Optimization for Business and Industry* (pp. 105–127). IGI Global.

Sampath, B., & Haribalaji, V. (2021). Influences of Welding Parameters on Friction Stir Welding of Aluminum and Magnesium: A Review. *Materials Research Proceedings*, *19*(1), 322–330.

Sampath, B., Pandian, M., Deepa, D., & Subbiah, R. (2022). Operating parameters prediction of liquefied petroleum gas refrigerator using simulated annealing algorithm. *AIP Conference Proceedings*, *2460*(1), 070003. doi:10.1063/5.0095601

Saravanan, M., Vasanth, M., Boopathi, S., Sureshkumar, M., & Haribalaji, V. (2022). Optimization of Quench Polish Quench (QPQ) Coating Process Using Taguchi Method. *Key Engineering Materials*, *935*, 83–91. doi:10.4028/p-z569vy

Sarode, P. U., Raul, J. G., Gaikwad, A. S., & Gawande, S. (2023). Metal Matrix Nanocomposites: A Brief Overview. *Nanoscience & Nanotechnology-Asia*, *13*(2), 24–42. doi:10.2174/2210681213666230301152349

Sathish, T., Sunagar, P., Singh, V., Boopathi, S., Al-Enizi, A. M., Pandit, B., Gupta, M., & Sehgal, S. S. (2023). Characteristics estimation of natural fibre reinforced plastic composites using deep multi-layer perceptron (MLP) technique. *Chemosphere*, *337*, 139346. doi:10.1016/j.chemosphere.2023.139346 PMID:37379988

Selvakumar, S., Adithe, S., Isaac, J. S., Pradhan, R., Venkatesh, V., & Sampath, B. (2023). A Study of the Printed Circuit Board (PCB) E-Waste Recycling Process. In Sustainable Approaches and Strategies for E-Waste Management and Utilization (pp. 159–184). IGI Global.

Senthil, T., Puviyarasan, M., Babu, S. R., Surakasi, R., Sampath, B., & Associates. (2023). Industrial Robot-Integrated Fused Deposition Modelling for the 3D Printing Process. In Development, Properties, and Industrial Applications of 3D Printed Polymer Composites (pp. 188–210). IGI Global.

Shoji, D., He, Z., Zhang, D., & Li, V. C. (2022). The greening of engineered cementitious composites (ECC): A review. *Construction & Building Materials*, *327*, 126701. doi:10.1016/j.conbuildmat.2022.126701

Shumuye, E. D., Li, W., Fang, G., Wang, Z., Liu, J., & Zerfu, K. (2023). Review on the durability of eco-friendly engineering cementitious composite (ECC). *Case Studies in Construction Materials*, *19*, e02324. doi:10.1016/j.cscm.2023.e02324

Singh, M., Saini, B., & Chalak, H. (2019). Performance and composition analysis of engineered cementitious composite (ECC)–A review. *Journal of Building Engineering*, *26*, 100851. doi:10.1016/j.jobe.2019.100851

Veeranjaneyulu, R., Boopathi, S., Kumari, R. K., Vidyarthi, A., Isaac, J. S., & Jaiganesh, V. (2023). Air Quality Improvement and Optimisation Using Machine Learning Technique. *IEEE Explore*, 1–6.

Vennila, T., Karuna, M., Srivastava, B. K., Venugopal, J., Surakasi, R., & Sampath, B. (2022). New Strategies in Treatment and Enzymatic Processes: Ethanol Production From Sugarcane Bagasse. In Human Agro-Energy Optimization for Business and Industry (pp. 219–240). IGI Global.

Yu, K., McGee, W., Ng, T. Y., Zhu, H., & Li, V. C. (2021). 3D-printable engineered cementitious composites (3DP-ECC): Fresh and hardened properties. *Cement and Concrete Research*, *143*, 106388. doi:10.1016/j.cemconres.2021.106388

Zhang, D., Yu, J., Wu, H., Jaworska, B., Ellis, B. R., & Li, V. C. (2020). Discontinuous micro-fibers as intrinsic reinforcement for ductile Engineered Cementitious Composites (ECC). *Composites. Part B, Engineering*, *184*, 107741. doi:10.1016/j.compositesb.2020.107741

Zhang, Z., Liu, S., Yang, F., Weng, Y., & Qian, S. (2021). Sustainable high strength, high ductility engineered cementitious composites (ECC) with substitution of cement by rice husk ash. *Journal of Cleaner Production*, *317*, 128379. doi:10.1016/j.jclepro.2021.128379

Ziada, M., & Erdem, S. (2023). Strain hardening green cementitious composites reinforced with nanoparticles: Mechanical and microstructural properties and high temperature effect. *Case Studies in Construction Materials*, *18*, e02033. doi:10.1016/j.cscm.2023.e02033

# Chapter 9
# Current Investigations and Field Performances of Engineered Cementitious Composites (ECC) in Transportation Infrastructure

**Chintada Chandrasekhar**
https://orcid.org/0000-0001-7502-8532
*Indian Institute of Technology Roorkee, Roorkee, India*

**G. D. R. N. Ransinchung**
*Indian Institute of Technology Roorkee, Roorkee, India*

## ABSTRACT

*Engineered cementitious composites (ECC) or strain-hardening cement-based composites (SHCC) were initially developed in the early 1990s, and later, many advancements were made in matrix composition. The development of ECC hinged on gaining a comprehensive understanding of how the materials integrated into the matrix performed in in-situ conditions. The field application provides an opportunity to capitalize on ECC's enhanced mechanical properties, particularly the structural performance caused by various loading conditions. This chapter briefly introduces concrete pavement cracking and fiber-reinforced concrete (FRC) classification and then introduces ECC and its mix design characteristics. Later, the developments and applications of ECC in transportation-related infrastructure were introduced. The innovative ideas related to highway or airport pavement with conventional or ultra-thin white topping overlay, field-related construction, and their current performance were discussed in separate sections.*

DOI: 10.4018/978-1-6684-8182-0.ch009

# 1. INTRODUCTION

Concrete pavements are generally subjected to heavy loads compared to flexible pavements due to their load-bearing capacity and less maintenance during service life. However, many concrete pavements and infrastructures have deteriorated due to the matrix's brittle behavior and cracking potential *J. Zhang et al.* (*2013*). Significant research is going on to improve the ductility of the concrete matrix. One of the efficient solutions is to incorporate the fibers in the matrix, called Fiber Reinforced Concrete (FRC) *V. C. Li et al.* (*2001*). Incorporating the fibers increased the toughness of the concrete matrix by resisting the propagation of cracks into major fractures due to the fiber bridging phenomenon *V. C. Li et al.* (*2001*), (*2002*). However, this modification in the matrix had a minimal impact on increasing ductility. FRC exhibits tension Softening (i.e., localization of failure mechanism due to fiber breakage, matrix cracking, or delamination) behavior under the tensile load after the first crack. It is advantageous in FRC in gradually losing strength and limiting the crack width due to the fiber bridging phenomenon, making FRC a quasi-brittle material. However, adding fiber increased the fracture toughness of concrete, but tensile strain capacity remains unchanged (or a slight improvement).

The current chapter reviewed the characteristics of FRC, the development of ECC, and field case studies of ECC in the context of transportation infrastructure. Figure 1 provides a detailed conceptual framework illustrating the structure and content of this chapter.

*Figure 1. Methodological framework of the review chapter*

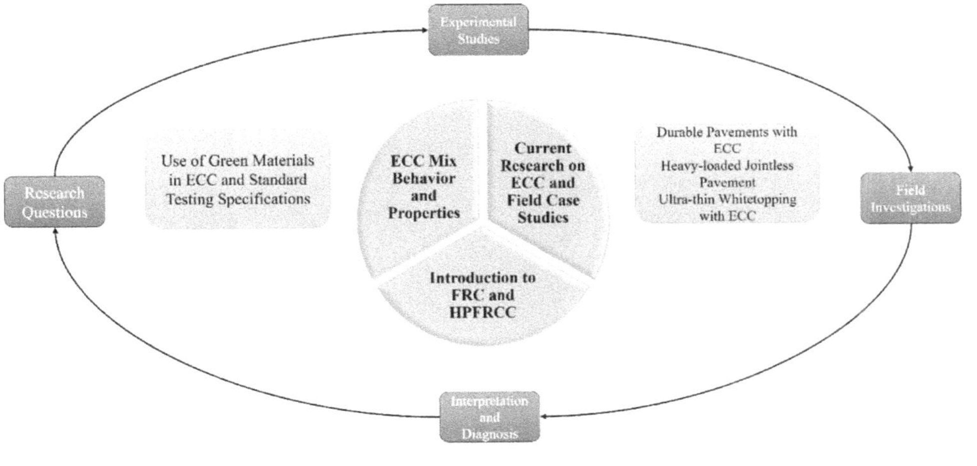

Few researchers suggested the behavior of FRC materials based on their tensile and bending characteristics during crack propagation *D. joo Kim et al. (2008)*; *Naaman & Reinhardt (2006)*. According to the model developed by *Naaman & Reinhardt (2006)*, after achieving the proportionality limit (point 1 in Figure 2), the tensile behavior of the particular material can be divided into either tensile strain hardening (curve (a) in Figure 2) or tensile strain softening (curve (b) in Figure 2). Similarly, the flexural behavior of FRC can also be classified as deflection softening (curve (a) in Figure 3) or deflection hardening (curve (b) in Figure 3). Also, it is important to mention that deflection hardening can occur in both tension strain softening and hardening FRCs. The FRCs showing strain-hardening behavior in tensile loading are High-Performance Fiber Reinforced Cementitious Composites (HPFRCC).

*Figure 2. Tensile stress-strain response of material in a) strain-hardening, b) strain-softening (Naaman & Reinhardt, 2006)*

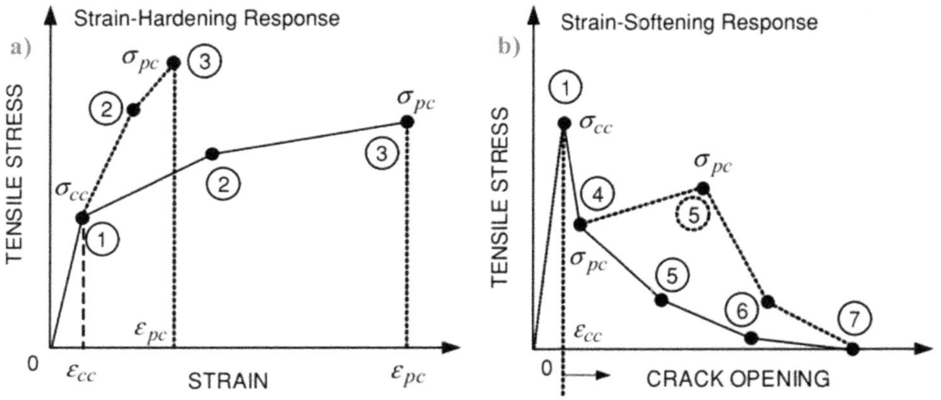

*Figure 3. Load-deflection response of FRCC in bending (D. Joo Kim et al., 2008)*

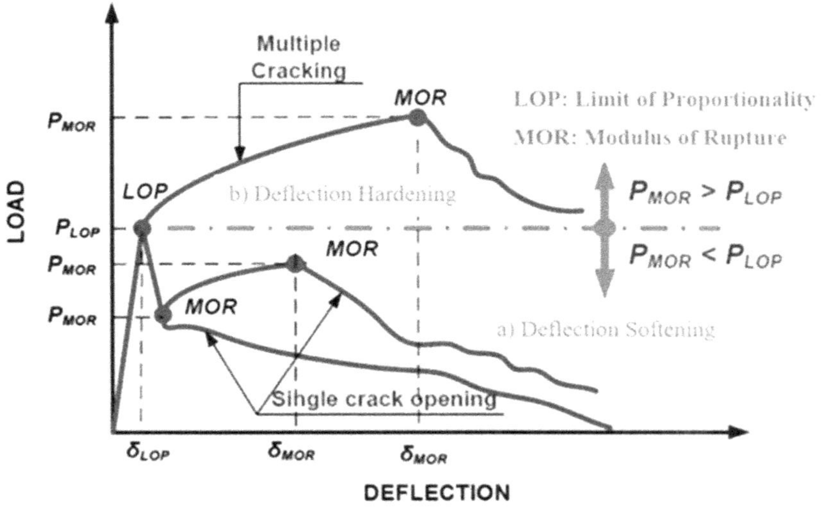

*Figure 4. Performance levels of FRC (Wille et al., 2014)*

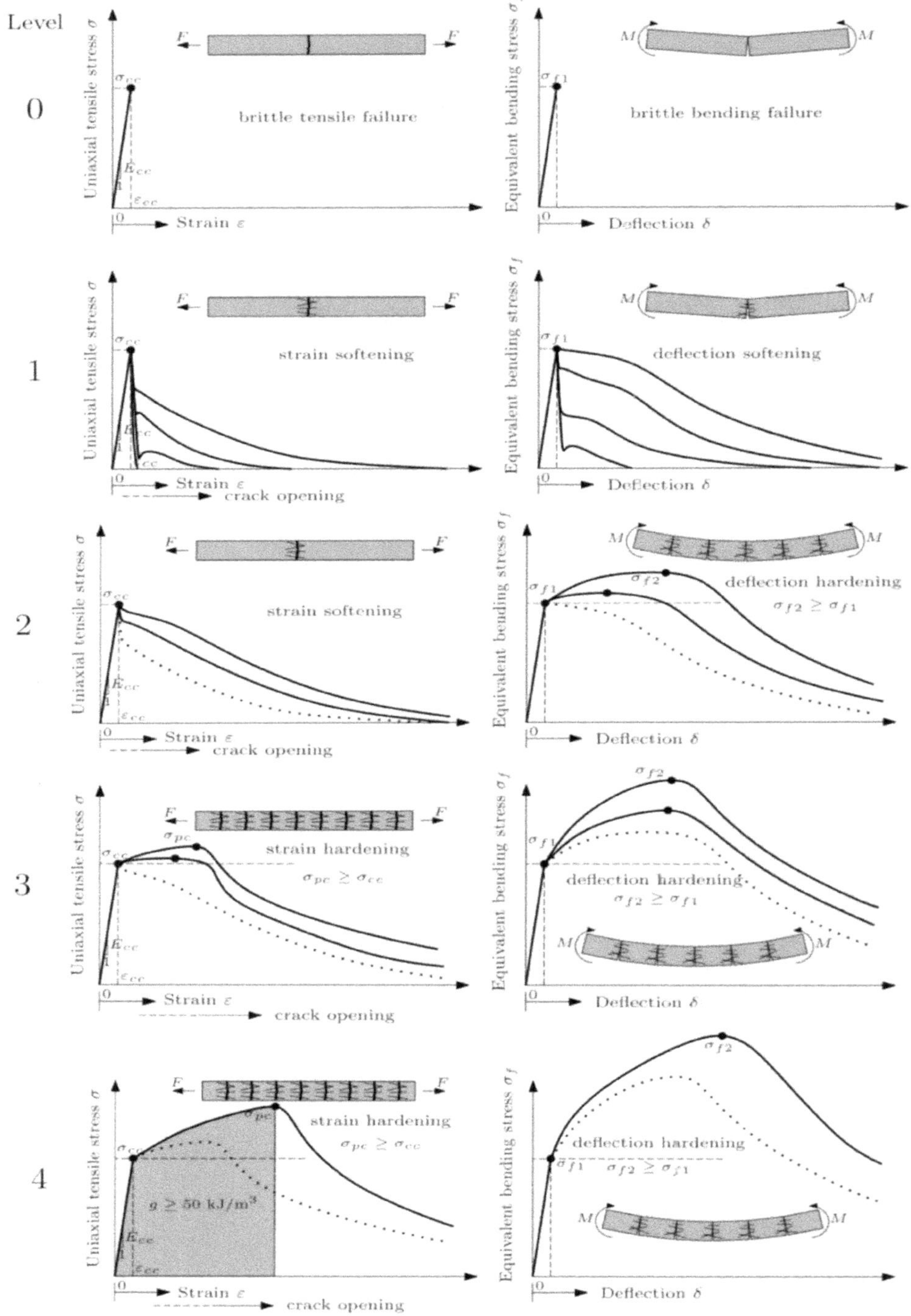

A study by *Wille et al.* (*2014*) categorized the classification of these HPFRCCs at different levels based on their tensile and bending behavior, as shown in Figure 3. The performance classification started with Level 0 to demonstrate the behavior of regular concrete (non-fiber reinforced) as a reference. Other categories are Level 1 – strain and deflection softening with a slight improvement in mechanical properties with the addition of fibers, Level 2 – only improvement in deflection hardening, Level 3 – both strain and deflection hardening, and Level 4 – with higher energy absorption capability. Figure 4 consists of the following parameters: cracking stress ($\sigma_{cc}$) and its corresponding strain ($\varepsilon_{cc}$), post-cracking strength ($\sigma_{pc}$), modulus of rupture ($\sigma_{f1}$) and its equivalent flexural strength($\sigma_{f2}$), and materials' elastic modulus ($E_{cc}$). It is to be highlighted that the stress-strain relationship is only valid till the matrix first cracks. The stress (vs) crack-opening usually describes the strain-softening behavior of the material. Similarly, the stress-strain relationship can be used assuming the smeared cracking approach during the strain-hardening or multiple cracking conditions *Wille et al.* (*2014*). According to the study by *V. C. Li* (*1998*), the crack pattern of SHCC in bending is noticeably different than that of regular FRC. During the loading, the first crack appears at the mid-span of the tensile face, and then the multiple cracks with the increase in load spread outside the mid-span area.

## 2. ECC BEHAVIOR AND MATERIALS

Engineered Cementitious Composite (ECC) is a special class of HPFRCC that Victor Li and his team developed in the early 1990s *V. C. Li* (*1994*). The design of ECC is based on the theory of micromechanics that relates matrix composite properties to the microstructure level *Lee et al.* (*2010*); *V. C. Li* (*1994*). The significant difference between FRC and ECC is the material behavior after the first crack at the flaw site. Instead of tension softening, the ECC shows the behavior of Strain-Hardening or Pseudo-Strain Hardening (i.e., the material increases tensile stress even after the first crack with the increase in tensile strain.). It possesses extreme tensile ductility of 3% to 5%, approximately 300 to 500 times that of conventional concrete or FRC *V. C. Li* (*2003*); *V. C. Li et al.* (*2001*). Also, few recent studies indicate the potential for ECC to achieve tensile ductility of up to 8%, nearly 800 times that of conventional concrete *Guan et al.* (*2019*); *K. Yu et al.* (*2017*). With an increase in load, the strain-hardening behavior reduces the localization failure of material by causing a more significant number of micro-cracks. The average crack width tends to stabilize at around 60μm after reaching a strain value of 1% *Weimann & LI* (*2003*); *Z. Zhang & Zhang* (*2017*), and numerous researchers have documented ECC's pseudo-strain-hardening *H. Li et al.* (*2009*); *V. C. Li et al.* (*1995*); *G. Yang et al.* (*2019*) and deflection-hardening *D. joo Kim et al.* (*2008*); *Pakravan et al.* (*2018*) behavior under tension and flexural loading.

ECCs can be considered a family of materials based on their wide range of strength and ductility properties depending on the structural applications. In addition to the tensile ductility and multiple cracking, ECC is also popular for different functionalities, such as Self-consolidating ECC for large-scale applications *Kong et al.* (*2003*); *Lepech & Li* (*2008*), High early-strength ECC, which is generally required for transportation infrastructure to early reopen the road *S. Wang & Li* (*2006*), Lightweight ECC where the structural dead load is required to be minimized *S. Wang & Li* (*2003*), Green ECC where the material greenness is required to sustainable infrastructure *Lepech et al.* (*2008*); *V. C. Li et al.* (*2004*), and Self-healing ECC is designed to recover the mechanical and transport properties after the damage *Y. Yang et al.* (*2005*).

A summary of the significant physical properties given by *V. C. Li* (*2008*) is shown in Figure 5. It should be highlighted that these physical properties are tailored through micromechanical modeling. Even a wide range of these properties has been observed in recent years. For example, a study by *Guan et al.* (*2019*) achieved a tensile strain of up to 10% with coarse river sand.

*Figure 5. Physical properties of ECC*

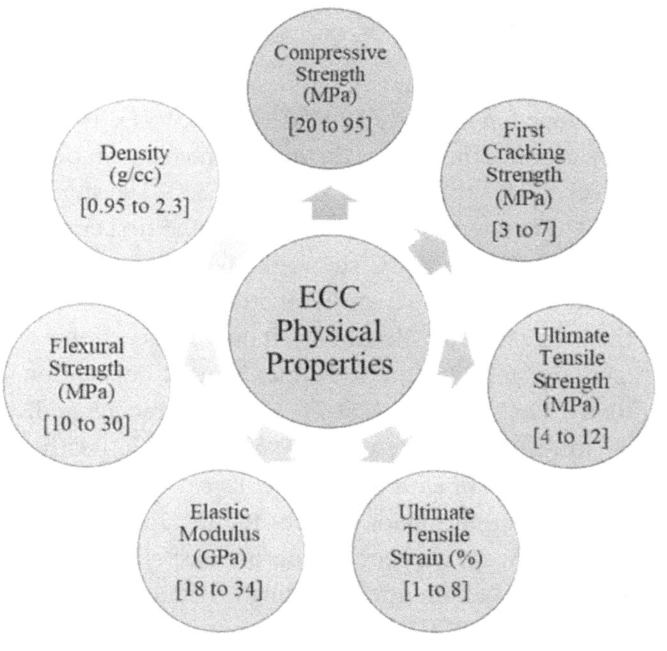

## 2.1. Green Materials in ECC

In recent years, while sustainability has been the main concern for the construction industry, the design of ECC with green binders is required. The research towards replacing cementitious materials with other green binders or fillers has increased over the years *Shoji et al.* (*2022*). In addition to the primary supplementary cementitious composites (SCMs) such as Fly ash (FA) *Turk & Nehdi* (*2021*); *E. H. Yang et al.* (*2007*); *J. Yu & Leung* (*2017*), Ground-granulated blast furnace slag (GGBS) *Iqbal Khan et al.* (*2017*); *S. Zhang et al.* (*2020*); *J. Zhou et al.* (*2010*), silica fume (SF) *Arivusudar & Babu* (*2020*); *Liu et al.* (*2018*); *J. Yu et al.* (*2017*), Rice husk ash (RHA) *Righi et al.* (*2017*); *Shoba & Asha* (*2019*); *Z. Zhang et al.* (*2021*), other OPC based binders that include, Glass pozzolana (GP) *Hisseine & Tagnit-Hamou* (*2020*); *Siad et al.* (*2017*), Ceramic waste *W. Wang et al.* (*2020*); *J. Zhao et al.* (*2020*) and small sized Iron ore tailings (IOTs) *X. Huang et al.* (*2013*); *X. Y. Huang et al.* (*2015*); *Xiao et al.* (*2023*) were also used in ECC design as a part of sustainable materials. The detailed classification of green materials used in ECC till the data is shown in Figure 6.

*Figure 6. Green materials adopted in ECC design*

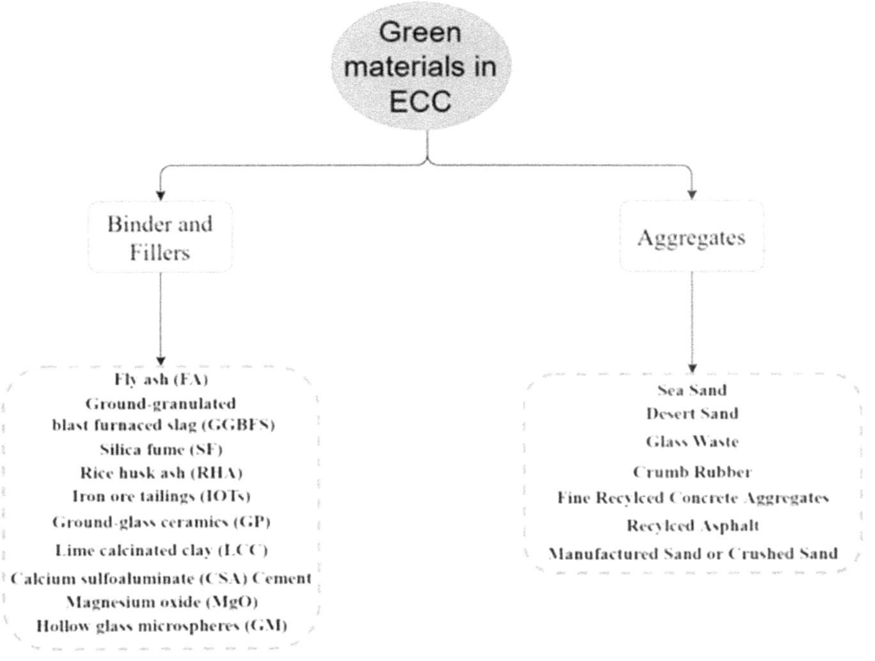

Moreover, traditional ECC was designed with silica sand, which incurs additional economic and environmental costs. As a part of sustainability, reducing embedded energy and emissions associated with the production and transport of sand is essential. Thus, more research towards replacing silica sand with regular sand or recycled fine aggregates produces properties similar to traditional ECC (ECC-M45) developed by *V. C. Li* (*2008*). The details of different sands used in various studies of ECC design with their characteristics are discussed in **Table 1.**

*Table 1. Characteristics of green aggregates in ECC*

| Sand Type | Characteristics | Reference Studies |
|---|---|---|
| Natural River Sand | Economical but reduces the compressive strength | *Guan et al.* (*2019*); *Tan et al.* (*2021*); *K. Yu et al.* (*2017*); *Zhu et al.* (*2020*) |
| Sea Sand | Increases the compressive strength but has a negative effect on tensile properties | *B. T. Huang et al.* (*2020*), (*2021*); *Yao et al.* (*2021*) |
| Desert Sand | Economical design but weakens the ductility and tensile strength | *An et al.* (*2021*); *Che et al.* (*2019*); *Jialing et al.* (*2017*); *Pavlu et al.* (*2019*) |
| Glass Waste (GW) | Improves strength and ductility but has more embodied energy and emissions than silica sand | *Adesina & Das* (*2020*)b, (*2020*)a; *Hossain et al.* (*2017*); *Letelier et al.* (*2019*) |
| Crumb Rubber (CR) | Increases the durability and flexural ductility but reduces the strength properties | *Adesina & Das* (*2021*)a; *Chen et al.* (*2021*); *Noorvand et al.* (*2019*); *Srikakulam & Khed* (*2020*) |
| Fine Recycled Concrete Aggregates (FRCA) | More economical and decreases the drying shrinkage but reduces the durability. | *J. Li & Yang* (*2017*); *Pavlu et al.* (*2019*); *Y. Zhou et al.* (*2021*) |
| Recycled Asphalt | Sustainable option by reducing the embodied carbon and cost. However, it compromises the performance of ECC in mechanical and durability characteristics. | *Adesina & Das* (*2020*)c, (*2021*)b; *Ibrahim et al.* (*2014*) |
| Manufactured sand (M-sand) | As a cost-effective alternative to river sand, it has become increasingly popular, primarily driven by concerns over the depletion of riverbed resources. Nevertheless, it's important to note that the decrease in tensile ductility can be attributed to factors like surface texture and particle shape. | *Ajith et al.* (*2021*); *Chandrasekhar & Ransinchung* (*2023*); *Sahmaran et al.* (*2009*); *Shanmugasundaram & Praveenkumar* (*2022*), (*2023*) |

## 2.2. Standard Specifications Related to ECC Testing

Numerous studies have extensively examined the mechanical and micro-mechanical properties of ECC. The mechanical behavior can be assessed through compressive, tensile, and flexural tests. On the other hand, micro-mechanical characteristics, such as toughness, fiber-bridging capacity, and bond characteristics, can be determined through tests like three-point bending, single crack tensile, or pull-out tests.

The compression testing of cylindrical or cubical specimens is the most common test to determine the compressive strength ($f_c$') and elastic modulus (E) of concrete or any other type of hydraulic mortars. The standard test methods for compressive strength are:

- ASTM C39/C39M - "Standard Test Method for Compressive Strength of Cylindrical Concrete Specimens" *ASTM C39/C39M* (*2020*)
- ASTM C109/C109M – "Standard Test Method for Compressive Strength of Hydraulic Cement Mortars (Using 2-in. or [50 mm] Cube Specimens)" *ASTM C109/C109M* (*2021*)
- IS 4031-6 (1988): "Methods of physical tests for hydraulic cement, Part 6: Determination of compressive strength of hydraulic cement (other than masonry cement)" *IS: 4031 (Part 6)* (*1988*)

The uni-axial tension test is a fundamental test to evaluate the tensile properties of any FRC, but ASTM or Indian Standard has not adopted any test procedure yet. Instead, a split tensile strength test was proposed to find the tensile properties of concrete. In this test, a compressive force is applied diametrically on two strips placed on two opposite sides of the cylindrical sample. However, the splitting tensile strength results are often assumed to be 30 to 40 percent of compressive strength, and they are usually 10 to 15 percent higher than the direct tension test. Thus, a standard test must be adopted to determine the uni-axial tensile properties as fiber incorporation improves the tensile properties. Some foreign institutions provided few recommendations for FRC testing, generally used for research purposes.

- RILEM TC 162-TDF: "Test and Design Methods for Steel Fibre Reinforced Concrete - Recommendations for uniaxial tension test" *Rilem TC 162-TDF* (*2001*)
- Japan Society of Civil Engineers (JSCE): "Recommendations for design and construction of high-performance fiber reinforced cement composites with multiple fine cracks (HPFRCC)" *Japan Society of Civil Engineers* (*2008*)

The beam flexural strength or modulus of rupture (MOR) test is another indirect tensile strength measurement. The flexural loading condition is necessary for concrete pavements, simulating vehicle loading in the field. The following are the standard test methods to determine the flexural strength:

- ASTM C1609/C1609M – "Standard Test Method for Flexural Performance of Fiber-Reinforced Concrete (Using Beam With Third-Point Loading)" *ASTM C1609/C1609M (2019)*
- ASTM C78/78M – "Standard Test Method for Flexural Strength of Concrete (Using Simple Beam with Third-Point Loading)" *ASTM C78/C78M (2021)*

Based on the J-integral analysis of steady-state crack propagation by *V. C. Li (1997)*, the authors have proposed strength and energy criteria to model the ECC based on the micro-mechanical analysis. When the material tailoring is made, certain material and interface properties are required to satisfy these requirements *Ma et al. (2015)*. These properties can be measured through the following tests:

The fiber-matrix interface parameters, which include the frictional bond ($\tau_0$), chemical bond ($G_d$), and slip-hardening coefficient (β), can be measured through the fiber-pullout test. Different test procedures and methods were adopted in various studies *Boughanem et al. (2014)*; *Jesson & Watts (2012)*; *V. C. Li et al. (1991)*; *Redon et al. (2001)*; *Saraireh et al. (2018)*; *Singh & Munjal (2020)*.

Single Crack Tensile Test (SCTT) is an alternative test to determine the ECC composites' fundamental tensile stress vs crack opening relationship. Based on the test results, the fiber bridging stress and complimentary energy can be calculated using the empirical relationships developed through the J-integral design methodology. In addition, the fracture toughness of ECC mortar is also an essential parameter to determine one of the design criteria. The SCTT test is identical to a direct tensile test using a dog-bone specimen with a notch at the middle, and the fracture toughness test is carried out on notched beam specimens. Past researchers *Hou et al. (2022)*; *Subedi et al. (2021)*; *Y. Wang et al. (2020)*; *J. Yu et al. (2019)*; *K. Yu et al. (2017)* explained both the tests specimen preparation and testing methodology.

In addition to the strength test and micro-mechanical analysis, the durability of the designed mixes (with locally accessible or green materials) is also a major concern. As the initial cost of the ECC composites is comparatively higher than the conventional concrete, the durability of the mixes should be relatively higher than the conventional concrete pavement so that the service life will be increased. A few researchers studied the durability properties of ECC mixes prepared with different materials *Adesina & Das (2020)*a; *Ammasi & Ragul (2018)*; *Righi et al. (2017)*; *Şahmaran & Li (2009)*. The following tests can test the durability behavior, as shown in **Table 2.**

*Table 2. Durability-related test procedures*

| Test | Methodology |
|---|---|
| Density, Porosity, Absorption Test | ASTM C642 – "Standard Test Method for Density, Absorption, and Voids in Hardened Concrete" *ASTM (2021)* |
| Rate of Water Absorption | ASTM C1585 – "Standard Test Method for Measurement of Rate of Absorption of Water by Hydraulic-Cement Concretes" *ASTM C1585 (2020)* |
| Alkali-Silica Reaction | ASTM C1260 – "Standard Test Method for Potential Alkali Reactivity of Aggregates (Mortar-Bar Method)" *ASTM C1260 (2022)* |
| Rapid Chloride Penetration | ASTM C1202 – "Standard Test Method for Electrical Indication of Concrete's Ability to Resist Chloride Ion Penetration" *ASTM C1202 (2022)* |
| Freeze-thaw Durability | ASTM C666 – "Standard Test Method for Resistance of Concrete to Rapid Freezing and Thawing" *Astm C666/C666M (2003)* |
| Bulk Electrical Resistivity | ASTM C1876 – "Standard Test Method for Bulk Electrical Resistivity or Bulk Conductivity of Concrete" *ASTM C1876 (2019)* |
| Ultra-Sonic Pulse Velocity | ASTM C597 – "Standard Test Method for Pulse Velocity Through Concrete" *ASTM C597 (2016)* |

## 3. DURABLE PAVEMENTS WITH ECC

Pavements are always essential to any country's transportation system and significantly contribute to the nation's economy. Therefore, better-designed and optimally constructed pavements are imperative in the transportation sector. Most developed countries have approximately 80% asphalt pavement in their nation's road network, primarily due to the advantages of low construction cost, less time for traffic diversion, and better surface quality *Pranav et al. (2022)*. On the other hand, concrete pavements also have several advantages, such as long service life, improved durability, and better night visibility. To this end, many countries are now taking steps toward constructing concrete pavements on the major routes. For example, at the end of the year 2017, Japan had 34,590 miles of concrete pavements (Road Statistics Annual Report, 2017). Also, many European countries have nearly 20% of major roads as rigid pavement, and it is consistently rising over time.

Even though the construction of rigid pavements is gaining popularity, there are a few challenges to improve: significantly low ability in tensile conditions, high susceptibility to temperature, and poor riding quality. Most pavements have reached their service life or severely deteriorated due to excess traffic loading or other environmental factors and need repair and rehabilitation. The most used rehabilitation technique is placing an overlay with asphalt or cement concrete, depending on the traffic condition. The overlay of either of these materials provides additional service life to the pavement, reduces deterioration, and protects the pavement structure *Owusu-Antwi et al. (1998)*. Past research showed that providing a 50 mm ultra-thin or 150 mm thick overlay gives good results depending on the application and traffic loading. However, these two methods are proving potential results; still, some problems, such as reflective cracking, appear in the overlay when the crack in existing pavement reflects due to traffic, shrinkage, and temperature loading. The developed cracking on overlay structures reduces the load-bearing capacity and pavement durability. Therefore, preventing reflective cracking and delamination is crucial to improving pavement design life *Yucel et al. (2013)*. The studies observed that the load-bearing capacity of ECC increased by 100% and deformability capacity by seven times when tested under quasi-static flexural loading.

Over the years, several applications of ECC in pavements have been studied by various researchers *G. A. Arce et al. (2021)*; *Bawono et al. (2019)*; *Hungria et al. (2020)*; *Lepech & Li (2010)*; *Sahmaran et al. (2014)*; *Yucel et al. (2013)*; *J. Zhang & Li (2002)*; *A. Zhao et al. (2015)*. Furthermore, ultra-thin white toppings (UTWs) are predominantly used to repair deteriorated asphalt pavements. The white topping is typically constructed on existing HMA pavements, which reached their service life and exhibited severe distresses, including fatigue cracking, rutting, and thermal cracking. Generally, the thickness of UTW ranges from 2 inches to 4 inches, and the slab size from 4 inches to 6 inches without joint sealant and load transfer elements *Mateos et al. (2015)*. Even though the reduced thickness of UTW slabs showed satisfactory performance, some distresses have been observed, and reduced joint spacing increased the workload during saw cutting compared to conventional concrete pavement construction *Gowripalan (2018)*; *D. Kim et al. (2008)*. ECC's enhanced mechanical and durable properties can sustain the load-related and thermal stresses developed in the UTW. (Zhang et al., 2015) developed a "Low Elastic Modulus Early Strength ECC material" for the application of UTW on existing concrete substrates. The developed mix achieved satisfactory compressive and flexural strength and tensile ductility up to 4% with an elastic modulus of 14 GPa.

Usually, Continuous Reinforced Concrete Pavements (CRCP) and Pre-Stressed Concrete Pavements (PCP) are used for jointless and crack-controlled structures. Even though CRCP increased the structural integrity but showed uncontrolled crack width, it shortens the service life. *Z. Zhang et al. (2017)* investigated the performance of ECC material for jointless concrete pavement. The self-healing behavior of ECC could also decrease continuous crack deterioration, which increases the service life. The self-healing ECC protects the base from water ingress and recovers its mechanical properties to some extent *Ma et al. (2014)*; *Z. Zhang et al. (2014)*. In addition, *Noorvand et al. (2021)* studied the performance of ECC on concrete pavement plasticity using experimental and numerical modeling analysis. The study established thickness vs. cycle-to-failure (T-N) relationships to optimize the thickness of the ECC layer for pavement applications. Likewise, the research on ECC material for various pavement applications prepares an optimized mix to improve the pavement infrastructure.

## 4. FIELD CASE STUDIES OF ECC IN TRANSPORTATION-RELATED APPLICATIONS

### 4.1. Construction of a Heavy-Loaded Jointless Pavement

Even though concrete pavements are more durable than asphalt pavements, the presence of joints produces the weakest links concerning durability and serviceability. In addition, plain concrete requires higher thickness to resist the current loading conditions, as existing guidelines suggest. Pavements hardly used ECC in the field yet. A plausible solution to strengthen the pavements is to eliminate the joints; thus, a material such as ECC containing higher deformable capacity shall be a viable option.

A full-scale demonstration project was completed in the year of 2015 involving the construction of heavy-loaded 80 m jointless pavement, with a width of 3 m and height of 15 cm in Wittstock/Dosse, Germany *Müller & Mechtcherine* (*2018*). The prepared test track represents the actual traffic conditions and is designed to examine the long-term performance under single-tire loading conditions up to 10 tons. The project also tested the feasibility of three paver technologies to construct the 80 m ECC pavement: 40 m with an asphalt paver, 20 m with a concrete roller paver, and another 20 m with a truss screed. The ECC layer is placed on the well-compacted gravel base, which is freeze-safe indeed. The ECC material mixing occurred using a local concrete plant consisting of a gravity mixture of 3 $m^3$ near the demonstration site. A total of 52 $m^3$ of ECC mix was produced and delivered using a concrete mixer truck with a rotated mixing drum. The project used three different paving techniques, and the construction took place in two days: on the first day, 30 mixes, and on the other day, 22 mixes, with each 1 $m^3$ volume on the second day. The construction activities involved in the project are illustrated in Figure 7.

Among the paving technologies, the prepared surface using truss screed gave satisfying smoothness compared to concrete roller paver. As asphalt pavers can handle a wide range of materials, the result of the current project showed that the consistency of SHCC should not vary too much, and the mixes with high stiffness are generally not recommended. The pavement test section has been continuously monitored using various sensor systems (strain gauges, temperature, and humidity sensors) installed during the construction *Mueller et al.* (*2019*). In the first sensor system, a total of six deformation-measuring gauges and four temperature and humidity sensors were installed in the pavement during the construction. Among these, three temperature sensors were installed within the pavement, and one was situated outside to monitor the air temperature. All these sensors were continuously assessing the pavement's condition at 30-minute intervals for a duration of 1.5 years. Based on these observations, it was determined that the initial shrinkage over the first three months due to the hydration of the cementitious composites was 0.25 mm/m. After this period, the deformation began to fluctuate with outside temperature changes, making it difficult to isolate the strains caused by the shrinkage from the derived data. The strain variations attributed to both temperature and humidity changes over the 15 months following the initial three months amounted to approximately 0.1 mm/min.

An additional set of five deformation sensor groups was installed in the pavement, each featuring three strain sensors stacked vertically to note the in-site load measurements. Two of these groups were positioned along the test track, while the other two were placed perpendicular beneath the wheel track. Further, the fifth deformation sensor was also placed along the test track but in the center of the road. For the loading, a wheel-type loader with a load of 29 tons was employed, in which the load was distributed as one front wheel bears a single load of 8.25 tons while one back wheel supports 6.25 tons. Figure 8 shows the loading condition, the schematic measurement setup of strains, and the deformation graph. Figure 8(c)) illustrates the bending response of the pavement, varying between 0.33 mm/min in compression at a depth of 25 mm below the pavement surface and 0.2 mm/min in tension recorded at a position 25 mm above the bottom of the pavement. A detailed investigation performed on the test track after three years of construction showed nearly no cracks on the surface except in a few areas where finely distributed cracks resulted due to the higher stress intensity.

*Figure 7. Construction activities of ECC pavement: a) material mixing, b) truss screed paving, c) paving through roller paver, d) paving with asphalt paver (Müller & Mechtcherine, 2018)*

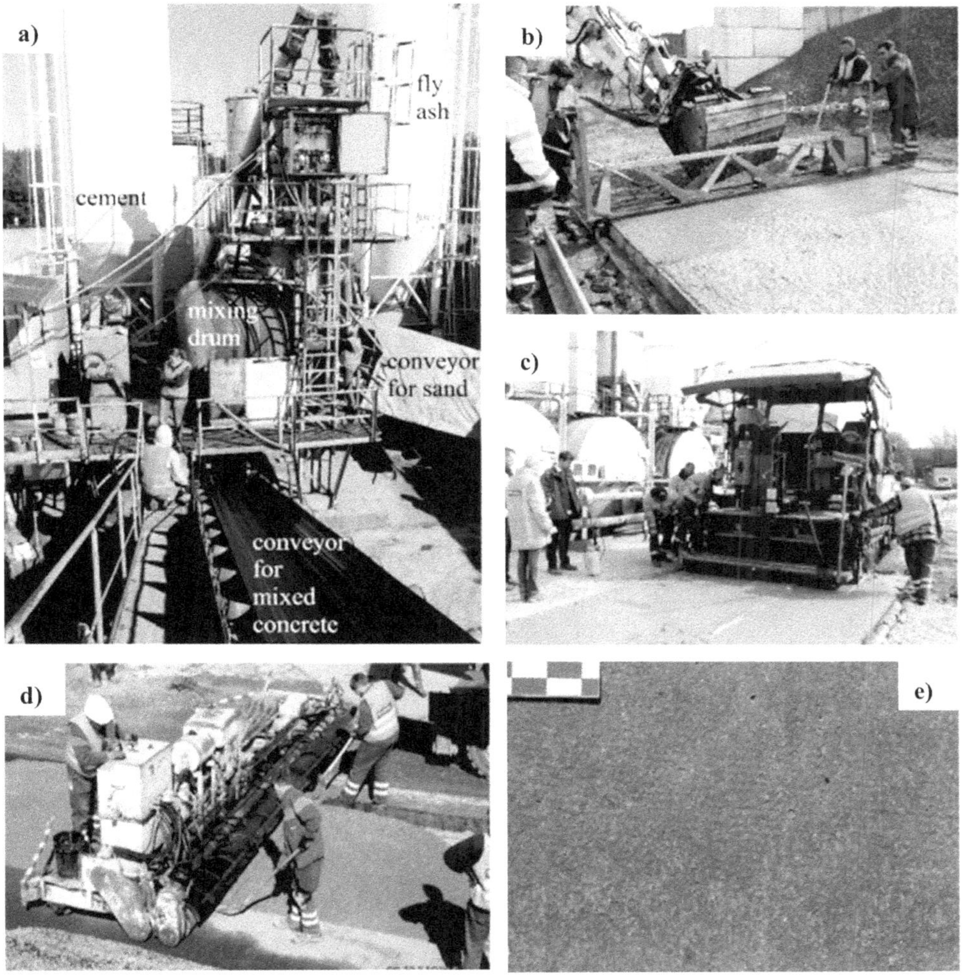

*Figure 8. Field testing a) load drive with wheel loader, b) schematic strain measurement set up, c) evaluating deformation changes (Müller & Mechtcherine, 2018)*

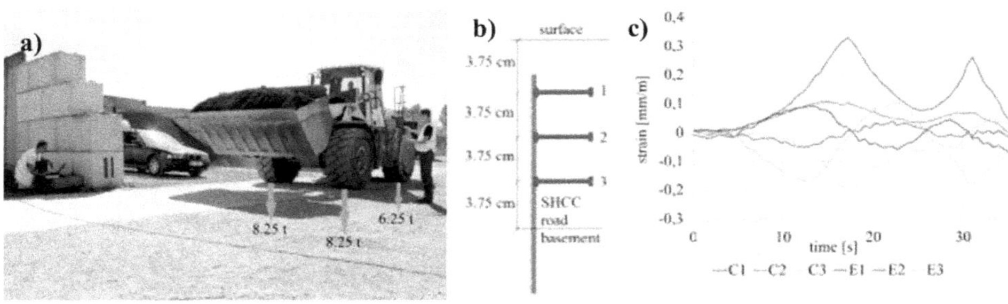


## 4.2. Ultrathin Whitetopping (UTW) Overlay

Ultrathin whitetoppings (UTWs) are an alternative solution for rehabilitating asphalt pavement (HMA) in deteriorated conditions. UTWs are mainly constructed on HMA pavements whose service life is about to end and have several distresses such as fatigue, rutting, thermal cracking, and other deteriorations *Winkelman* (*2005*). Compared to conventional concrete overlay, the main advantage of UTWs is their reduced thickness, which is around 2 inches to 4 inches (50.8 mm to 101.6 mm) with a slab size of 4 feet by 4 feet to 6 feet by 6 feet (1.22 m x 1.22 m to 1.83 x to 1..83 m) *Mateos et al.* (*2015*). Figure 9 shows the schematic diagram of UTWs on existing deteriorated HMA pavement.

*Figure 9. Schematic diagram of UTWs*

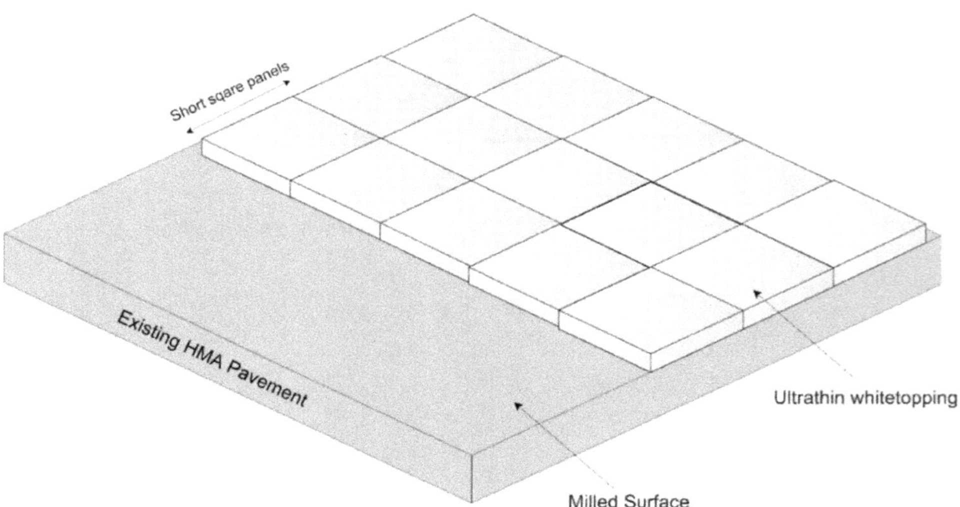

Although the UTWs with short slab size showed better performance, several distresses, such as corner breaking near joints and cracking due to poor pavement structural condition and debonding between the overlay and existing layer, have been observed during its service life. In addition, concrete shrinkage and joint sawing also cause distress in the overlay. The shrinkage develops the tensile stresses at the interface between the overlay and existing pavement, leading to debonding, thus weakening the load-carrying capacity of the pavement. The improper joint sawing accelerates the spalling or joint raveling in the panels. Also, the reduced joint spacing increases the workload compared to conventional concrete and, in turn, increases the cost.

### 4.2.1. Jointless Ultrathin Whitetopping (UTW) With ECC

A full-scale demonstration project with ECC-UTW is being evaluated at the "Louisiana Department of Transportation and Development (LADOTD) Pavement Research Facility (PRF)" in Port Allen, Louisiana *Hungria et al.* (*2020*). The project has constructed three UTW sections with a total length of 54.87 m. Among which the first section (from 0 to 18.29 m) was a jointless ECC-UTW with a thickness of 63.5 mm, the second section (from 18.29 to 36.58 m) was a jointless ECC-UTM with a thickness of 101.6 mm and the final section (from 36.58 to 54.87 m) of 101.6 mm thick with conventional concrete UTW as a control. Figure 10 shows the cross-sectional view of the existing pavement and overlay with ECC and regular concrete.

The project followed the ECC mixture M3-1.5%, which is comparatively low-cost ECC material using less fiber percentage and locally accessible materials for constructing UTW from the previous study by *G. Arce et al.* (*2019*). In the other case of concrete UTW, a regular Type-D concrete pavement mix (according to LADOTD specifications) with a target compressive strength of 28 MPa was used. The mix proportions of ECC and concrete and the mechanical properties of the selected ECC mix have been reported in Table 3.

After the successful completion of the milling process, two preliminary assessments were carried out. The evenness of the milled asphalt layer was evaluated using "Topcom Optical Level and Sokkia GRX1 Real Time Kinematic (RTK) GPS." A Falling Weight Deflectometer (FWD) was conducted to evaluate the structural integrity of the milled asphalt layer to place the strain transducers effectively. Based on these two functional and structural assessment results, three possible locations were selected to place the strain-measuring instruments. In all the test sections, three replicas of strain gauges were placed in longitudinal and transverse directions. Figure 11 shows the location of strain gauge placements for each test section. A portion of transition existed at the end of section 1 (from 18.29 m to 19.81 m), which is around 1.53 m, to eliminate the overlay thickness difference between the two sections (63.5 mm to 101.6 mm). Three ready-mix trucks of volume 6.1 $m^3$, 6.9

m³, and 8.0 m³ were used during the construction of these UTWs. The first truck was used for section 1 and the transition at the end of section 1; the remaining two were for section 2 and section 3 of UTW. It is important to mention that large clumps were observed during the construction of the second section with truck 2. Due to this, additional mixing was performed for 10 minutes at 15 rpm, then the process resumed. All three test sections were constructed in a single day (6:00 AM to 3:00 PM) to avoid variation in environmental changes.

*Figure 10. a) Overall cross-section of existing pavement, b) and c) newly laid overlay with ECC after milling the existing HMA layer, d) overlay with regular concrete*

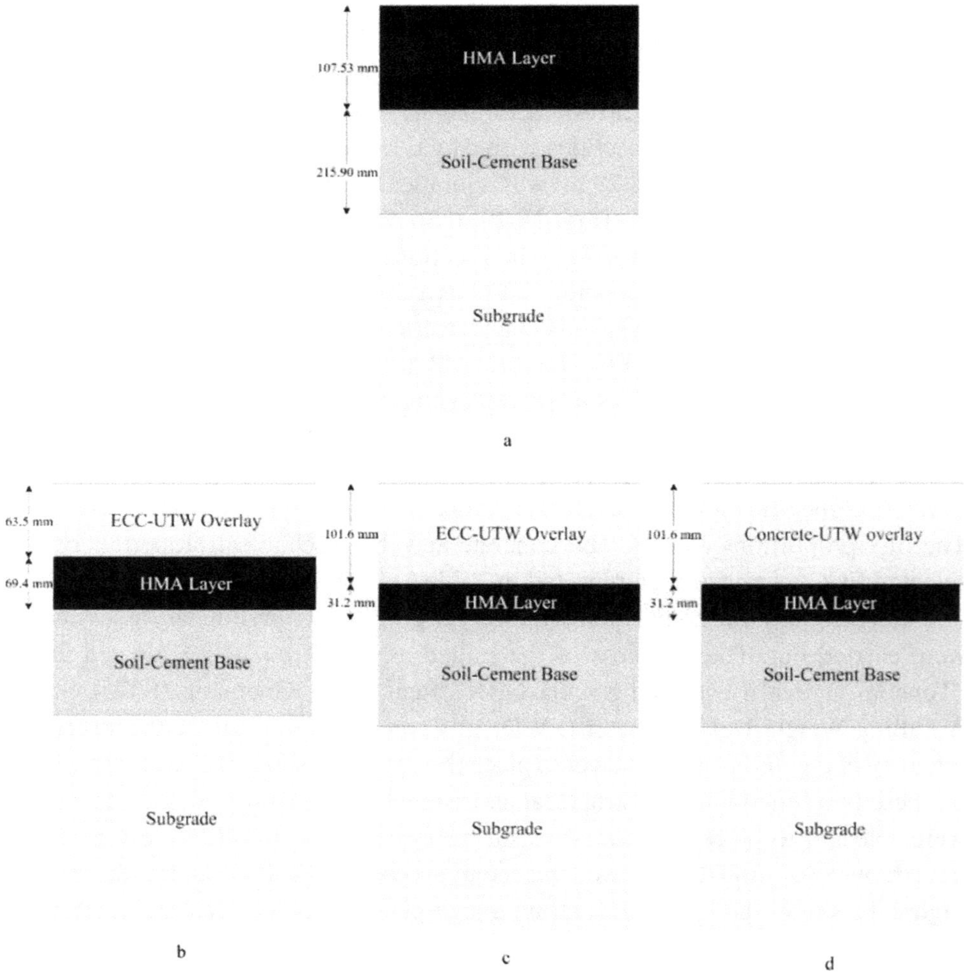

*Table 3. Properties of ECC and mix proportion of ECC and regular concrete (Hungria et al., 2020)*

| **M3-1.5% ECC Mix Proportions** | | | | | |
|---|---|---|---|---|---|
| Cement (Kg/$m^3$) | Fly Ash (Kg/ $m^3$) | Sand (Kg/ $m^3$) | Water (Kg/$m^3$) | Fiber (Kg/$m^3$) | HRWRA (ml/$m^3$) |
| 300.7 | 902.0 | 449.6 | 314.1 | 19.5 | 390.7 |

| **Properties of M3-1.5%** | | | | |
|---|---|---|---|---|
| Compressive Strength (MPa) | Tensile Strength (MPa) | Tensile Strain capacity (%) | Flexural Strength (MPa) | Deflection Capacity (mm) |
| 37.6 | 4.07 | 2.61 | 9.58 | 1.93 |

| **Type -D Concrete Mix Proportions** | | | | | | | |
|---|---|---|---|---|---|---|---|
| Cement (Kg/$m^3$) | Fly Ash (Kg/$m^3$) | Gravel 1 (Kg/$m^3$) | Gravel 2 (Kg/$m^3$) | Sand (Kg/$m^3$) | Water (Kg/$m^3$) | WR (ml/$m^3$) | AE (ml/$m^3$) |
| 253.9 | 63.7 | 805.1 | 278.3 | 745.8 | 133.3 | 688.5 | 61.9 |

The post-crack development behavior was analyzed at regular intervals, which immediately started after one day of construction. The regular UTW section developed two cracks throughout the joints after one day, whereas cracks developed after six days in the 101.6 mm ECC-UTW section and continued up to 9 days. By the end of 9 days, a total of 6 cracks of size ranging from 0.25 mm to 2.54 mm were developed in the ECC-UTW section 2. In the case of 63.5 mm ECC-UTW, no cracks were observed within the first two weeks of the post-construction period. All three UTW sections were continuously monitored for seven months (May 2019 to December 2019) to examine the new crack growth. Also, crack growth was negligible after two weeks of construction in all the sections. Despite that, when the winter season arrived in November, new cracks formed, and crack growth resumed in all the sections, including the 63.5 mm ECC-UTW section (which was not the case before the winter). But whatever the cracks developed in the 63.5 mm ECC section were microcracks of size <0.25 mm, whereas the 101.6 mm ECC section developed macrocracks (0.25 mm to 2.54 mm).

*Figure 11. Strain gauge placements at different test sections after asphalt surface milling, a) milled asphalt surface before construction (Section 1), b) milled asphalt surface before construction (Section 2), and c) milled asphalt surface before construction (Section 3)*

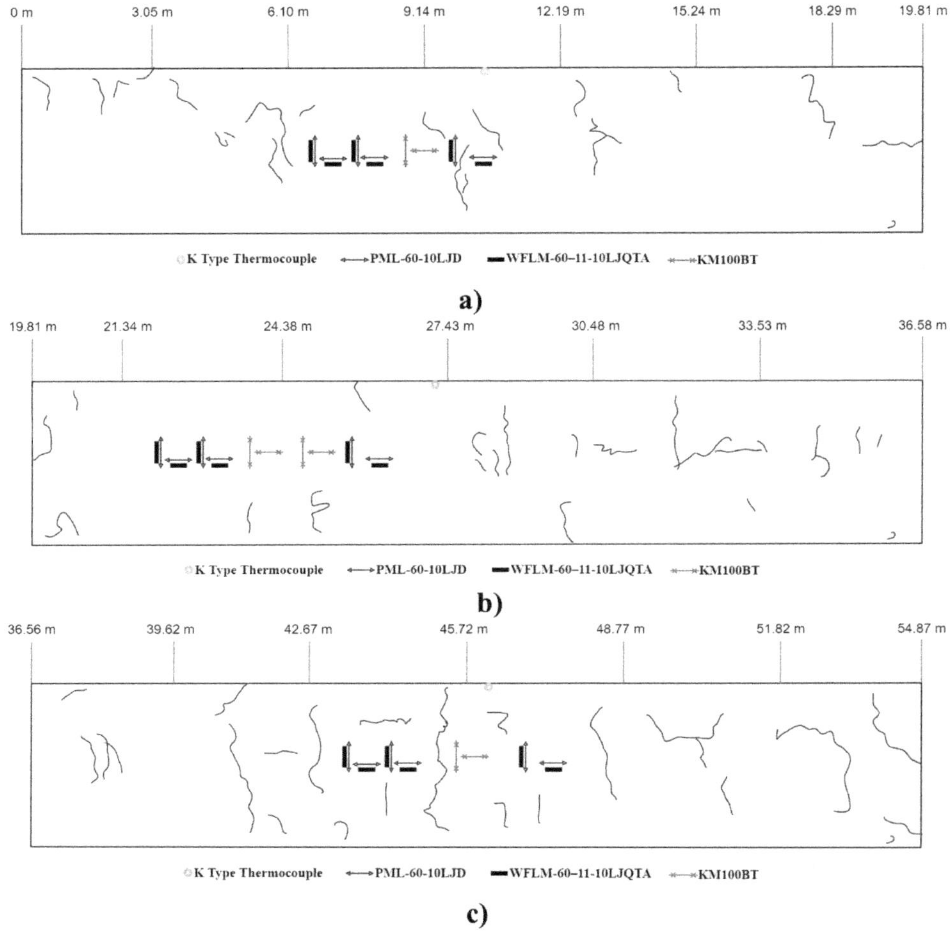


Furthermore, the regular concrete UTW developed cracks at the joints, similar to the pristine condition. The observed cracking of the regular concrete section was expected, as the joints control the cracking due to shrinkage and thermal conditions. However, the same was not expected in the 101.6 mm ECC-UTW due to its remarkable deformability compared to regular concrete. There were several reasons attributed to the macro-cracking in the ECC-UTW. Notably, the formation of fiber or mortar clumping during the pour of the second ready-mix truck, even though the additional mixing time could not improve the mixing properties.

However, in the case of the 63.5 mm ECC-UTW section, there were no cracks in the first two weeks, which emphasized the superior performance of the ECC mix against the early shrinkage and thermal stresses caused by temperature variations. Besides, the cracks developed in the winter season were micro-cracks (<0.25 mm) and can be attributed to the pseudo-hardening (PSH) behavior of the ECC material. In addition, detailed monitoring is required for section 1 when the accelerated load and the environmental stresses act together. The current findings from the project are relevant as they suggest that jointless or widely spaced thin overlays with ECC materials might be possible.

## 4.3. Other Transportation-Related Applications Using ECC

The various other field applications using ECC have been summarized in **Table 4.**

*Table 4. Applications of ECC in various other infrastructures*

| Place of Application | Method of Processing | The Benefit of Using ECC |
|---|---|---|
| Bridge Deck Link Slab - Michigan, USA. *Y. Y. Kim et al. (2004)*; *Lepech & Li (2009)* | Precast and Cast-in-place | The problem of regular maintenance and traffic noise was reduced. Higher fatigue resistance and thermal deformability due to the tensile ductility property of ECC. |
| Patch Repair of Bridge Deck *Kunieda & Rokugo (2006)*; *M. Li & Li (2009), (2011)* | Cast-in-place construction | Lowered the frequency of maintenance and cost. Resistance against wheel loads and chloride environment. |
| Composite Bridge Desk - Mihara Bridge, Japan. *Rokugo & Kanda (2013)* | Cast-in-place Construction | Reduced the risk of steel deck fatigue failure and dead weight. Higher tensile ductility and fatigue resistance. |
| Railway Bridge Beam-Column Joint *R. Zhang et al. (2015)* | Precast slabs at the plant | Increased the shear capacity of frames even without steel reinforcement. |
| Tunnel Lining and Retrofitting – Japan *Rokugo (2018)* | Spraying process | Reduced maintenance and labor costs. Suppressed the water infiltration and increased the resistance against carbonation and freeze-thaw durability. |
| High-Speed Rail Line (Tokaido Shinkansen, Japan) *Rokugo (2018)* | Cast-in-place | Crack resistance with tight crack width and reduced steel corrosion. |
| Viaduct Retrofit – Seisho By-pass (Japan) | Precast at plant | Reduced the risk of collapse during seismic loading due to high energy absorption capacity. Resistance against chloride penetration. |

## 5. SUMMARY AND CONCLUSION

In this chapter, the authors looked back into the studies on fiber-reinforced concrete and the development of HPFRCC, their characteristics, and classification. The chapter also described fiery attempts and observations for the promising use of ECC in various applications. While developing a new class of high-performance cementitious composites, it is recommended that the designed mix should meet the following criteria to study further behavior in the field:

- The mix should perform competitively with conventional materials, and the future cost should be comparable.
- The material characteristics and design must be understood and easy to incorporate.
- A construction method to derive the material performance and its characteristics in the field during the service life.
- Finally, the designed material should deal with the existing problems.

In addition to the conventional material usage in ECC design, the continuous development of green ECC composites has emerged as a prominent research area to reduce the carbon footprint of the infrastructure. The chapter included some research papers that studied the usage of green binders, fillers, and aggregates. A more detailed study of green materials and their discriminating combinations serves to develop emerging greener ECC.

Other significant obstructions to ECC applications in infrastructure construction mentioned by experts were the lack of field experience or performances about these materials. The field demonstrations are the starting point for collecting the experience and would provide the information to publish the collected field data to the engineering community. The chapter provided an in-depth analysis of two field applications. The first application involved an 80-meter-long pavement layer that utilized different paving technologies, while the second study focused on jointless ultrathin whitetoppings (UTW) of varying thicknesses. The long-term monitoring of these ECC field studies holds significant value for confirming performance and incentivizes employing ECC to construct diverse transportation infrastructure projects.

The chapter also delved into several other ECC field applications, summarized towards the end. As illustrated, ECC exhibits versatile applications in transportation infrastructure, buildings, and water-related construction projects. It is foreseeable that ECC's exceptional structural and functional attributes will lead to its broader adoption in various other domains in the near future.

## REFERENCES


Adesina, A., & Das, S. (2020a). Durability Evaluation of Green-Engineered Cementitious Composite Incorporating Glass as Aggregate. *Journal of Materials in Civil Engineering*, *32*(12), 1–13. doi:10.1061/(ASCE)MT.1943-5533.0003444

Adesina, A., & Das, S. (2020b). Mechanical performance of engineered cementitious composite incorporating glass as aggregates. *Journal of Cleaner Production*, *260*, 121113. doi:10.1016/j.jclepro.2020.121113

Adesina, A., & Das, S. (2020c). Performance of green fibre-reinforced composite made with sodium-carbonate-activated slag as a binder. *Innovative Infrastructure Solutions*, *5*(2), 1–9. doi:10.1007/s41062-020-00296-w

Adesina, A., & Das, S. (2021a). Performance of engineered cementitious composites incorporating crumb rubber as aggregate. *Construction & Building Materials*, *274*, 122033. doi:10.1016/j.conbuildmat.2020.122033

Adesina, A., & Das, S. (2021b). Sustainable utilization of recycled asphalt as aggregates in engineered cementitious composites. *Construction & Building Materials*, *283*, 122727. doi:10.1016/j.conbuildmat.2021.122727

Ajith, G., Shanmugasundaram, N., & Praveenkumar, S. (2021). Effect of mineral admixtures and manufactured sand on compressive strength of engineered cementitious composite. *Journal of Building Pathology and Rehabilitation*, *6*(1), 38. Advance online publication. doi:10.1007/s41024-021-00137-y

Ammasi, A. K., & Ragul. (2018). Strength and Durability of High Volume Fly Ash in Engineered Cementitious Composites. *Materials Today: Proceedings*, *5*(11), 24050–24058. doi:10.1016/j.matpr.2018.10.198

An, X., Che, J., Liu, H., Yang, S., & Doh, S. I. (2021). Study on freeze-thaw resistance with NaCl of desert sand engineering cement composites. *Physics and Chemistry of the Earth, 121*, 102954. doi:10.1016/j.pce.2020.102954

Arce, G., Noorvand, H., Hassan, M., Rupnow, T., & Hungria, R. (2019). Cost-Effective ECC with Low Fiber Content for Pavement Application. *MATEC Web of Conferences, 271*, 07001. 10.1051/matecconf/201927107001

Arce, G. A., Noorvand, H., Hassan, M. M., Rupnow, T., & Dhakal, N. (2021). Feasibility of low fiber content PVA-ECC for jointless pavement application. *Construction & Building Materials*, *268*, 121131. doi:10.1016/j.conbuildmat.2020.121131

Arivusudar, N., & Babu, S. S. (2020). Mechanical properties of engineered cementitious composites developed with silica fume. *Cement, Wapno, Beton, 2020*(4), 282–291. doi:10.32047/cwb.2020.25.4.3

ASTM. (2021). Standard Test Method for Density, Absorption, and Voids in Hardened Concrete C642-97. In ASTM C642. https://doi.org/ doi:10.1520/C0642-21.2

ASTM C1202. (2022). Standard Test Method for Electrical Indication of Concrete's Ability to Resist Chloride Ion Penetration. In *ASTM International*. https://doi.org/ doi:10.1520/C1202-22E01.2

ASTM C1260. (2022). Standard Test Method for Potential Alkali Reactivity of Aggregates (Mortar-Bar Method). In *ASTM International*. https://doi.org/ doi:10.1520/C1260-22.2

ASTM C1585. (2020). Standard Test Method for Measurement of Rate of Absorption of Water by Hydraulic Cement Concretes. In *ASTM International*. https://doi.org/ doi:10.1520/C1585-20.2

ASTM C1876. (2019). Standard Test Method for Bulk Electrical Conductivity of Hardened Concrete. In *ASTM International*. doi:10.1520/C1876-19

ASTM C597. (2016). Standard Test Method for Pulse Velocity Through Concrete. In *ASTM International*. https://doi.org/ doi:10.1520/C0597-16.2

Bawono, A. A., Lechner, B., & Yang, E. H. (2019). Skid resistance and surface water drainage performance of engineered cementitious composites for pavement applications. *Cement and Concrete Composites*, *104*(August), 103387. doi:10.1016/j.cemconcomp.2019.103387

Boughanem, S., Jesson, D. A., Mulheron, M. J., Smith, P. A., Eddie, C., Psomas, S., & Rimes, M. (2014). Tensile characterisation of thick sections of Engineered Cement Composite (ECC) materials. *Journal of Materials Science*, *50*(2), 882–897. doi:10.1007/s10853-014-8649-6

Chandrasekhar, C., & Ransinchung, G. D. R. N. (2023). Engineered Cementitious Composites (ECC) with Manufactured Sand (M- sand) for Pavement Applications. *Composites Communications*, *41*(March), 101657. doi:10.1016/j.coco.2023.101657

Che, J., Wang, D., Liu, H., & Zhang, Y. (2019). Mechanical properties of desert sand-based fiber reinforced concrete (DS-FRC). *Applied Sciences (Basel, Switzerland)*, *9*(9), 1857. Advance online publication. doi:10.3390/app9091857

Chen, M., Zhong, H., Chen, L., Zhang, Y., & Zhang, M. (2021). Engineering properties and sustainability assessment of recycled fibre reinforced rubberised cementitious composite. *Journal of Cleaner Production*, *278*, 123996. doi:10.1016/j.jclepro.2020.123996

Gowripalan, N. (2018). Autogenous Shrinkage of Concrete at Early Ages. *25th Australasian Conference on Mechanics of Structures and Materials, December*, 269–276. 10.1007/978-981-13-7603-0_27

Guan, X., Li, Y., Liu, T., Zhang, C., Li, H., & Ou, J. (2019). An economical ultra-high ductile engineered cementitious composite with large amount of coarse river sand. *Construction & Building Materials*, *201*, 461–472. doi:10.1016/j.conbuildmat.2018.12.207

Hisseine, O. A., & Tagnit-Hamou, A. (2020). Nanocellulose for ecological nanoengineered strain-hardening cementitious composites incorporating high-volume ground-glass pozzolans. *Cement and Concrete Composites*, *112*(March), 103662. doi:10.1016/j.cemconcomp.2020.103662

Hossain, M. U., Poon, C. S., Lo, I. M. C., & Cheng, J. C. P. (2017). Comparative LCA on using waste materials in the cement industry: A Hong Kong case study. *Resources, Conservation and Recycling*, *120*, 199–208. doi:10.1016/j.resconrec.2016.12.012

Hou, M., Zhang, D., & Li, V. C. (2022). Material processing, microstructure, and composite properties of low carbon Engineered Cementitious Composites (ECC). *Cement and Concrete Composites*, *134*(March), 104790. doi:10.1016/j.cemconcomp.2022.104790

Huang, B. T., Wu, J. Q., Yu, J., Dai, J. G., Leung, C. K. Y., & Li, V. C. (2021). Seawater sea-sand engineered/strain-hardening cementitious composites (ECC/SHCC): Assessment and modeling of crack characteristics. *Cement and Concrete Research, 140*, 106292. doi:10.1016/j.cemconres.2020.106292

Huang, B. T., Yu, J., Wu, J. Q., Dai, J. G., & Leung, C. K. (2020). Seawater sea-sand Engineered Cementitious Composites (SS-ECC) for marine and coastal applications. *Composites Communications*, *20*(February), 100353. doi:10.1016/j.coco.2020.04.019

Huang, X., Ranade, R., & Li, V. C. (2013). Feasibility Study of Developing Green ECC Using Iron Ore Tailings Powder as Cement Replacement. *Journal of Materials in Civil Engineering*, *25*(7), 923–931. doi:10.1061/(ASCE)MT.1943-5533.0000674

Huang, X. Y., Ni, W., & Li, K. Q. (2015). Development of engineered cementitious composites containing iron ore tailing powders. *Gongcheng Kexue Xuebao. Chinese Journal of Engineering*, *37*(11), 1491–1497. doi:10.13374/j.issn2095-9389.2015.11.015

Hungria, R., Arce, G., Hassan, M., Anderson, M., Mahdi, M., Rupnow, T., & Bogus, S. M. (2020). Evaluation of novel jointless engineered cementitious composite ultrathin whitetopping (ECC-UTW) overlay. *Construction & Building Materials*, *265*, 120659. doi:10.1016/j.conbuildmat.2020.120659

Ibrahim, A., Mahmoud, E., Khodair, Y., & Patibandla, V. C. (2014). Fresh, Mechanical, and Durability Characteristics of Self-Consolidating Concrete Incorporating Recycled Asphalt Pavements. *Journal of Materials in Civil Engineering*, *26*(4), 668–675. doi:10.1061/(ASCE)MT.1943-5533.0000832

Iqbal Khan, M., Fares, G., & Mourad, S. (2017). Optimized Fresh and Hardened Properties of Strain Hardening Cementitious Composites: Effect of Mineral Admixtures, Cementitious Composition, Size, and Type of Aggregates. *Journal of Materials in Civil Engineering*, *29*(10), 1–16. doi:10.1061/(ASCE)MT.1943-5533.0002039

IS 4031 (Part 6). (1988). Methods Of Physical Tests For Hydraulic Cement, Part 6: Determination Of Compressive Strength Of Hydraulic Cement (other than masonry cement). Bureau of Indian Standards.

Japan Society of Civil Engineers. (2008). Recommendations for Design and Construction of High Performance Fiber Reinforced Cement Composites with Multiple Fine Cracks (HPFRCC). In Concrete Engineereing Series (Vol. 82). Author.

Jesson, D. A., & Watts, J. F. (2012). The interface and interphase in polymer matrix composites: Effect on mechanical properties and methods for identification. *Polymer Reviews (Philadelphia, Pa.)*, *52*(3–4), 321–354. doi:10.1080/15583724.2012.710288

Jialing, C., Quanwei, L., Minggin, L., & Dan, W. (2017). Experimental research on mechanical properties of desert sand steel-PVA fiber engineered cementitious composites. *Functional Materials*, *24*(4), 584–592. doi:10.15407/fm24.04.584

Kim, D., Naaman, A. E., & El-Tawil, S. (2008). Comparative flexural behavior of four fiber reinforced cementitious composites. *Cement and Concrete Composites*, *30*(10), 917–928. doi:10.1016/j.cemconcomp.2008.08.002

Kim, D., Suh, C., & Won, M. (2008). Development of the Thickness Design for Concrete Pavement Overlays Over Existing Asphalt Pavement Structures. *Center for Transportation Research at The University of Texas at Austin, 7*(21), 1–121. www.utexas.edu/research/ctr

Kim, Y. Y., Fischer, G., & Li, V. C. (2004). Performance of bridge deck link slabs designed with ductile engineered cementitious composite. *ACI Structural Journal, 101*(6), 792–801. doi:10.14359/13454

Kong, H. J., Bike, S. G., & Li, V. C. (2003). Development of a self-consolidating engineered cementitious composite employing electrosteric dispersion/stabilization. *Cement and Concrete Composites, 25*(3), 301–309. doi:10.1016/S0958-9465(02)00057-4

Kunieda, M., & Rokugo, K. (2006). Recent progress on HPFRCC. *Journal of Advanced Concrete Technology, 4*(1), 19–33. doi:10.3151/jact.4.19

Lee, B. Y., Lee, Y., Kim, J. K., & Kim, Y. Y. (2010). Micromechanics-Based fiber-bridging analysis of strain-hardening cementitious composite accounting for fiber distribution. *CMES -. Computer Modeling in Engineering & Sciences, 61*(2), 111–132.

Lepech, M. D., & Li, V. C. (2008). Large-Scale Processing of Engineered Cementitious. *ACI Materials Journal, 105*(4), 358–366. doi:10.1016/j.cemconcomp.2009.07.002

Lepech, M. D., & Li, V. C. (2009). Application of ECC for bridge deck link slabs. *Materials and Structures/Materiaux et Constructions, 42*(9), 1185–1195. doi:10.1617/s11527-009-9544-5

Lepech, M. D., & Li, V. C. (2010). Sustainable pavement overlays using Engineered Cementitious Composites. *International Journal of Pavement Research and Technology, 3*(5), 241–250.

Lepech, M. D., Li, V. C., Robertson, R. E., & Keoleian, G. A. (2008). Design of green engineered cementitious composites for improved sustainability. *ACI Materials Journal, 105*(6), 567–575. doi:10.14359/20198

Letelier, V., Henríquez-Jara, B. I., Manosalva, M., Parodi, C., & Ortega, J. M. (2019). Use of waste glass as a replacement for raw materials in mortars with a lower environmental impact. *Energies, 12*(10), 1974. Advance online publication. doi:10.3390/en12101974

Li, H., Leung, C. K. Y., Xu, S., & Cao, Q. (2009). Potential use of strain hardening ECC in permanent formwork with small scale flexural beams. *Journal of Wuhan University of Technology. Materials Science Edition*, *24*(3), 482–487. doi:10.1007/s11595-009-3482-5

Li, J., & Yang, E. H. (2017). Macroscopic and microstructural properties of engineered cementitious composites incorporating recycled concrete fines. *Cement and Concrete Composites*, *78*, 33–42. doi:10.1016/j.cemconcomp.2016.12.013

Li, M., & Li, V. C. (2009). Influence of material ductility on performance of concrete repair. *ACI Materials Journal*, *106*(5), 419–428. doi:10.14359/51663142

Li, M., & Li, V. C. (2011). High-early-strength engineered cementitious composites for fast, durable concrete repair-material properties. *ACI Materials Journal*, *108*(1), 3–12. doi:10.14359/51664210

Li, V. C. (1994). From micromechanics to structural engineering -the design of cementitious composites for civil engineering applications. *Structural Engineering / Earthquake Engineering*, *10*(2), 1–34.

Li, V. C. (1997). Engineered Cementitious Composites (Ecc) – Tailored Composites through Micromechanical Modeling. Canadian Society for Civil Engineering.

Li, V. C. (1998). Engineered Cementitious Composites (ECC) – Tailored Composites through Micromechanical Modeling. In Fiber Reinforced Concrete: Present and the Future (pp. 64–97). Academic Press.

Li, V. C. (2003). On Engineered Cementitious Composites (ECC). *Journal of Advanced Concrete Technology*, *1*(3), 215–230. doi:10.3151/jact.1.215

Li, V. C. (2008). Engineered cementitious composite (ecc): Material, structural, and durability performance. Concrete Construction Engineering Handbook, 1001–1048.

Li, V. C., Lepech, M., & Wang, S. (2004). Development of green engineered cementitious composites for sustainable infrastructure systems. *International Workshop on Sustainable Development and Concrete Technology, 1*(September), 181–191. http://core.kmi.open.ac.uk/download/pdf/11346106.pdf#page=192

Li, V. C., Mishra, D. K., & Wu, H.-C. (1995). Matrix design for pseudo-strain-hardening fibre reinforced cementitious composites. *Materials and Structures*, *28*(10), 586–595. doi:10.1007/BF02473191

Li, V. C., Wang, S., & Wu, C. (2001). Tensile strain-hardening behavior or polyvinyl alcohol engineered cementitious composite (PVA-ECC). *ACI Materials Journal, 98*(6), 483–492. doi:10.14359/10851 PMID:34682615

Li, V. C., Wang, Y., & Backer, S. (1991). A micromechanical model of tension-softening and bridging toughening of short random fiber reinforced brittle matrix composites. *Journal of the Mechanics and Physics of Solids, 39*(5), 607–625. doi:10.1016/0022-5096(91)90043-N

Li, V. C., Wu, C., Wang, S., Ogawa, A., & Saito, T. (2002). Interface tailoring for strain-hardening polyvinyl alcohol-engineered cementitious composite (PVA-ECC). *ACI Materials Journal, 99*(5), 463–472. doi:10.14359/12325 PMID:28338189

Liu, Y., Zhou, X., Lv, C., Yang, Y., & Liu, T. (2018). Use of Silica Fume and GGBS to Improve Frost Resistance of ECC with High-Volume Fly Ash. *Advances in Civil Engineering, 2018*, 1–11. Advance online publication. doi:10.1155/2018/7987589

Ma, H., Qian, S., & Zhang, Z. (2014). Effect of self-healing on water permeability and mechanical property of Medium-Early-Strength Engineered Cementitious Composites. *Construction & Building Materials, 68*, 92–101. doi:10.1016/j.conbuildmat.2014.05.065

Ma, H., Qian, S., Zhang, Z., Lin, Z., & Li, V. C. (2015). Tailoring Engineered Cementitious Composites with local ingredients. *Construction & Building Materials, 101*(Part 1), 584–595. doi:10.1016/j.conbuildmat.2015.10.146

Mateos, A. J. H., Paniagua, J. C., & Paniagua, F. (2015). Development of Improved Guidelines and Designs for Thin Whitetopping : Literature Review. *University of California Pavement Research Center,* 1–31. www.its.ucdavis.edu

Mueller, S., Ranjbarian, M., & Mechtcherine, V. (2019). Fatigue behavior of strain-hardening cement-based composites—From the single fiber level to real-scale application. *Structural Concrete, 20*(4), 1231–1242. doi:10.1002/suco.201900032

Müller, S., & Mechtcherine, V. (2018). *Use of Strain-Hardening Cement-Based Composites (SHCC) in Real Scale Applications BT*. In V. Mechtcherine, V. Slowik, & P. Kabele (Eds.), *Strain-Hardening Cement-Based Composites* (pp. 690–700). Springer Netherlands.

Naaman, A. E., & Reinhardt, H. W. (2006). Proposed classification of HPFRC composites based on their tensile response. *Materials and Structures/Materiaux et Constructions, 39*(5), 547–555. doi:10.1617/s11527-006-9103-2

Noorvand, H., Arce, G., Hassan, M., Rupnow, T., & Mohammad, L. N. (2019). Investigation of the Mechanical Properties of Engineered Cementitious Composites with Low Fiber Content and with Crumb Rubber and High Fly Ash Content. *Transportation Research Record: Journal of the Transportation Research Board*, *2673*(5), 418–428. doi:10.1177/0361198119837510

Noorvand, H., Arce, G. A., & Hassan, M. M. (2021). Evaluation of the effects of engineered cementitious composites (ECC) plasticity on concrete pavement performance. *The International Journal of Pavement Engineering*, *0*(0), 1–13. doi:10.1080/10298436.2021.1954180

Owusu-Antwi, E. B., Khazanovich, L., & Titus-Glover, L. (1998). Mechanistic-based model for predicting reflective cracking in asphalt concrete-overlaid pavements. *Transportation Research Record: Journal of the Transportation Research Board*, *1629*(1), 234–241. doi:10.3141/1629-26

Pakravan, H. R., Jamshidi, M., & Latifi, M. (2018). The effect of hydrophilic (polyvinyl alcohol) fiber content on the flexural behavior of engineered cementitious composites (ECC). *Journal of the Textile Institute*, *109*(1), 79–84. doi:10.1080/00405000.2017.1329132

Pavlu, T., Kocí, V., & Hájek, P. (2019). Environmental assessment of two use cycles of recycled aggregate concrete. *Sustainability (Basel)*, *11*(21), 6185. Advance online publication. doi:10.3390/su11216185

Pranav, S., Lahoti, M., Shan, X., Yang, E. H., & Muthukumar, G. (2022). Economic input-output LCA of precast corundum-blended ECC overlay pavement. *Resources, Conservation and Recycling*, *184*(September), 106385. doi:10.1016/j.resconrec.2022.106385

Redon, C., Li, V. C., Wu, C., Hoshiro, H., Saito, T., & Ogawa, A. (2001). Measuring and Modifying Interface Properties of PVA Fibers in ECC Matrix. *Journal of Materials in Civil Engineering*, *13*(6), 399–406. doi:10.1061/(ASCE)0899-1561(2001)13:6(399)

RighiD. P.da CostaF. B. P.GraeffÂ. G.da Silva FilhoL. C. P. (2017). Tensile behaviour and durability issues of Engineered Cementitious Composites with Rice Husk Ash. *Revista Materia, 22*(2). https://doi.org/ doi:10.1590/s1517-707620170002.0182

Rilem, T. C. (2001). Test and design methods for steel fibre reinforced concrete: Uni-axial tension test for steel fibre reinforced concrete. In Materials and Structures/ Materiaux et Constructions (Vol. 34, pp. 3–6). Academic Press.

Rokugo, K. (2018). Strain-hardening cement-based composites. In SHCC44 (Vol. 15, pp. 671–680). doi:10.1007/978-94-024-1194-2

Rokugo, K., & Kanda, T. (2013). Strain Hardening Cement Composites: Structural Design and Performance. In Rilem Tac (Vol. 6). Academic Press.

Sahmaran, M., Lachemi, M., Hossain, K. M. A., Ranade, R., & Li, V. C. (2009). Influence of aggregate type and size on ductility and mechanical properties of engineered cementitious composites. *ACI Materials Journal*, *106*(3), 308–316. doi:10.14359/56556

Şahmaran, M., & Li, V. C. (2009). Durability properties of micro-cracked ECC containing high volumes fly ash. *Cement and Concrete Research*, *39*(11), 1033–1043. doi:10.1016/j.cemconres.2009.07.009

Sahmaran, M., Yücel, H. E., Yildirim, G., Al-Emam, M., & Lachemi, M. (2014). Investigation of the Bond between Concrete Substrate and ECC Overlays. *Journal of Materials in Civil Engineering*, *26*(1), 167–174. doi:10.1061/(ASCE)MT.1943-5533.0000805

Saraireh, D., Walls, S., Suryanto, B., Starrs, G., & McCarter, W. J. (2018). The influence of multiple micro-cracking on the electrical impedance of an engineered cementitious composite. In RILEM Bookseries (Vol. 15). doi:10.1007/978-94-024-1194-2_34

Shanmugasundaram, N., & Praveenkumar, S. (2022). Mechanical properties of engineered cementitious composites (ECC) incorporating different mineral admixtures and fibre: A review. *Journal of Building Pathology and Rehabilitation*, *7*(1), 40. Advance online publication. doi:10.1007/s41024-022-00182-1

Shanmugasundaram, N., & Praveenkumar, S. (2023). Adhesive characteristics of novel greener engineered cementitious composite with conventional concrete substrate. *Construction & Building Materials*, *407*(September), 133591. doi:10.1016/j.conbuildmat.2023.133591

Shoba, K. B., & Asha, P. (2019). Relationship between compressive strength and ultrasonic pulse velocity of engineered cementitious composites inducing rice husk ash. *International Journal of Recent Technology and Engineering*, *7*(6), 682–685.

Shoji, D., He, Z., Zhang, D., & Li, V. C. (2022). The greening of engineered cementitious composites (ECC): A review. *Construction and Building Materials*, *327*, 126701. doi:10.1016/j.conbuildmat.2022.126701

Siad, H., Lachemi, M., Sahmaran, M., & Hossain, K. M. A. (2017). Mechanical, Physical, and Self-Healing Behaviors of Engineered Cementitious Composites with Glass Powder. *Journal of Materials in Civil Engineering*, *29*(6), 1–12. doi:10.1061/(ASCE)MT.1943-5533.0001864

Singh, S. B., & Munjal, P. (2020). Engineered cementitious composite and its applications. *Materials Today: Proceedings, 32*, 797–802. doi:10.1016/j.matpr.2020.03.743

Srikakulam, L. M., & Khed, V. C. (2020). Experimental investigation on the strength parameters of rubberized engineered cementitious composite with M sand. *Materials Today: Proceedings, 27*, 1230–1234. doi:10.1016/j.matpr.2020.02.146

Subedi, S., Arce, G. A., Hassan, M. M., Barbato, M., & Mohammad, L. N. (2021). Effect of raw sugarcane bagasse ash as sand replacement on the fiber-bridging properties of engineered cementitious composites. *Transportation Research Record: Journal of the Transportation Research Board, 2675*(11), 1028–1042. doi:10.1177/03611981211023762

Tan, G., Zhu, Z., Wang, W., Wu, C., Ou, J., Cui, G., & Zhang, D. (2021). Flexural ductility and crack-controlling capacity of polypropylene fiber reinforced ECC thin sheet with waste superfine river sand based on acoustic emission analysis. *Construction & Building Materials, 277*, 122321. Advance online publication. doi:10.1016/j.conbuildmat.2021.122321

Turk, K., & Nehdi, M. L. (2021). Flexural toughness of sustainable ECC with high-volume substitution of cement and silica sand. *Construction & Building Materials, 270*, 121438. Advance online publication. doi:10.1016/j.conbuildmat.2020.121438

Wang, S., & Li, V. (2003). Light Weigth Engineered Cementitious Composites (ECC). In *Fourth int'l RILEM workshop on high performance fiber reinforced cement composites (HPFRCC 4)* (pp. 379–390).

Wang, S., & Li, V. C. (2006). High-early-strength engineered cementitious composites. *ACI Materials Journal, 103*(2), 97–105. doi:10.14359/15260

Wang, W., Wu, H., Ma, Z., & Wu, R. (2020). Using eco-friendly recycled powder from CDW to prepare strain hardening cementitious composites (SHCC) and properties determination. *Materials (Basel), 13*(5), 1143. Advance online publication. doi:10.3390/ma13051143 PMID:32143483

Wang, Y., Liu, F., Yu, J., Dong, F., & Ye, J. (2020). Effect of polyethylene fiber content on physical and mechanical properties of engineered cementitious composites. *Construction & Building Materials, 251*, 118917. doi:10.1016/j.conbuildmat.2020.118917

Weimann, M. B., & Li, V. C. (2003). Drying Shrinkage and Crack Width of Engineered Cementitious Composites (Ecc). In *Brittle Matrix Composites 7*. Institute of Fundamental Technological Research. doi:10.1533/9780857093103.37

Wille, K., El-Tawil, S., & Naaman, A. E. (2014). Properties of strain hardening ultra high performance fiber reinforced concrete (UHP-FRC) under direct tensile loading. *Cement and Concrete Composites*, *48*, 53–66. doi:10.1016/j.cemconcomp.2013.12.015

Winkelman, T. J. (2005). Whitetopping Performance in Illinois. Illinois Department of Transportation.

Xiao, H., Zhang, N., Li, G., Zhang, Y., Wang, Y., Wang, Y., & Zhang, Y. (2023). Graphene-Iron Ore Tailings–Based Cementitious Composites with High Early Flexural Strength. *Materials (Basel)*, *16*(1), 327. Advance online publication. doi:10.3390/ma16010327 PMID:36614666

Yang, E. H., Yang, Y., & Li, V. C. (2007). Use of high volumes of fly ash to improve ECC mechanical properties and material greenness. *ACI Materials Journal*, *104*(6), 620–628. doi:10.14359/18966

Yang, G., Yu, J., & Luo, Y. (2019). Development and Mechanical Performance of Fire-Resistive Engineered Cementitious Composites. *Journal of Materials in Civil Engineering*, *31*(5), 1–10. doi:10.1061/(ASCE)MT.1943-5533.0002666

Yang, Y., Lepech, M. D., & Li, V. C. (2005). Self-Healing of Engineered Cementitious Composites Under Cyclic Wetting and Drying. In *International Workshop on Durability of Reinforced Concrete Under Combined Mechanical and Climatic Loads* (*Vol. 1*, Issue 1, pp. 231–242). Academic Press.

Yao, Q., Li, Z., Lu, C., Peng, L., Luo, Y., & Teng, X. (2021). Development of Engineered Cementitious Composites Using Sea Sand and Metakaolin. *Frontiers in Materials*, *8*(August), 1–13. doi:10.3389/fmats.2021.711872

Yu, J., Chen, Y., & Leung, C. K. Y. (2019). Mechanical performance of Strain-Hardening Cementitious Composites (SHCC) with hybrid polyvinyl alcohol and steel fibers. *Composite Structures*, *226*(July), 111198. Advance online publication. doi:10.1016/j.compstruct.2019.111198

Yu, J., & Leung, C. K. Y. (2017). Strength Improvement of Strain-Hardening Cementitious Composites with Ultrahigh-Volume Fly Ash. *Journal of Materials in Civil Engineering*, *29*(9), 05017003. Advance online publication. doi:10.1061/(ASCE)MT.1943-5533.0001987

Yu, J., Lu, C., Leung, C. K. Y., & Li, G. (2017). Mechanical properties of green structural concrete with ultrahigh-volume fly ash. *Construction & Building Materials*, *147*, 510–518. doi:10.1016/j.conbuildmat.2017.04.188

Yu, K., Wang, Y., Yu, J., & Xu, S. (2017). A strain-hardening cementitious composites with the tensile capacity up to 8%. *Construction & Building Materials*, *137*, 410–419. doi:10.1016/j.conbuildmat.2017.01.060

Yucel, H. E., Jashami, H., Sahmaran, M., Guler, M., & Yaman, I. O. (2013). Thin ECC overlay systems for rehabilitation of rigid concrete pavements. *Magazine of Concrete Research*, *65*(2), 108–120. doi:10.1680/macr.12.00022

Zhang, J., & Li, V. C. (2002). Monotonic and fatigue performance in bending of fiber-reinforced engineered cementitious composite in overlay system. *Cement and Concrete Research*, *32*(3), 415–423. doi:10.1016/S0008-8846(01)00695-0

Zhang, J., Wang, Z., & Ju, X. (2013). Application of ductile fiber reinforced cementitious composite in jointless concrete pavements. *Composites. Part B, Engineering*, *50*, 224–231. doi:10.1016/j.compositesb.2013.02.007

Zhang, R., Matsumoto, K., Hirata, T., Ishizeki, Y., & Niwa, J. (2015). Application of PP-ECC in beam-column joint connections of rigid-framed railway bridges to reduce transverse reinforcements. *Engineering Structures*, *86*, 146–156. doi:10.1016/j.engstruct.2015.01.005

Zhang, S., Li, V. C., & Ye, G. (2020). Micromechanics-guided development of a slag/fly ash-based strain-hardening geopolymer composite. *Cement and Concrete Composites*, *109*, 103510. doi:10.1016/j.cemconcomp.2020.103510

Zhang, Z., Liu, S., Yang, F., Weng, Y., & Qian, S. (2021). Sustainable high strength, high ductility engineered cementitious composites (ECC) with substitution of cement by rice husk ash. *Journal of Cleaner Production*, *317*(March), 128379. doi:10.1016/j.jclepro.2021.128379

Zhang, Z., Qian, S., Liu, H., & Li, V. C. (2017). Ductile concrete material with self-healing capacity for jointless concrete pavement use. *Transportation Research Record: Journal of the Transportation Research Board*, *2640*(1), 78–83. doi:10.3141/2640-09

Zhang, Z., Qian, S., & Ma, H. (2014). Investigating mechanical properties and self-healing behavior of micro-cracked ECC with different volume of fly ash. *Construction & Building Materials*, *52*, 17–23. doi:10.1016/j.conbuildmat.2013.11.001

Zhang, Z., & Zhang, Q. (2017). Self-healing ability of Engineered Cementitious Composites (ECC) under different exposure environments. *Construction & Building Materials*, *156*, 142–151. doi:10.1016/j.conbuildmat.2017.08.166

Zhang, Z., Zhang, Q., Qian, S., & Li, V. C. (2015). Low E modulus early strength engineered cementitious composites material development for ultrathin whitetopping overlay. *Transportation Research Record: Journal of the Transportation Research Board*, *2481*(1), 41–47. doi:10.3141/2481-06

Zhao, A., Yang, J., & Yang, E. H. (2015). Self-cleaning engineered cementitious composites. *Cement and Concrete Composites*, *64*, 74–83. doi:10.1016/j.cemconcomp.2015.09.007

Zhao, J., Yan, C., Liu, S., Zhang, J., Li, S., & Yan, Y. (2020). Effect of solid waste ceramic on uniaxial tensile properties and thin plate bending properties of polyvinyl alcohol engineered cementitious composite. *Journal of Cleaner Production*, *268*, 122329. doi:10.1016/j.jclepro.2020.122329

Zhou, J., Qian, S., Beltran, M. G. S., Ye, G., Van Breugel, K., & Li, V. C. (2010). Development of engineered cementitious composites with limestone powder and blast furnace slag. *Materials and Structures/Materiaux et Constructions, 43*(6), 803–814. doi:10.1617/s11527-009-9549-0

Zhou, Y., Gong, G., Huang, Y., Chen, C., Huang, D., Chen, Z., & Guo, M. (2021). Feasibility of incorporating recycled fine aggregate in high performance green lightweight engineered cementitious composites. *Journal of Cleaner Production*, *280*, 124445. doi:10.1016/j.jclepro.2020.124445

Zhu, Z., Tan, G., Zhang, W., & Wu, C. (2020). Preliminary analysis of the ductility and crack-control ability of engineered cementitious composite with superfine sand and polypropylene fiber (SSPP-ECC). *Materials (Basel)*, *13*(11), 1–15. doi:10.3390/ma13112609 PMID:32521673

# Chapter 10
# Engineered Cementitious Composites for Diverse Applications

**K. R. Dhanya**
*Northamps, India*

## ABSTRACT

*Engineered cementitious composites (ECC) are developed for the wide applications of construction industry and it is considered as a special form of high-performance fiber-reinforced cementitious composites. This material exhibits excellent shear capacity, capable of controlling crack-width and better damage tolerance capacity. ECC enhances the basic properties of materials thereby increasing the structural performance, exhibits some unique properties, that cannot be replaced by conventional concrete and on implementing ECC structures for desired applications, shows superior performance. The chapter describes the importance and benefits of ECC materials in the field of construction, current uses, and future scope. Conventional concrete material shows low tensile strength as well as strain capacity. ECC is a potential material to exhibit superior durability performances as well as structural viability.*

## INTRODUCTION

Engineered Cementitious Composites (ECC) are also known as high-performance fiber reinforced cementitious composites (HRFRCC). ECC materials are considered as high ductile and medium tensile strength composites. The composite shows remarkable properties such as damage tolerance, shear capacity, capable to control crack width and synergistic interaction with reinforcing bars. On comparison with

DOI: 10.4018/978-1-6684-8182-0.ch010

regular concrete, the initial rate of ECC is higher but the life-cycle cost is very low and offer better safety due to larger ductility. These can be applicable to several sectors to enhance their service life, minimizing maintenance and repair-cost. ECC materials contains fiber content around 2% which comprises of polyethylene fibers as well as polyvinyl alcohol fibers (Gupta, 2017). Engineered Cementitious Composites (ECC) are indeed known for their fiber-reinforced nature, but the specific fiber content can vary depending on the ECC mix and the requirements of the application. The statement that ECC materials typically contain fiber content around 2% is a generalization and may not be universally accurate. ECC can be designed with varying fiber contents depending on the desired performance characteristics. While some ECC mixes may contain fibers in the range of 2%, others may have higher or lower fiber contents. The types of fibers used can also vary, and ECC may include a combination of different fiber materials to achieve specific mechanical properties and durability. The most common fibers used in ECC include polyethylene (PE) fibers and polyvinyl alcohol (PVA) fibers. These fibers are selected for their high tensile strain capacity and their ability to enhance the ductility and crack control of the concrete matrix. However, the precise composition of ECC can be tailored to meet the requirements of a given application, and the fiber content can be adjusted accordingly.

Therefore, it is essential to consult the specific mix design and engineering requirements for an ECC project to determine the exact fiber content and types of fibers used, as they can vary depending on the project's goals and specifications.

Engineered Cementitious Composites have a history of research and applications dating back several decades. ECC was initially developed by researchers in the 1980s and 1990s, with pioneering work by Professor Victor Li at the University of Michigan. This marked the beginning of ECC research and development. Early studies focused on ECC's exceptional properties, such as its high ductility and tensile strain capacity, which were found to be superior to conventional concrete. Researchers conducted extensive material testing to characterize these properties. ECC's ability to withstand blast loading led to applications in the construction of blast-resistant structures, particularly in military and security-related projects are noted widely. It was explored as a material for retrofitting existing buildings in earthquake-prone regions and then researchers began to investigate self-healing ECC, which involved incorporating microcapsules with healing agents into the material. When cracks formed, these capsules ruptured, releasing healing agents to repair the damage. In 2010, researchers explored robotic construction techniques to create intricate architectural forms and structures.

ECC exhibit several beneficial properties such as durability in hot and humid environments (Li et al., 2007) permeability and crack resistance (Li & Yang, 2005), self-healing characteristics (Li et al., 2001), mechanical enhancements (Li & Maalej,

1996), etc. Several research studies focused on the residual strength, stiffness and strain hardening properties of ECC and either in natural conditions to very mild conditions self-healing abilities are carried out at an elevated temperature of 200$^0$C to 800$^0$C (Hertbert, 2012; Yu et al., 2015).

The recent research studies shows that ECC have been considered as a material for providing environmental sustainability. The wide use of these materials on commercial level and resulting benefits are more compared to conventional concrete material. This material improves the durability, safety but decreases the cost happened for maintenance of structures. The outstanding features of this material proven as highly predominant material in commercial sectors in current time. For the site construction, ECC materials with self-combing capacity are considered as a good choice, in transport sector, high-strength ECC are generally applicable and in building with dead load light weight ECC are suitable (Manish & Kewalramani, 2016; Michael, 2008). The choice between high-strength and lightweight ECC depends on a range of factors, including the intended use of the material, structural design, and construction needs. High-strength ECC is typically characterized by its enhanced mechanical properties, including higher compressive and tensile strength. This type of ECC is often used in applications where structural strength, durability, and crack resistance are critical. Examples include earthquake-resistant structures, bridges, and infrastructure projects. Lightweight ECC may contain lightweight aggregates or other additives to reduce its density while maintaining other desirable properties like ductility and durability. It can be used in construction projects where reducing the overall weight of the structure is important, such as in some residential and lightweight construction applications.

Shang and Lu clarified FRP fabric materials or steel bars on combination with ECC enhance the strength of structure member and ECC can be widely used in structure repairs and seismic activity resistance. These materials possess impacts as well as blasts because of the strain hardening property, high strain capacity and ultra-ductile behaviour. The influence of high temperature up to 800$^0$C on compressive strength of ECC was analysed by Shang and Lu (2014).

Nowadays, ECC have gained much attention in civil infrastructures such as building, transportation, water and energy industry fields. In large building, the ECC coupling beams with high tensile ductility enhances seismic protection and exhibits durability by energy absorption during an earthquake (Kanda, n.d.). Some research studies depict that on combining ECC with polyvinyl alcohol and polyethylene fibers, the flexural fatigue properties remarkably changes (Suthiwarapirak et al., 2004) and the study was reported by Suthiwarapirak et al. in 2004. When PVA-ECC is combined, the material shows extremely high fiber/matrix bond strength and PE-ECC combinations shows higher tensile strength. Hybrid fiber of ECC of steel/ polyethylene obtain stable ultimate strength and strain capacity for blast resistance as well as impact structures and this was reported by Zhang et al (2007). Zollo, 1997, Atahan et al. 2012; Maalej et al. 2012; Soe et al. 2013a; 2013b; Ali et al. 2017; Kai et al. 2017; Nehdi and Ali 2019 stated that the results achieved explains the use of ECC in protective structures (Ali et al., 2017; Atahan, 2012; Kai et al., 2017; Maalei, 2012; Nehdi & Ali, 2019; Soe et al., 2013a; Soe et al., 2013b; Zollo, 1997). ECC materials absorb and resist impact energy compared to conventional concrete and FRC materials.

One of the most striking properties of ECC material is its toughness, long-term existence as well as self-healing robustness in contact with marine environment. This was mentioned by Li and Li (2011) and under severe marine conditions, ECC preventing normal failure mechanism in concrete and holds the tensile ductility nature. ECC mixtures contain polyvinyl alcohol fibers, steel fibers and polyethylene fibers and shows remarkable properties for specific applications. When ECC is mixed with slag, it typically enhances the entire ductile nature and also reduces the residual crack width, mentioned by Ozbay et al in 2013 (Ozbay, 2013).

## Properties

The flexibility of ECC materials is such that it can be easily fabricated, extruded and sprayed and can be measured over viscosity parameter which allows the sprayable ECC structures. There are various studies related to the workability of ECC with self-compacting nature. Several researchers investigated the self-healing property of ECC material and Qian et al (2009) mentioned that pre-cracked strain hardening fiber reinforced cementitious composites shows self-healing property. The crack width and tensile strain ductility of ECC can be selectively enhanced by using recycled materials and this study was reported by Yang et al (2007). Khan et al (2016), studied that white sand when incorporating into ECC mixture, creates remarkable effects. The sustainability of ECC was further analysed by Kewalramani et al (2017) in hot arid climates. The natural fibers flax and cottonised flax were used with ECC by Snoeck and De Belie (2012). ECC is nowadays widely used for pavement repair application and was studied by Li (2008). The studies proves that ECC overlay can provide an extended service life and minimised maintenance efforts. This is one of the major applications of ECC structures. The extremely high ductile nature of ECC make it for use in link-slabs and can be replaced by conventional expansion joints in bridges and this will provide smooth ride conditions on bridges. ECC can be successfully used in existing concrete structures and self-healing property is most important property as a repair material. The life-cycle, GHG emission and energy consumption cost of ECC were very low compared to concrete overlay as well as in traditional expansion joints. The low maintenance capability is considered as the lower life-cycle costs and environmental impact of ECC materials compared to others and it is one of the beneficial properties of these structures. Various studies show that during the inelastic deformation, ECC exhibit tensile strain hardening properties. After the first crack, ECC structures exhibit extremely high ductile nature therefore these types of composites are also known as Ductile Fibers Reinforced Composite. ECC is known for its exceptional tensile strain capacity, which is significantly higher than that of traditional concrete. This means that ECC can deform and stretch significantly before experiencing failure. As a result, ECC can undergo extensive strain, accommodating tensile loads even after the formation of the first crack. This composite exhibits a multiple cracking behavior, where numerous microcracks form within the material. These microcracks are distributed throughout the ECC matrix and help to dissipate energy and redistribute stresses. This behavior allows ECC to continue carrying loads even after cracks have developed. The multiple cracking behavior and high tensile strain capacity result in ECC structures remaining highly ductile. Ductility is the ability of a material or structure to undergo large deformations without catastrophic failure. ECC's enhanced ductility is a key feature, and it is this property that allows it to be described as a Ductile Fiber Reinforced Composite. The ability of ECC to control crack widths is another significant aspect of its ductility. Even as cracks form, they tend to be very fine, which helps maintain the structural integrity of the material and prevents the rapid propagation of cracks. ECC's ductility contributes to its durability. It can better withstand environmental stresses, such as freeze-thaw cycles, chemical exposure, and cyclic loading, due to its ability to distribute and absorb strains and stresses without significant loss of structural capacity.

Compared to traditional fiber reinforced composites, these structures show pseudo strain hardening nature under tensile loading. ECC material has got much attention on vast applications such as structure repair, new constructions, seismic activity resistance, high strain capacity, impact and blast resistance. Several kinds of structural applications are reported in various Countries.

ECC beams are widely used in construction fields and the aim of using these structures are to defend it from earthquakes. The main feature of these materials is damage tolerance capacity, high energy absorption and extremely large ductile nature. In the construction of road beds of Mihara bridge at Japan, ECC materials are largely used and during the construction, there is a sharp reduce around 40%, of other materials. The interstate highway of USA was used ECC slab and due to its remarkable nature, this slab spent less material than other traditional concrete slabs (Li, 2006).

ECC materials shows extraordinary mechanical properties and the use of these structures in bridges providing excellent compression strength. These materials can also be applicable for the construction of dams. One of the case studies illustrate that in Japan, one of the oldest dams was repaired using ECC. The badly scratched surface was replaced by incorporation of ECC materials. Another study reported that

ECC is widely applied in earth retaining wall in Gifu at Japan. The cracks found in the wall had been easily eliminated and repaired using ECC. Generally, it is found that ECC is a promising material for strengthening purpose due to its exceptionally high mechanical property. ECC's extraordinary mechanical properties make it a highly suitable and promising material for strengthening and enhancing the durability of various construction projects, including bridges and dams. Its use in real-world applications, such as the repair of old dams and retaining walls, underscores its effectiveness in improving the performance and longevity of infrastructure. The application of ECC materials in decks (Lepech & Li, 2009), high-rise buildings and repair applications (Dehghani, 2015; Dehghani et al., 2015) are more preferable areas of ECC. Several research analyses shows that the corrosion resistance and self-healing characteristics of ECC is due to the microcracking behaviour (Li, 2011; Minard, n.d.; Sahmaran, 2007). These materials provide large damage tolerance and exceptionally good durability and can be used in several environments for infrastructure. Nowadays sustainable ECC materials can be fabricated using pozzolan.

Research studies shows that PVA/ ECC can be fabricated with local ingredients and the manufacturing method of mixture is cost effective (Zhang & Zhang, 2018). They also reported that PVA materials largely affects the mechanical properties of ECC such as tensile strength, post-peak ductility and compressive strength. ECC can be manufactured using locally sourced ingredients, which contributes to its cost-effectiveness. The primary constituents of ECC typically include portland cement, fine aggregates, water, superplasticizers, and, notably, PVA (polyvinyl alcohol) fibers. These ingredients are often readily available and can be sourced locally, reducing material transportation costs. Its mix design can be tailored to suit local availability, making it adaptable to a wide range of construction projects. PVA fibers are highly ductile and have a high tensile strain capacity. When incorporated into ECC, they enhance the tensile strength and post-peak ductility of the material. PVA fibers contribute to the multiple cracking behavior and strain-hardening characteristics of ECC. The tensile strength of ECC is greatly improved by the addition of PVA fibers. ECC exhibits exceptional tensile strain capacity and can undergo extensive deformation before failure, which is attributed to the presence of PVA fibers. This property is highly beneficial in applications where tensile strength is crucial, such as earthquake-resistant structures. Post-peak ductility refers to the material's ability to deform and absorb energy after reaching its peak load. ECC's post-peak ductility is valuable in preventing brittle failures and ensuring the safety of structures, particularly in seismic events. In summary, research and industry findings support the notion that PVA/ECC can be fabricated with local ingredients in a cost-effective manner. Furthermore, PVA fibers play a crucial role in enhancing the mechanical properties of ECC, especially tensile strength and post-peak ductility, making it a valuable material for a wide range of construction projects.

Polyethylene can be effectively added to ECC for superior mechanical behaviours. It will increase the tensile property and reduces the brittleness. The shear strength, flexural strength, deflection strength, ductility, toughness, tensile strength and energy dissipation factor also increases enormously. Another research study mentioned that addition of fly ash to ECC reduced the chemical interactive force between fibre and matrix and finally enhances the deformation ability of ECC. Reports based on effective use of ECC mentioned that ultrahigh molecular weight polyethylene was used for the production of calcium sulfoaluminate/ECC material and it was successfully used as a coating material for concrete element reinforcing and repairing purpose. To improve the ductile nature of ECC, some materials should be added and crumb rubber when incorporated with ECC, micro-cracks were more easily activated and thereby facilitates the multiple cracking (Taha, 2008; Zhang, 2013; Zhang, Ma, & Qian, 2015; Zhang & Zhang, 2018; Zhang, Zhang, Qian et al, 2015).

Research reports shown that the manufacture of ECC is also with the addition of sea sand and this must save cost, time and as a result, reduces the construction period. Numerous studies are reported for the application of ECC and this material is widely targeted due to long life cycle, high strength, low-cost, long-term existence, low maintenance and eco-friendly nature. Most significant use of ECC is in highway structures and in highway bridges. Highway structures are always affected with adverse environmental loads and most probably requiring repairing and replacements due to micro-cracks, steel corrosion etc. ECC has higher ultimate strain compared to other materials and most important property over others are its modulus of rupture and fracture energy. The weather conditions and environmental effects such as hot and humid weather provide improved resistance in ECC structures. These materials when immersing in alkaline environment shows high resistance and does not have any particular expansion after 30 days dipping in alkaline solution. Research shows that the crack developed pattern also different from fiber reinforced composites. With the help of micromechanics action, ECC forms bonds and attain the enhanced ductile nature. But on analysing the production cost, it is much higher than the traditional concrete by 300%. Neglecting the initial cost, this material will always show long-term financial security. The goal of sustainability of ECC can be improved by minimizing the cement component and polyvinyl alcohol component or eliminating these materials with industrial green-byproducts (Lepech, 2008; Li, 2004).

ECC is increasingly used to retrofit existing structures in earthquake-prone regions. Its ductile nature allows ECC to absorb and distribute seismic forces, improving the structural resilience of buildings. Its thixotropic properties make it suitable for 3D printing applications. It has been employed to create complex architectural elements and structures using robotic construction techniques, enabling innovative and cost-effective designs. ECC can be used for the construction of high-performance, energy-efficient building facades. Its thin and lightweight properties make it suitable for cladding applications, where it can enhance a building's aesthetics and energy efficiency. Its properties make it a good choice for precast and modular construction, where elements can be fabricated offsite and transported for assembly. This approach improves construction speed and quality. Engineered cementitious composites can be integrated with sensors and monitoring systems to create smart infrastructure. It allows for real-time monitoring of structural health and early detection of any issues, improving maintenance and safety. This can be used in sustainable urban infrastructure projects, including permeable pavements, which allow for stormwater infiltration and reduce the urban heat island effect. The unique properties of ECC enable architectural innovations, such as the creation of thin, curved, and freeform structures that challenge traditional construction techniques.

On analysing some of the properties of ECC materials over others, exhibit better performance. The shear property of reinforced ECC material always show excellent result on comparing with reinforced concrete material which consists of ductility, cracks control and load capacity and can be applicable in shear elements (46, 47). Bridge applications are one of the promising applications of ECC materials and it is also introduced in jointless bridges by incorporating link slab instead of mechanical expandable joint. The fracture energy of ECC is greater than fibre reinforced composites and formation of cracks are also different. The ultimate strength and flexure strength are remarkably high. The numerous applications of ECC in highway bridges and other construction sectors are important. The use of these materials in jointless bridges by adding as link slab gained much attention. The link slab made of ECC provide the bridge as higher life cycle guarantee and minimized maintenance work in future. This will solve most of the issues related with bridge construction. Less energy consumption, less production of solid waste and less use of raw materials make ECC as an outstanding construction material and resolve problems associated with the commercial sector. ECC materials with low fibre volume content provide smaller crack width and larger tensile strength.

Some of the case studies are illustrated here, ECC materials have been used in a variety of real-world applications to improve the performance and durability of structures. It was used to retrofit several aging concrete bridge structures in California. By applying ECC layers to the bridge columns, the ductile nature of ECC allowed the columns to absorb and distribute seismic forces, significantly improving the bridges' earthquake resistance. An ECC-based self-healing concrete was applied in a tunnel project in Norway. When cracks appeared due to stress or environmental factors, the capsules ruptured, releasing the healing agent to seal the cracks and maintain the structural integrity of the tunnel. ECC's thixotropic properties were employed in 3D printing applications for the construction of complex architectural elements in China. ECC was used for the construction of high-performance building facades in Japan. The thin, lightweight, and durable properties of ECC made it suitable for cladding applications, contributing to the aesthetics and energy efficiency of the buildings. Engineered cementitious composites was also applied to permeable pavements in urban areas of the United States. The use of ECC allowed for stormwater infiltration and helped mitigate the urban heat island effect, contributing to sustainable urban infrastructure. ECC was integrated into a green building project in Australia, where its durability and reduced maintenance needs aligned with the project's sustainability goals. It contributed to the long-term environmental sustainability of the structure.

ECC can be widely applicable to fire-resistive and blast resistive uses. The most common use is related to bridge construction, jointless bridges as link slab, highway structures etc. The existing uses of ECC materials create attention and increase its use in other sectors also. The most significant criteria for choosing ECC in different sectors are its less energy consumption, limited solid waste creation and very low amounts of raw material utilization for the entire production.

## REFERENCES

Ali, M. S., Soliman, A. M., & Nehdi, M. L. (2017). Hybrid fiber reinforced engineered cementitious composites under tensile and impact loading. *Materials & Design, 17*, 139–149. doi:10.1016/j.matdes.2016.12.047

Atahan, H. P. (2012). Behaviour of PVA fiber-reinforced cementitious composites under static and impact flexural effects. *Composite Structures, 104*, 320–330.

Bhat, V. C. (2012). Effect of elevated temperature on strain-hardening engineered cementitious composites. *Construction & Building Materials*, (69), 370–380.

Dehghani, A. F. (2015). Structural behaviour of small-scale masonry elements strengthened with engineered cementitious composites. *WCEE Conference*, 28800-28810.

Dehghani, A. F., Fischer, G., & Nateghi Alahi, F. (2015). Strengthening masonry infill panels using engineered cementitious composites. *Materials and Structures, 48*(1-2), 185–204. doi:10.1617/s11527-013-0176-4

Gupta, A. (2017). *All about fexibile concrete or bendable concrete, engineered cementitious composites.* Retrieved from https://civildigital.com/all-about-flexible-concrete-bendabl e-concreteengineered-cementitious-composite-ecc

Hertbert, V. L. (2012). Self-healing of enginereed cementitious composites in the natural environment. *High Performance Fiber Reinforced Cement Composites, 2*, 155-162.

Kai, M. X., Xiao, Y., Shuai, X. L., & Ye, G. (2017). Compressive behaviour of engineered cementitious composites under high strain-rate loading. *Journal of Materials in Civil Engineering*, *29*(4), 04016254. doi:10.1061/(ASCE)MT.1943-5533.0001781

Kanda, T. S. N. (n.d.). New high-rise R/C structure using ECC coupling beams. *Second International conference on strain-hardening cementitious composites*, 289-296.

Kanda, V. L. (2009). Engineered cementitious composites for structural applications. *Journal of Materials in Civil Engineering*, *10*(2), 66–69.

Kewalramani, Mohamed, & Syed. (2017). *Engineered cementitious composites for modern civil engineering structures in hot arid coastal climatic conditions*. Academic Press.

Khan, M. I., Fares, G., Mourad, S., & Abbass, W. (2016). Optimized fresh and hardened properties of strain hardening cementitious composites: Effect of sand size and workability. *Journal of Materials in Civil Engineering*, *28*(12), 1943–5533. doi:10.1061/(ASCE)MT.1943-5533.0001665

Lepech, V. L. (2008). Design of green engineered cementitious composites for improved sustsinability. *ACI Materials Journal*, *105*(6), 567–575.

Lepech, M. D., & Li, V. C. (2009). Application of ECC bridge deck link slabs. *Materials and Structures*, *42*(9), 1185–1195. doi:10.1617/s11527-009-9544-5

Li, M. a. (2011). Cracking and healing of engineered cementitious composites under chloride environment. *ACI Materials Journal*, *108*(3), 333–340.

Li, V. (2006). Article. *Structure Magazine*, 45-48.

Li, V. (2008). *Engineered cementitious composites (ECC)- Material, Structural and durability performance*. Concrete Construction Engineering Handbook.

Li, V. C., Li, M., & Wu, C. (2001). Tensile strain-hardening behavior of PVA-ECC. *Journal of Materials in Civil Engineering*, *13*(6), 419–428.

Li, V. C., & Maalej, M. (1996). Tensile strain-hardening behavior of PVA/ECC. In *Proceedings of the First International RILEM Symposium on Ultra High Performance Concrete and High-Performance Construction Materials* (pp. 183-192). Academic Press.

Li, V. C., Wu, H. C., & Wang, S. (2007). Tensile strain-hardening behavior of polyvinyl alcohol-engineered cementitious composites (PVA-ECC). *ACI Materials Journal*, *104*(3), 245–252.

Li, V. C., & Yang, E. H. (2005). Building blocks for self-healing structural composites. *ACI Materials Journal, 102*(6), 405–412.

Li, V. C. M. L., S. W. (2004). Development of green engineered cementitious composites for sustainable infrastructure systems. Beijing-China development of green ECC for sustainable iinfrastructure systems. *Proc. International workshop on sustainable development and concrete technology.*

Lu., X. S. (2014). Impact of high temperature on the compressive strength of ECC. *Advances in Materials Science and Engineering.*

Maalei, M. Q. (2012). Review of potential structural applications of hybrid fiber engineered cementitious composites. *Construction & Building Materials, 36*, 216–227. doi:10.1016/j.conbuildmat.2012.04.010

Manish, A., & Kewalramani, O. A. (2016). Engineered cementitious composites for modern civil engineering structures in hot arid coastal climatic conditions. *International high-performance built environment conference- A sustainable built environment conference.*

Michael, D. V. (2008). Large-scale processing of engineered cementitious composites. *ACI Materials Journals, 105.*

Minard, A. (n.d.). Bendable concrete heals itself-just ass water. *National Geographic News.*

Nehdi, M. L., & Ali, M. (2019). Experimental and numerical study of engineered cementitious composites with strain recovery under impact loading. *Applied Sciences (Basel, Switzerland), 9*(5), 994. doi:10.3390/app9050994

Ozbay, E. K. (2013). Dual effectiveness of freezing-thawing and sulfate attack on high volume slag incorporated ECC. *Composites B Engineering, 45*(1), 1384-1390.

Qian, S. Z., Zhou, J., de Rooij, M. R., Schlangen, E., Ye, G., & van Breugel, K. (2009). Self-healing behaviour of strain hardening cementitious composites incorporating local waste materials. *Cement and Concrete Composites, 31*(9), 613–621. doi:10.1016/j.cemconcomp.2009.03.003

Sahmaran, M. L. (2007). Transport properties of engineered cementitious composites under chloride exposure. *ACI Materials Journal, 104*, 604–611.

Shimizu, K. T. K. (2006). Shear behaviour of PVA-ECC beams. *International RILEM Workshop on high performance fiber reinforced cementitious composites in structural applications.*

Snoeck, D., & De Belie, N. (2012). Mechanical and self-healing properties of cementitious composites reinforced with flax and cottonised flax and compared with polyvinyl alcohol fibres. *Biosystems Engineering*, *111*(4), 325–335. doi:10.1016/j.biosystemseng.2011.12.005

Soe, K. Z., Zhang, Y. X., & Zhang, L. C. (2013a). Material properties of a new hybrid fiber-reinforced engineered cementitious composite. *Construction & Building Materials*, *43*, 399–407. doi:10.1016/j.conbuildmat.2013.02.021

Soe, K. Z., Zhang, Y. X., & Zhang, L. C. (2013b). Impact resistance of hybrid-fiber engineered cementitious composite panels. *Composite Structures*, *104*, 320–330. doi:10.1016/j.compstruct.2013.01.029

Suthiwarapirak, P. M., Matsumoto, T., & Kanda, T. (2004). Multiple cracking and fiber bridging characteristics of engineered cementitious composites under fatigue flexure. *Journal of Materials in Civil Engineering*, *16*(5), 433–443. doi:10.1061/(ASCE)0899-1561(2004)16:5(433)

Taha, M. M. (2008). Mechanical, fracture and microstructural investigations of rubber concrete. *Journal of Materials in Civil Engineering*, *20*(10), 640–649. doi:10.1061/(ASCE)0899-1561(2008)20:10(640)

Yang, E. Y. (2007). Use of high volume of fly ash to improve ECC mechanical properties and material greenness. *ACI Materials Journal*, *104*(6), 620–628.

Yu, K., Dai, J.-G., Lu, Z.-D., & Leung, C. K. Y. (2015). Mechanical properties of engineered cementitious composites subjected to elevated temperature. *Journal of Materials in Civil Engineering*, *27*(10), 04014268. doi:10.1061/(ASCE)MT.1943-5533.0001241

Zhang, J. M., Maalej, M., & Quek, S. T. (2007). Performance of hybrid-fiber ECC blast/shelter panels subjected to drop weight impact. *Journal of Materials in Civil Engineering*, *19*(10), 855–863. doi:10.1061/(ASCE)0899-1561(2007)19:10(855)

Zhang, Z., & Zhang, Q. (2018). Matrix tailoring of Engineered Cementitious Composites (ECC) with non-oil-coated, low tensile strength PVA fiber. *Construction & Building Materials*, *161*, 420–431. doi:10.1016/j.conbuildmat.2017.11.072

Zhang, Z. G. (2013). Influence of crumb rubber on the mechanical behaviour of engineered cementitious composites. *Proceedings of the VIII International Conference on fracture mechanics of concrete structures*.

Zhang, Z. G., Ma, H., & Qian, S. (2015). Investigation on properties of ECC incorporating crumb rubber of different sizes. *Journal of Advanced Concrete Technology*, *31*(5), 241–251. doi:10.3151/jact.13.241

Zhang, Z. G., & Zhang, Q. (2018). Matrix tailoring of engineered cementitious composites (ECC) with non-oil-coated, low tensile strength PVA fiber. *Construction & Building Materials*, *161*, 420–431. doi:10.1016/j.conbuildmat.2017.11.072

Zhang, Z. G., Zhang, Q., Qian, S., & Li, V. C. (2015). Low e-modulus early strength engineered cementitious composites material: Development for ultrathin white topping overlay. *Transportation Research Record: Journal of the Transportation Research Board*, *2481*(1), 41–47. doi:10.3141/2481-06

Zollo, R. (1997). Fiber-reinforced concrete: An overview after 30 years of development. *Cement and Concrete Composites*, *19*(2), 107–122. doi:10.1016/S0958-9465(96)00046-7

# Compilation of References


AASHTO. (2004). *LRFD Bridge Design Specifications* (3rd ed.). AASHTO.

Abdulhussein, Z. A., & Kopecskó, K. (2012). Effect of Supplementary Cementitious Materials on Properties of Concrete. *ACI Materials Journal*, *109*(1), 71–80.

Abdulkadir, I., Mohammed, B. S., Montasir, O. A. A., & Liew, M. S. (2022). Effects of Graphene Oxide and Crumb Rubber on the Fresh Properties of Self-Compacting Engineered Cementitious Composite Using Response Surface Methodology. *Materials (Basel)*, *15*(7), 2519. doi:10.3390/ma15072519 PMID:35407853

Abdul-Rahman, M., Al-Attar, A. A., Hamada, H. M., & Tayeh, B. (2020). Microstructure and structural analysis of polypropylene fibre reinforced reactive powder concrete beams exposed to elevated temperature. *Journal of Building Engineering*, *29*, 101167. doi:10.1016/j.jobe.2019.101167

Abid, S. R., Hilo, A. N., & Daek, Y. H. (2018). Experimental tests on the underwater abrasion of engineered cementitious composites. *Construction & Building Materials*, *171*, 779–792. doi:10.1016/j.conbuildmat.2018.03.213

ACI Committee 318. (2008). *Building Code Requirements for Structural Concrete (ACI 318 - 08) and Commentary*. American Concrete Institute.

Adesina, A. (2020). Nanomaterials in Cementitious Composites: Review of Durability Performance. *Journal of Building Pathology and Rehabilitation*, *5*(1), 21. doi:10.1007/s41024-020-00089-9

Adesina, A., & Das, S. (2020). Mechanical performance of engineered cementitious composite incorporating glass as aggregates. *Journal of Cleaner Production*, *260*, 121113. doi:10.1016/j.jclepro.2020.121113

Adesina, A., & Das, S. (2020a). Durability Evaluation of Green-Engineered Cementitious Composite Incorporating Glass as Aggregate. *Journal of Materials in Civil Engineering*, *32*(12), 1–13. doi:10.1061/(ASCE)MT.1943-5533.0003444

Adesina, A., & Das, S. (2020c). Performance of green fibre-reinforced composite made with sodium-carbonate-activated slag as a binder. *Innovative Infrastructure Solutions*, *5*(2), 1–9. doi:10.1007/s41062-020-00296-w

Adesina, A., & Das, S. (2021). Development of eco-friendly engineered cementitious composites using glass aggregates: Shrinkage properties. *Cleaner Engineering and Technology*, *5*, 100299. doi:10.1016/j.clet.2021.100299

Adesina, A., & Das, S. (2021a). Performance of engineered cementitious composites incorporating crumb rubber as aggregate. *Construction & Building Materials*, *274*, 122033. doi:10.1016/j.conbuildmat.2020.122033

Adesina, A., & Das, S. (2021b). Sustainable utilization of recycled asphalt as aggregates in engineered cementitious composites. *Construction & Building Materials*, *283*, 122727. doi:10.1016/j.conbuildmat.2021.122727

Ahmad, S. H., & Shah, S. P. (1985). Structural properties of high strength concrete and its implications for precast prestressed concrete. *PCI Journal*, *30*(6), 92–119. doi:10.15554/pcij.11011985.92.119

AIJ. (2008). Recommendations for Practice of Thermal Cracking Control of Massive Concrete in Building. *Architectural Institute of Japan, AIJ*, *2008*, 69.

Ajith, G., Shanmugasundaram, N., & Praveenkumar, S. (2021). Effect of Mineral Admixtures and Manufactured Sand on Compressive Strength of Engineered Cementitious Composite. *Journal of Building Pathology and Rehabilitation*, *6*(1), 1–9. doi:10.1007/s41024-021-00137-y

Al-Fakih, A., Al-Shugaa, M. A., Al-Osta, M. A., & Thomas, B. S. (2023). Mechanical, environmental, and economic performance of engineered cementitious composite incorporated limestone calcined clay cement: A review. *Journal of Building Engineering*, 107901. doi:10.1016/j.jobe.2023.107901

Al-Fakih, A., Mohammed, B. S., & Liew, M. (2021a). On rubberized engineered cementitious composites (R-ECC): A review of the constituent material. *Case Studies in Construction Materials*, *14*, e00536. doi:10.1016/j.cscm.2021.e00536

Ali, A. K., Ali, M. E., Younes, A. A., Farag, A., & ... (2021). Proton exchange membrane based on graphene oxide/polysulfone hybrid nano-composite for simultaneous generation of electricity and wastewater treatment. *Journal of Hazardous Materials*, *419*, 126420. doi:10.1016/j.jhazmat.2021.126420 PMID:34166952

Ali, M. A. E. M., Soliman, A. M., & Nehdi, M. L. (2017). Hybrid-fiber reinforced engineered cementitious composite under tensile and impact loading. *Materials & Design*, *117*, 139–149. doi:10.1016/j.matdes.2016.12.047

Allujami, H. M., Abdulkareem, M., Jassam, T. M., Al-Mansob, R. A., Ng, J. L., & Ibrahim, A. (2022). Nanomaterials in Recycled Aggregates Concrete Applications: Mechanical Properties and Durability. A Review. *Cogent Engineering*, *9*(1), 2122885. doi:10.1080/23311916.2022.2122885

Alnaggar, M., Abdelrahman, A. M., & Shamseldin, S. S. (2019, August). Influence of coarse aggregate type and size on the chloride penetration resistance of ultra-high performance concrete. *Construction & Building Materials*, *235*, 117466.

Alraeeini, A. S., & Nikbakht, E. (2022). Corrosion effect on the flexural behaviour of concrete filled steel tubulars with single and double skins using engineered cementitious composite. *Structures*, *44*, 1680–1694. doi:10.1016/j.istruc.2022.08.095

Altayeb, M., Wang, X., & Musa, T. H. (2021). An Ensemble Method for Predicting the Mechanical Properties of Strain Hardening Cementitious Composites. *Construction & Building Materials*, *286*, 122807. doi:10.1016/j.conbuildmat.2021.122807

Al-Thomali, S., Maslehuddin, S., & Shameem, S. (2021, March). Influence of pozzolanic materials on the properties of ultra-high performance concrete. *Construction & Building Materials*, *283*, 122730.

Altwair, N. M., & Abu Jarir, S. J. (2018).. . *Influence of Surface Roughness on Adhesion Between the Existing and New Plain Concretes.*, (September), 581–591.

Alvi, I. H., Li, Q., Hou, Y., Onyekwena, C. C., Zhang, M., & Ghaffar, A. (2023). A Critical Review of Cement Composites containing Recycled Aggregates with Graphene Oxide Nanomaterials. *Journal of Building Engineering*, *69*, 105989. doi:10.1016/j.jobe.2023.105989

Amini, K., Ghasemi, A., Amiri, S. S., Mirvalad, S., & Korayem, A. A. (2021). The Synergic Effects of Metakaolin and Polycarboxylate-Ether on Dispersion of Graphene Oxide in Cementitious Environments and Macro-Level Properties of Graphene Oxide Modified Cement Composites. *Construction & Building Materials*, *270*, 121462. doi:10.1016/j.conbuildmat.2020.121462

Amin, M., Zeyad, A. M., Tayeh, B. A., & Agwa, I. S. (2021). Engineering properties of self-cured normal and high strength concrete produced using polyethylene glycol and porous ceramic waste as coarse aggregate. *Construction & Building Materials*, *299*, 124243. doi:10.1016/j.conbuildmat.2021.124243

Ammasi, A. K., & Ragul. (2018). Strength and Durability of High Volume Fly Ash in Engineered Cementitious Composites. *Materials Today: Proceedings*, *5*(11), 24050–24058. doi:10.1016/j.matpr.2018.10.198

An, X., Che, J., Liu, H., Yang, S., & Doh, S. I. (2021). Study on freeze-thaw resistance with NaCl of desert sand engineering cement composites. *Physics and Chemistry of the Earth, 121*, 102954. doi:10.1016/j.pce.2020.102954

Anderson, M., Bogus, S. M., Arce, G., Hungria, R., & Hassan, M. (2020). Advantageous Construction Techniques for ECC Overlays. In *Tran-SET 2020* (pp. 247–257). American Society of Civil Engineers Reston.

Arce, G., Noorvand, H., Hassan, M., Rupnow, T., & Hungria, R. (2019). Cost-Effective ECC with Low Fiber Content for Pavement Application. *MATEC Web of Conferences, 271*, 07001. 10.1051/matecconf/201927107001

Arce, G. A., Noorvand, H., Hassan, M. M., Rupnow, T., & Dhakal, N. (2021). Feasibility of low fiber content PVA-ECC for jointless pavement application. *Construction & Building Materials*, *268*, 121131. doi:10.1016/j.conbuildmat.2020.121131

Arioglu. (1995). Discussion of relationship between splitting tensile strength and the compressive strength. *IMO Technical Journal, 4*, 1059-1062.

Arivusudar, N., & Babu, S. S. (2020). Mechanical properties of engineered cementitious composites developed with silica fume. *Cement, Wapno, Beton*, *2020*(4), 282–291. doi:10.32047/cwb.2020.25.4.3

AS 3600, Handbook of Reinforced Concrete Deign, Cement Concrete & Aggregates Australia (2009)

Aslani, F., & Wang, L. (2019). Fabrication and characterization of an engineered cementitious composite with enhanced fire resistance performance. *Journal of Cleaner Production*, *221*, 202–214. doi:10.1016/j.jclepro.2019.02.241

Aslani, F., Wang, L., & Zheng, M. (2019). The Effect of Carbon Nanofibers on Fresh and Mechanical Properties of Lightweight Engineered Cementitious Composite Using Hollow Glass Microspheres. *Journal of Composite Materials*, *53*(17), 2447–2464. doi:10.1177/0021998319827078

ASTM C1202. (2022). Standard Test Method for Electrical Indication of Concrete's Ability to Resist Chloride Ion Penetration. In *ASTM International*. https://doi.org/ doi:10.1520/C1202-22E01.2

ASTM C1260. (2022). Standard Test Method for Potential Alkali Reactivity of Aggregates (Mortar-Bar Method). In *ASTM International*. https://doi.org/ doi:10.1520/C1260-22.2

ASTM C1585. (2020). Standard Test Method for Measurement of Rate of Absorption of Water by Hydraulic Cement Concretes. In *ASTM International*. https://doi.org/ doi:10.1520/C1585-20.2

ASTM C1876. (2019). Standard Test Method for Bulk Electrical Conductivity of Hardened Concrete. In *ASTM International*. doi:10.1520/C1876-19

ASTM C597. (2016). Standard Test Method for Pulse Velocity Through Concrete. In *ASTM International*. https://doi.org/ doi:10.1520/C0597-16.2

ASTM C666-15. (2015). *Standard test method for resistance of concrete to rapid freezing and thawing*. American Society for Testing and Materials.

ASTM. (2021). Standard Test Method for Density, Absorption, and Voids in Hardened Concrete C642-97. In ASTM C642. https://doi.org/ doi:10.1520/C0642-21.2

Atahan, H. P. (2012). Behaviour of PVA fiber-reinforced cementitious composites under static and impact flexural effects. *Composite Structures*, *104*, 320–330.

Babu, B. S., Kamalakannan, J., Meenatchi, N., Karthik, S., & Boopathi, S. (2022). Economic impacts and reliability evaluation of battery by adopting Electric Vehicle. *IEEE Explore*, 1–6.

Bahari, A., Sadeghi-Nik, A., Shaikh, F. U. A., Sadeghi-Nik, A., Cerro-Prada, E., Mirshafiei, E., & Roodbari, M. (2022). Experimental Studies on Rheological, Mechanical, and Microstructure Properties of Self-Compacting Concrete containing Perovskite Nanomaterial. *Structural Concrete*, *23*(1), 564–578. doi:10.1002/suco.202000548

Bahraq, A. A., Maslehuddin, M., & Al-Dulaijan, S. U. (2020). Macro-and micro-properties of engineered cementitious composites (ECCs) incorporating industrial waste materials: A review. *Arabian Journal for Science and Engineering*, *45*(10), 7869–7895. doi:10.1007/s13369-020-04729-7

Bai, R., Liu, S., Yin, L., Yan, C., Zhou, J., Lu, L., & Li, L. (2023). Freeze-thaw damage assessment of engineered cementitious composites using the electrochemical impedance spectroscopy method. *Materials & Design*, *230*, 111965. doi:10.1016/j.matdes.2023.111965

Bakri, M. K. B., Rahman, M. R., Taib, S. N. L., Khui, P. L. N., Kakar, A., & Jayamani, E. (2021). Nano-reinforcement in sustainable polymer composites. In *Advances in sustainable polymer composites* (pp. 231–243). Elsevier. doi:10.1016/B978-0-12-820338-5.00010-2

Bao, Y., & Li, V. C. (2020). Feasibility study of lego-inspired construction with bendable concrete. *Automation in Construction*, *113*, 103161. doi:10.1016/j.autcon.2020.103161

Barhemat, R., Mahjoubi, S., Li, V. C., & Bao, Y. (2022). Lego-inspired reconfigurable modular blocks for automated construction of engineering structures. *Automation in Construction*, *139*, 104323. doi:10.1016/j.autcon.2022.104323

Basha, S. I., Rehman, A. U., Aziz, M. A., & Kim, J.-H. (2023). Cement Composite with Carbon-based Nanomaterials for 3D Concrete Printing Applications – A Review. *Chemical Record (New York, N.Y.)*, *23*(4), e202200293. doi:10.1002/tcr.202200293 PMID:36782098

Bawono, A. A., Lechner, B., & Yang, E. H. (2019). Skid resistance and surface water drainage performance of engineered cementitious composites for pavement applications. *Cement and Concrete Composites*, *104*(August), 103387. doi:10.1016/j.cemconcomp.2019.103387

Bhat, P. S., Chang, V., & Li, M. (2014). Effect of elevated temperature on strain-hardening engineered cementitious composites. *Construction & Building Materials*, *69*, 370–380. doi:10.1016/j.conbuildmat.2014.07.052

Bhushan, B., Khamparia, A., Sagayam, K. M., Sharma, S. K., Ahad, M. A., & Debnath, N. C. (2020). Blockchain for smart cities: A review of architectures, integration trends and future research directions. *Sustainable Cities and Society*, *61*, 102360. doi:10.1016/j.scs.2020.102360

Boopathi, S. (2013). *Experimental study and multi-objective optimization of near-dry wire-cut electrical discharge machining process* [PhD Thesis]. http://hdl.handle.net/10603/16933

Boopathi, S. (2021). Improving of Green Sand-Mould Quality using Taguchi Technique. *Journal of Engineering Research.*

Boopathi, S. (2022b). Performance Improvement of Eco-Friendly Near-Dry wire-Cut Electrical Discharge Machining Process Using Coconut Oil-Mist Dielectric Fluid. *World Scientific: Journal of Advanced Manufacturing Systems.*

Boopathi, S. (2024). Minimization of Manufacturing Industry Wastes Through the Green Lean Sigma Principle. *Sustainable Machining and Green Manufacturing*, 249–270.

Boopathi, S. (2024b). Energy Cascade Conversion System and Energy-Efficient Infrastructure. In Sustainable Development in AI, Blockchain, and E-Governance Applications (pp. 47–71). IGI Global.

Boopathi, S., & Kumar, P. (2024). Advanced bioprinting processes using additive manufacturing technologies: Revolutionizing tissue engineering. *3D Printing Technologies: Digital Manufacturing, Artificial Intelligence, Industry 4.0*, 95.

Boopathi, S., Khare, R., KG, J. C., Muni, T. V., & Khare, S. (2023). Additive Manufacturing Developments in the Medical Engineering Field. In Development, Properties, and Industrial Applications of 3D Printed Polymer Composites (pp. 86–106). IGI Global.

Boopathi, S., Thillaivanan, A., Mohammed, A. A., Shanmugam, P., & VR, P. (2022). Experimental investigation on Abrasive Water Jet Machining of Neem Wood Plastic Composite. *IOP: Functional Composites and Structures, 4*, 025001.

Boopathi, S. (2019). Experimental investigation and parameter analysis of LPG refrigeration system using Taguchi method. *SN Applied Sciences*, *1*(8), 892. doi:10.1007/s42452-019-0925-2

Boopathi, S. (2022). Experimental investigation and multi-objective optimization of cryogenic Friction-stir-welding of AA2014 and AZ31B alloys using MOORA technique. *Materials Today. Communications*, *33*, 104937. doi:10.1016/j.mtcomm.2022.104937

Boopathi, S. (2022a). An investigation on gas emission concentration and relative emission rate of the near-dry wire-cut electrical discharge machining process. *Environmental Science and Pollution Research International*, *29*(57), 86237–86246. doi:10.1007/s11356-021-17658-1 PMID:34837614

Boopathi, S. (2023). An Investigation on Friction Stir Processing of Aluminum Alloy-Boron Carbide Surface Composite. In *Springer:Advances in Processing of Lightweight Metal Alloys and Composites* (pp. 249–257). Springer. doi:10.1007/978-981-19-7146-4_14

Boopathi, S. (2023). Impact analysis of ceramic tile powder aggregates on self-compacting concrete. *Engineering Research Express*.

Boopathi, S. (2024a). Advancements in Machine Learning and AI for Intelligent Systems in Drone Applications for Smart City Developments. In *Futuristic e-Governance Security With Deep Learning Applications* (pp. 15–45). IGI Global. doi:10.4018/978-1-6684-9596-4.ch002

Boopathi, S., Balasubramani, V., & Kumar, R. S. (2023). Influences of various natural fibers on the mechanical and drilling characteristics of coir-fiber-based hybrid epoxy composites. *Engineering Research Express*, *5*(1), 015002. doi:10.1088/2631-8695/acb132

Boopathi, S., Balasubramani, V., Kumar, R. S., & Singh, G. R. (2021). The influence of human hair on kenaf and Grewia fiber-based hybrid natural composite material: An experimental study. *Functional Composites and Structures*, *3*(4), 045011. doi:10.1088/2631-6331/ac3afc

Boopathi, S., & Davim, J. P. (2023a). Applications of Nanoparticles in Various Manufacturing Processes. In *Sustainable Utilization of Nanoparticles and Nanofluids in Engineering Applications* (pp. 1–31). IGI Global. doi:10.4018/978-1-6684-9135-5.ch001

Boopathi, S., & Davim, J. P. (2023b). *Sustainable Utilization of Nanoparticles and Nanofluids in Engineering Applications*. IGI Global. doi:10.4018/978-1-6684-9135-5

Boopathi, S., Jeyakumar, M., Singh, G. R., King, F. L., Pandian, M., Subbiah, R., & Haribalaji, V. (2022). An experimental study on friction stir processing of aluminium alloy (AA-2024) and boron nitride (BNp) surface composite. *Materials Today: Proceedings*, *59*(1), 1094–1099. doi:10.1016/j.matpr.2022.02.435

Boopathi, S., & Kanike, U. K. (2023). Applications of Artificial Intelligent and Machine Learning Techniques in Image Processing. In *Handbook of Research on Thrust Technologies' Effect on Image Processing* (pp. 151–173). IGI Global. doi:10.4018/978-1-6684-8618-4.ch010

Boopathi, S., Myilsamy, S., & Sukkasamy, S. (2021). *Experimental Investigation and Multi-Objective Optimization of Cryogenically Cooled Near-Dry Wire-Cut EDM Using TOPSIS Technique*. IJAMT.

Boopathi, S., & Sivakumar, K. (2013). Experimental investigation and parameter optimization of near-dry wire-cut electrical discharge machining using multi-objective evolutionary algorithm. *International Journal of Advanced Manufacturing Technology*, *67*(9–12), 2639–2655. doi:10.1007/s00170-012-4680-4

Boopathi, S., & Sivakumar, K. (2016). Optimal parameter prediction of oxygen-mist near-dry wire-cut EDM. *Inderscience: International Journal of Manufacturing Technology and Management*, *30*(3–4), 164–178. doi:10.1504/IJMTM.2016.077812

Boopathi, S., Sureshkumar, M., Jeyakumar, M., Kumar, R. S., & Subbiah, R. (2022). Influences of Fabrication Parameters on Natural Fiber Reinforced Polymer Composite (NFRPC) Material: A Review. *Materials Science Forum*, *1075*, 115–124. doi:10.4028/p-095f0t

Boopathi, S., Sureshkumar, M., & Sathiskumar, S. (2022). Parametric Optimization of LPG Refrigeration System Using Artificial Bee Colony Algorithm. *International Conference on Recent Advances in Mechanical Engineering Research and Development*, 97–105.

Boopathi, S., Thillaivanan, A., Pandian, M., Subbiah, R., & Shanmugam, P. (2022). Friction stir processing of boron carbide reinforced aluminium surface (Al-B4C) composite: Mechanical characteristics analysis. *Materials Today: Proceedings*, *50*(5), 2430–2435. doi:10.1016/j.matpr.2021.10.261

Boopathi, S., Umareddy, M., & Elangovan, M. (2023). Applications of Nano-Cutting Fluids in Advanced Machining Processes. In *Sustainable Utilization of Nanoparticles and Nanofluids in Engineering Applications* (pp. 211–234). IGI Global. doi:10.4018/978-1-6684-9135-5.ch009

Boopathi, S., Venkatesan, G., & Anton Savio Lewise, K. (2023). Mechanical Properties Analysis of Kenaf–Grewia–Hair Fiber-Reinforced Composite. In *Springer: Recent Advances in Mechanical Engineering* (pp. 101–110). Springer. doi:10.1007/978-981-16-9057-0_11

Bo-Tao. (2021). Tensile and compressive performance of high-strength engineered cementitious composites (ECC) with seawater and sea-sand. *RILEM Bookseries, 30*, 1034–1041. . doi:10.1007/978-3-030-58482-5

Boughanem, S., Jesson, D. A., Mulheron, M. J., Smith, P. A., Eddie, C., Psomas, S., & Rimes, M. (2014). Tensile characterisation of thick sections of Engineered Cement Composite (ECC) materials. *Journal of Materials Science*, *50*(2), 882–897. doi:10.1007/s10853-014-8649-6

Broome, J. C., & Hayes, R. W. (1997). A comparison of the clarity of traditional construction contracts and of the New Engineering Contract. *International Journal of Project Management*, *15*(4), 255–261. doi:10.1016/S0263-7863(96)00078-6

BS8007:1987:British Standard Design of Concrete structures for retaining aqueous liquids", British Standards, 1987.

Carnio, N. J., & Lew, H. S. (1982). Re-Examination of the relation between splitting tensile and compressive strength of normal weight concrete. *ACI Journal Proceedings*, *79*(3), 214–219.

CEB. (1992). *Durable concrete structures. Comite Euro-International Beton*. Thomas Telford Ltd.

Chandrasekhar, C., & Ransinchung, G. D. R. N. (2023). Engineered Cementitious Composites (ECC) with Manufactured Sand (M- sand) for Pavement Applications. *Composites Communications*, *41*(March), 101657. doi:10.1016/j.coco.2023.101657

Chan, N., Rojanschi, C. Y., & Li, S. (2018). Effect of water-to-cement ratio and curing method on the strength, shrinkage and slump of the biosand filter concrete body. *Water Science and Technology*, *77*(6), 1744–1750. doi:10.2166/wst.2018.063 PMID:29595177

Che, J., Wang, D., Liu, H., & Zhang, Y. (2019). Mechanical properties of desert sand-based fiber reinforced concrete (DS-FRC). *Applied Sciences (Basel, Switzerland)*, *9*(9), 1857. Advance online publication. doi:10.3390/app9091857

Chen, J., Zhang, Y., Li, J., & Chen, D. (2017, July). Durability of ultra-high performance concrete under coupling attack of freeze-thaw and chloride. *Construction & Building Materials*, *154*, 470–477.

Chen, M., Zhong, H., Chen, L., Zhang, Y., & Zhang, M. (2021). Engineering properties and sustainability assessment of recycled fibre reinforced rubberised cementitious composite. *Journal of Cleaner Production*, *278*, 123996. doi:10.1016/j.jclepro.2020.123996

Chen, R., Jin, L., & Zhang, Y. (2018, November). Corrosion resistance of polyvinyl alcohol fiber reinforced ultra-high-performance concrete. *Construction & Building Materials*, *165*, 1067–1076.

Chen, Z., Liu, Y., Wu, J., Guo, X., Chen, W., & Gao, Y. (2023). Graphene Oxide-Coated Fly Ash for High Performance and Low-Carbon Cementitious Composites. *Journal of Materials Research and Technology*, *25*, 6710–6724. doi:10.1016/j.jmrt.2023.07.141

Cho, B. H., Chung, W., & Nam, B. H. (2020). Molecular dynamics simulation of calcium-silicate-hydrate for nano-engineered cement composites—A review. *Nanomaterials (Basel, Switzerland)*, *10*(11), 2158. doi:10.3390/nano10112158 PMID:33138107

Choi, Y., & Yuan, R. L. (2005). Experimental relationship between splitting tensile strength and compressive strength of GFRC and PFRC. *Cement and Concrete Research*, *35*(8), 1587–1591. doi:10.1016/j.cemconres.2004.09.010

Chuah, S., Pan, Z., Sanjayan, J. G., Wang, C. M., & Duan, W. H. (2014). Nano Reinforced Cement and Concrete Composites and New Perspective from Graphene Oxide. *Construction & Building Materials*, *73*, 113–124. doi:10.1016/j.conbuildmat.2014.09.040

Cosentino, I., Ferro, G. A., Restuccia, L., Bensaid, S., Deorsola, F., & Liendo, F. (2020). Nearly Zero $CO_2$ Cementitious Composites. *Material Design & Processing Communications*, *2*(6), e125. doi:10.1002/mdp2.125

Da Costa, F., Righi, D., Graeff, A., & da Silva Filho, L. (2019). Experimental study of some durability properties of ECC with a more environmentally sustainable rice husk ash and high tenacity polypropylene fibers. *Construction & Building Materials*, *213*, 505–513. doi:10.1016/j.conbuildmat.2019.04.092

Das, S., Lekhya, G., Shreya, K., Shekinah, K. L., Babu, K. K., & Boopathi, S. (2024). Fostering Sustainability Education Through Cross-Disciplinary Collaborations and Research Partnerships: Interdisciplinary Synergy. In Facilitating Global Collaboration and Knowledge Sharing in Higher Education With Generative AI (pp. 60–88). IGI Global.

Dehghani, A. F. (2015). Structural behaviour of small-scale masonry elements strengthened with engineered cementitious composites. *WCEE Conference*, 28800-28810.

Dehghani, A. F., Fischer, G., & Nateghi Alahi, F. (2015). Strengthening masonry infill panels using engineered cementitious composites. *Materials and Structures*, *48*(1-2), 185–204. doi:10.1617/s11527-013-0176-4

Deng, B.-Y., Li, L.-Z., Tan, D., Uddin, M. N., Cai, Z.-W., & Yu, K.-Q. (2023). Sustainable and cost-effective ultra-lightweight engineered cementitious composite: Design and material characterization. *Cement and Concrete Composites*, *136*, 104895. doi:10.1016/j.cemconcomp.2022.104895

Deng, H., & Li, H. (2018). Assessment of self-sensing capability of Carbon Black Engineered Cementitious Composites. *Construction & Building Materials*, *173*, 1–9. doi:10.1016/j.conbuildmat.2018.04.031

Ding, Y., Yu, J., Yu, K.-Q., & Xu, S. (2018). Basic mechanical properties of ultra-high ductility cementitious composites: From 40 MPa to 120 MPa. *Composite Structures*, *185*, 634–645. doi:10.1016/j.compstruct.2017.11.034

Ding, Y., Yu, K., & Li, M. (2022). A review on high-strength engineered cementitious composites (HS-ECC): Design, mechanical property and structural application. *Structures*, *35*, 903–921. doi:10.1016/j.istruc.2021.10.036

Ding, Y., Yu, K., & Mao, W. (2020). Compressive performance of all-grade engineered cementitious composites: Experiment and theoretical model. *Construction & Building Materials*, *244*, 118357. doi:10.1016/j.conbuildmat.2020.118357

Ding, Y., Yu, K.-Q., Yu, J., & Xu, S. (2018). Structural behaviors of ultra-high performance engineered cementitious composites (UHP-ECC) beams subjected to bending-experimental study. *Construction & Building Materials*, *177*, 102–115. doi:10.1016/j.conbuildmat.2018.05.122

Divya, S. (2023). Nucleation Effect of Graphene in Mechanical and Microstructure Properties of Mortar with Copper Slag and Quarry Dust. *Fullerenes, Nanotubes, and Carbon Nanostructures*, *0*(0), 1–20. doi:10.1080/1536383X.2023.2255319

Divya, S., Praveenkumar, S., & Bassam, A. (2022). Performance of Modified Nano Carbon Blended with Supplementary Materials in Cement Composite – An Interpretive Review. *Construction & Building Materials*, *346*(June), 128452. doi:10.1016/j.conbuildmat.2022.128452

Divya, S., Saranya, R., & Praveenkumar, S. (2023, September). 2024. "Pozzolanic Attributes of Hydraulic Cement Paste Hybridized with Agricultural By-Product and Nano-Carbon.". *Construction & Building Materials*, *411*, 134119. doi:10.1016/j.conbuildmat.2023.134119

Dongre, R. S. (2020). *Reinforce Fabricated Nano-Composite Matrixes for Modernization of S & T in New Millennium.* Composite and Nanocomposite Materials-from Knowledge to Industrial Applications.

Durairaj, M., Jayakumar, S., Karpagavalli, V., Maheswari, B. U., Boopathi, S., & ... (2023). Utilization of Digital Tools in the Indian Higher Education System During Health Crises. In *Multidisciplinary Approaches to Organizational Governance During Health Crises* (pp. 1–21). IGI Global. doi:10.4018/978-1-7998-9213-7.ch001

Elsener, B. (2002). Macrocell corrosion of steel in concrete–implications for corrosion monitoring. *Cement and Concrete Composites*, *24*(1), 65–72. doi:10.1016/S0958-9465(01)00027-0

Engineered Cementitious Composites. (n.d.-a). https://en.wikipedia.org/wiki/Engineered_cementitious_composite

Engineered Cementitious Composites. (n.d.-b). https://www.engineeringcivil.com/engineered-cementitious-composite.html

Feng, Y., Kero, J., Yang, Q., Chen, Q., Engström, F., Samuelsson, C., & Qi, C. (2019). Mechanical Activation of Granulated Copper Slag and Its Influence on Hydration Heat and Compressive Strength of Blended Cement. *Materials (Basel)*, *12*(5), 772. doi:10.3390/ma12050772 PMID:30845730

Feng, Y., Zhang, B., Xie, J., Xue, Z., Huang, K., & Tan, J. (2023). Effects of Recycled Sand and Nanomaterials on Ultra-High Performance Concrete: Workability, Compressive Strength and Microstructure. *Construction & Building Materials*, *378*, 131180. doi:10.1016/j.conbuildmat.2023.131180

Fowziya, S., Sivaranjani, S., Devi, N. L., Boopathi, S., Thakur, S., & Sailaja, J. M. (2023). Influences of nano-green lubricants in the friction-stir process of TiAlN coated alloys. *Materials Today: Proceedings*. Advance online publication. doi:10.1016/j.matpr.2023.06.446

Fukuyama, H., Sato, Y., Li, V. C., Matsuzaki, Y., & Mihashi, H. (2000). Ductile Engineered Cementitious Composite Elements for Seismic Structural Application. WCEE, 12, 1–8.

Gao, S., Wang, Z., Wang, W., & Qiu, H. (2018). Effect of Shrinkage-Reducing Admixture and Expansive Agent on Mechanical Properties and Drying Shrinkage of Engineered Cementitious Composite (ECC). *Construction & Building Materials*, *179*, 172–185. doi:10.1016/j.conbuildmat.2018.05.203

Gao, S., Zhao, X., Qiao, J., Guo, Y., & Hu, G. (2019). Study on the bonding properties of Engineered Cementitious Composites (ECC) and existing concrete exposed to high temperature. *Construction & Building Materials*, *196*, 330–344. doi:10.1016/j.conbuildmat.2018.11.136

Gao, X., Zhang, Y., Li, J., & Chen, D. (2016, November). Corrosion resistance of ultra-high-performance concrete reinforced with different types of fibers. *Cement and Concrete Research*, *89*, 19–31.

Gardner, N. J. (1990). Effect of temperature on the early-age properties of type I, type III and type I/fly ash concretes. *ACI Materials Journal*, *87*(1), 68–78.

Gardner, N. J., Sau, P. L., & Cheung, M. S. (1988). Strength development and durability of concretes cast and cured at 0°C. *ACI Materials Journal*, *85*(6).

Gayathiri, K., & Praveenkumar, S. (2022). *Silicon Influence of Nano Silica on Fresh and Hardened Properties of Cement-Based Materials – A Review*. Springer Netherlands. doi:10.1007/s12633-021-01598-z

Ghose, S., Hsieh, K., Boroumand, A., Ausavarungnirun, R., & Mutlu, O. (2018). Enabling the adoption of processing-in-memory: Challenges, mechanisms, future research directions. *arXiv Preprint arXiv:1802.00320*.

Gogotsi, Y., & Huang, Q. (2021). MXenes: Two-Dimensional Building Blocks for Future Materials and Devices. *ACS Nano*, *15*(4), 5775–5780. doi:10.1021/acsnano.1c03161 PMID:33906288

Goud, E. N., Rajeswari, A., Sharma, J., & Gupta, A. (2021). An Overall Review on Types and Preparation of Nano Biocomposites. *International Journal of Research in Engineering. Science and Management*, *4*(12), 69–74.

Gowri, N. V., Dwivedi, J. N., Krishnaveni, K., Boopathi, S., Palaniappan, M., & Medikondu, N. R. (2023). Experimental investigation and multi-objective optimization of eco-friendly near-dry electrical discharge machining of shape memory alloy using Cu/SiC/Gr composite electrode. *Environmental Science and Pollution Research International*, *30*(49), 1–19. doi:10.1007/s11356-023-26983-6 PMID:37126160

Gowripalan, N. (2018). Autogenous Shrinkage of Concrete at Early Ages. *25th Australasian Conference on Mechanics of Structures and Materials, December*, 269–276. 10.1007/978-981-13-7603-0_27

Guan, X., Li, Y., Liu, T., Zhang, C., Li, H., & Ou, J. (2019). An economical ultra-high ductile engineered cementitious composite with large amount of coarse river sand. *Construction & Building Materials*, *201*, 461–472. doi:10.1016/j.conbuildmat.2018.12.207

Gu, D., Zhao, W., Xie, Y., Wang, X., Su, K., & Zolotarev, O. V. (2021). A personalized medical decision support system based on explainable machine learning algorithms and ecc features: Data from the real world. *Diagnostics (Basel)*, *11*(9), 1677. doi:10.3390/diagnostics11091677 PMID:34574018

Gunasekara, C., Setunge, S., & Law, D. W. (2017). Correlations between mechanical properties of low-calcium fly ash geopolymer concretes. *Journal of Materials in Civil Engineering*, *29*(9), 1–9. doi:10.1061/(ASCE)MT.1943-5533.0001916

Guo, M., Zhong, Q., Zhou, Y., Hu, B., Huang, Z., & Yue, Y. (2020). Influence of flexural loading and chloride exposure on the fatigue behavior of high-performance lightweight engineered cementitious composites. *Construction & Building Materials*, *249*, 118512. doi:10.1016/j.conbuildmat.2020.118512

Guo, P., Meng, W., Xu, M., Li, V. C., & Bao, Y. (2021). Predicting Mechanical Properties of High-Performance Fiber-Reinforced Cementitious Composites by Integrating Micromechanics and Machine Learning. *Materials (Basel)*, *14*(12), 3143. doi:10.3390/ma14123143 PMID:34201068

Gupta, A. (2017). *All about fexibile concrete or bendable concrete, engineered cementitious composites*. Retrieved from https://civildigital.com/all-about-flexible-concrete-bendable-concreteengineered-cementitious-composite-ecc

Gürbüz, E., & Erdem, S. (2020). Development and thermo-mechanical analysis of high-performance hybrid fibre engineered cementitious composites with microencapsulated phase change materials. *Construction & Building Materials*, *263*, 120139. doi:10.1016/j.conbuildmat.2020.120139

Hanumanthakari, S., Gift, M. M., Kanimozhi, K., Bhavani, M. D., Bamane, K. D., & Boopathi, S. (2023). Biomining Method to Extract Metal Components Using Computer-Printed Circuit Board E-Waste. In *Handbook of Research on Safe Disposal Methods of Municipal Solid Wastes for a Sustainable Environment* (pp. 123–141). IGI Global. doi:10.4018/978-1-6684-8117-2.ch010

Haribalaji, V., Boopathi, S., Asif, M. M., Jeyakumar, M., Subbiah, R., & Lewise, K. A. S. (2022). Influences of Friction stir tool parameters for joining two similar AZ61A alloy plates. *Materials Today: Proceedings*, *50*(5), 2547–2553. doi:10.1016/j.matpr.2021.12.074

Haribalaji, V., Venkatesan, G., Pandian, M., Subbiah, R., & Boopathi, S. (2022). Investigation on corrosion and tensile Characteristics: Friction stir welding of AA7075 and AA2014. *Materials Today: Proceedings*, *66*(3), 743–748. doi:10.1016/j.matpr.2022.04.037

Harikaran, M., Boopathi, S., Gokulakannan, S., & Poonguzhali, M. (2023). Study on the Source of E-Waste Management and Disposal Methods. In *Sustainable Approaches and Strategies for E-Waste Management and Utilization* (pp. 39–60). IGI Global. doi:10.4018/978-1-6684-7573-7.ch003

Hertbert, V. L. (2012). Self-healing of enginereed cementitious composites in the natural environment. *High Performance Fiber Reinforced Cement Composites, 2*, 155-162.

He, Y., Huang, W., & Zhang, Y. (2019, April). Performance of ultra-high performance concrete under the coupled effects of sulfate attack and freeze-thaw cycles. *Construction & Building Materials*, *191*, 523–531.

He, Y., Ma, L., Zhou, L., Liu, G., Jiang, Y., & Gao, J. (2020). Preparation and application of bismuth/MXene nano-composite as electrochemical sensor for heavy metal ions detection. *Nanomaterials (Basel, Switzerland)*, *10*(5), 866. doi:10.3390/nano10050866 PMID:32365912

Himasha, M. W. C., Gamage, J. C. P. H., De Silva, G. I. P., & Attanayaka, V. (2020). Modification of Engineered Cementitious Composite Mortar to use as an adhesive for CFRP / Concrete bond. *11th Int. Conf. Sustain. Built Environ, 1*.

Hime, W. G. (1993). Corrosion of Steel—Random Thoughts and Wishful Thinking. *Concrete International*, *15*(10), 54–57.

Hisseine, O. A., & Tagnit-Hamou, A. (2020). Nanocellulose for ecological nanoengineered strain-hardening cementitious composites incorporating high-volume ground-glass pozzolans. *Cement and Concrete Composites*, *112*(March), 103662. doi:10.1016/j.cemconcomp.2020.103662

Hong, D. L. H. (2020). Deformation Properties of Rubberized Ecc Incorporating Nano Graphene Using Response Surface Methodology. *Materials (Basel)*, *13*(12), 1–14. doi:10.3390/ma13122831 PMID:32545443

Hooshmand, A., Kianoush, R., Siad, H., Lachemi, M., Booya, E., & Moslemi, M. (2021). Volumetric strain behaviour and self-healing of large scale engineered cementitious composite and normal concrete panels under natural conditions. *Construction & Building Materials*, *308*, 125078. doi:10.1016/j.conbuildmat.2021.125078

Hossain, M. U., Poon, C. S., Lo, I. M. C., & Cheng, J. C. P. (2017). Comparative LCA on using waste materials in the cement industry: A Hong Kong case study. *Resources, Conservation and Recycling*, *120*, 199–208. doi:10.1016/j.resconrec.2016.12.012

Hosseini, K., Atrian, M. A., Mirvalad, S., Korayem, A. H., & Ebrahimi, M. (2023). Influence of Ground Granulated Blast Furnace Slag on Mechanical Properties and Durability of Graphene Oxide-reinforced Cementitious Mortars. *Structural Concrete*, *24*(3), 6270–6282. doi:10.1002/suco.202200888

Hou, M., Zhang, D., & Li, V. C. (2022). Material processing, microstructure, and composite properties of low carbon Engineered Cementitious Composites (ECC). *Cement and Concrete Composites*, *134*(March), 104790. doi:10.1016/j.cemconcomp.2022.104790

Ho, V. D., Ng, C.-T., Coghlan, C. J., Goodwin, A., Mc Guckin, C., & Dusan, L. T. O. (2020). Electrochemically Produced Graphene with Ultra Large Particles Enhances Mechanical Properties of Portland Cement Mortar. *Construction & Building Materials*, *234*, 117403. doi:10.1016/j.conbuildmat.2019.117403

Huang, B. T., Wu, J. Q., Yu, J., Dai, J. G., Leung, C. K. Y., & Li, V. C. (2021). Seawater sea-sand engineered/strain-hardening cementitious composites (ECC/SHCC): Assessment and modeling of crack characteristics. *Cement and Concrete Research*, *140*, 106292. Advance online publication. doi:10.1016/j.cemconres.2020.106292

Huang, B. T., Yu, J., Wu, J. Q., Dai, J. G., & Leung, C. K. (2020). Seawater sea-sand Engineered Cementitious Composites (SS-ECC) for marine and coastal applications. *Composites Communications*, *20*(February), 100353. doi:10.1016/j.coco.2020.04.019

Huang, H., Han, L., & Zhang, Y. (2019, March). Durability of ultra-high performance concrete under sulfate and freeze-thaw coupling attack. *Construction & Building Materials*, *223*, 1–9.

Huang, H., Hu, Z., Han, L., Sun, H., & Liu, X. (2016, November). Performance evaluation of ultra-high performance concrete exposed to seawater environment for 15 years. *Materials & Design*, *109*, 1–9.

Huang, R., Li, C., & Zhu, Y. (2022, February). A comprehensive study on the durability of ultra-high performance concrete under coupled carbonation and chloride attack. *Construction & Building Materials*, *311*, 125008.

Huang, W., Zhang, Y., & Li, H. (2020, October). Durability of ultra-high-performance fiber-reinforced concrete exposed to freezing-thawing and sulfate attack. *Construction & Building Materials*, *241*, 1–10.

Huang, X. Y., Ni, W., & Li, K. Q. (2015). Development of engineered cementitious composites containing iron ore tailing powders. *Gongcheng Kexue Xuebao. Chinese Journal of Engineering*, *37*(11), 1491–1497. doi:10.13374/j.issn2095-9389.2015.11.015

Huang, X., Ranade, R., & Li, V. C. (2013). Feasibility Study of Developing Green ECC Using Iron Ore Tailings Powder as Cement Replacement. *Journal of Materials in Civil Engineering*, *25*(7), 923–931. doi:10.1061/(ASCE)MT.1943-5533.0000674

Hu, D., Ren, X., Fu, H., Wang, Y., Feng, X., & Li, H. (2022). Constructing highly rough skin layer of thin film (nano) composite polyamide membranes to enhance separation performance: A review. *Journal of Applied Polymer Science*, *139*(30), e52692. doi:10.1002/app.52692

Hueste, M. B. D., Chompreda, P., Trejo, D., Cline, D. B. H., & Keating, P. B. (2004). Mechanical properties of high-strength concrete for prestressed members. *Structural Journal.*, *101*, 457–465.

Hu, H., Liu, H., & He, H. (2021, July). Investigation of the durability of ultra-high performance concrete under combined freeze-thaw and carbonation conditions. *Construction & Building Materials*, *291*, 123387.

Hungria, R., Arce, G., Hassan, M., Anderson, M., Mahdi, M., Rupnow, T., & Bogus, S. M. (2020). Evaluation of novel jointless engineered cementitious composite ultrathin whitetopping (ECC-UTW) overlay. *Construction & Building Materials*, *265*, 120659. doi:10.1016/j.conbuildmat.2020.120659

Hu, W., Zhang, D., Ftwi, E., Ellis, B. R., & Li, V. C. (2023). Development of sustainable low carbon Engineered Cementitious Composites with waste polyethylene fiber, sisal fiber and carbonation curing. *Resources, Conservation and Recycling*, *197*, 107096. doi:10.1016/j.resconrec.2023.107096

Ibrahim, A., Mahmoud, E., Khodair, Y., & Patibandla, V. C. (2014). Fresh, Mechanical, and Durability Characteristics of Self-Consolidating Concrete Incorporating Recycled Asphalt Pavements. *Journal of Materials in Civil Engineering*, *26*(4), 668–675. doi:10.1061/(ASCE)MT.1943-5533.0000832

Ingle, R. B., Swathi, S., Mahendran, G., Senthil, T., Muralidharan, N., & Boopathi, S. (2023). Sustainability and Optimization of Green and Lean Manufacturing Processes Using Machine Learning Techniques. In *Circular Economy Implementation for Sustainability in the Built Environment* (pp. 261–285). IGI Global. doi:10.4018/978-1-6684-8238-4.ch012

Iqbal Khan, M., Fares, G., & Mourad, S. (2017). Optimized Fresh and Hardened Properties of Strain Hardening Cementitious Composites: Effect of Mineral Admixtures, Cementitious Composition, Size, and Type of Aggregates. *Journal of Materials in Civil Engineering*, *29*(10), 1–16. doi:10.1061/(ASCE)MT.1943-5533.0002039

Irfan, M. S., Ali, M. A., Khan, K. A., Umer, R., Kanellopoulos, A., & Samad, Y. A. (2021). An Electronic Textile Embedded Smart Cementitious Composite. *Engineering Reports*, *4*(3), e12468. doi:10.1002/eng2.12468

IS 4031 (Part 6). (1988). Methods Of Physical Tests For Hydraulic Cement, Part 6: Determination Of Compressive Strength Of Hydraulic Cement (other than masonry cement). Bureau of Indian Standards.

Istiaque Ahmed, K., Tahir, M., Hadi Habaebi, M., Lun Lau, S., & Ahad, A. (2021). Machine learning for authentication and authorization in iot: Taxonomy, challenges and future research direction. *Sensors (Basel)*, *21*(15), 5122. doi:10.3390/s21155122 PMID:34372360

Janardhana, K., Singh, V., Singh, S. N., Babu, T. R., Bano, S., & Boopathi, S. (2023). Utilization Process for Electronic Waste in Eco-Friendly Concrete: Experimental Study. In Sustainable Approaches and Strategies for E-Waste Management and Utilization (pp. 204–223). IGI Global.

Japan Society of Civil Engineers. (2008). Recommendations for Design and Construction of High Performance Fiber Reinforced Cement Composites with Multiple Fine Cracks (HPFRCC). In Concrete Engineereing Series (Vol. 82). Author.

Jesson, D. A., & Watts, J. F. (2012). The interface and interphase in polymer matrix composites: Effect on mechanical properties and methods for identification. *Polymer Reviews (Philadelphia, Pa.)*, *52*(3–4), 321–354. doi:10.1080/15583724.2012.710288

Jialing, C., Quanwei, L., Minggin, L., & Dan, W. (2017). Experimental research on mechanical properties of desert sand steel-PVA fiber engineered cementitious composites. *Functional Materials, 24*(4), 584–592. doi:10.15407/fm24.04.584

Jiang, B., Qian, Z., Gu, D., & Pan, J. (2023). Repair concrete structures with high-early-strength engineered cementitious composites (HES-ECC): Material design and interfacial behavior. *Journal of Building Engineering, 68*, 106060. doi:10.1016/j.jobe.2023.106060

Jing, G., Ye, Z., Lu, X., & Hou, P. (2017). Effect of Graphene Nanoplatelets on Hydration Behaviour of Portland Cement by Thermal Analysis. *Advances in Cement Research, 29*(2), 63–70. doi:10.1680/jadcr.16.00087

Jin, L., Zhang, Y., & Li, J. (2019, June). Durability of ultra-high performance concrete with hybrid fibers under coupling action of freeze-thaw cycles and chloride salt. *Construction & Building Materials, 232*, 117255.

JSCE. (2002). Standard specifications for concrete structures-2002. Materials and Construction, Japan Society of Civil Engineers, JSCE.

Kai, M. X., Xiao, Y., Shuai, X. L., & Ye, G. (2017). Compressive behaviour of engineered cementitious composites under high strain-rate loading. *Journal of Materials in Civil Engineering, 29*(4), 04016254. doi:10.1061/(ASCE)MT.1943-5533.0001781

Kanda, T. S. N. (n.d.). New high-rise R/C structure using ECC coupling beams. *Second International conference on strain-hardening cementitious composites*, 289-296.

Kanda, V. L. (2009). Engineered cementitious composites for structural applications. *Journal of Materials in Civil Engineering, 10*(2), 66–69.

Karthik, S., Hemalatha, R., Aruna, R., Deivakani, M., Reddy, R. V. K., & Boopathi, S. (2023). Study on Healthcare Security System-Integrated Internet of Things (IoT). In Perspectives and Considerations on the Evolution of Smart Systems (pp. 342–362). IGI Global.

Karthik, K., Teferi, A. B., Sathish, R., Gandhi, A. M., Padhi, S., Boopathi, S., & Sasikala, G. (2023). Analysis of delamination and its effect on polymer matrix composites. *Materials Today: Proceedings*. Advance online publication. doi:10.1016/j.matpr.2023.07.199

Kavitha, C., Varalatchoumy, M., Mithuna, H., Bharathi, K., Geethalakshmi, N., & Boopathi, S. (2023). Energy Monitoring and Control in the Smart Grid: Integrated Intelligent IoT and ANFIS. In Applications of Synthetic Biology in Health, Energy, and Environment (pp. 290–316). IGI Global.

Kavitha, C., Malini, P. G., Kantumuchu, V. C., Kumar, N. M., Verma, A., & Boopathi, S. (2023). An experimental study on the hardness and wear rate of carbonitride coated stainless steel. *Materials Today: Proceedings, 74*, 595–601. doi:10.1016/j.matpr.2022.09.524

Kewalramani, Mohamed, & Syed. (2017). *Engineered cementitious composites for modern civil engineering structures in hot arid coastal climatic conditions*. Academic Press.

Kewalramani, Mohamed, & Syed. (2017). Engineered cementitious composites for modern civil engineering structures in hot arid coastal climatic conditions. *Procedia Eng., 180*, 767–774. https://doi.org/.proeng.2017.04.237. doi:10.1016/j

Kewalramani, Mohamed, & Syed. (2017). Engineered Cementitious Composites for Modern Civil Engineering Structures in Hot Arid Coastal Climatic Conditions. *Procedia Engineering, 180*, 767-774. doi:10.1016/j.proeng.2017.04.237

Khan, M. I., Fares, G., Mourad, S., & Abbass, W. (2016). Optimized fresh and hardened properties of strain hardening cementitious composites: Effect of sand size and workability. *Journal of Materials in Civil Engineering, 28*(12), 1943–5533. doi:10.1061/(ASCE)MT.1943-5533.0001665

Khayat, K. H., Khayat, H. M. A., Chia, S., & Hwang, B. (2016, September). Permeability of strain-hardening fiber-reinforced cementitious composites under different loading conditions. *Journal of Materials in Civil Engineering, 28*(9).

Khokhar, S. A., Ahmed, T., Khushnood, R. A., Basit, M. U., & Javed, S. (2024). Development of low carbon engineered cementitious composite (ECC) using nano lime calcined clay cement (nLC3) based matrix. *Case Studies in Construction Materials, 20*, e02669. doi:10.1016/j.cscm.2023.e02669

Kim, D., Suh, C., & Won, M. (2008). Development of the Thickness Design for Concrete Pavement Overlays Over Existing Asphalt Pavement Structures. *Center for Transportation Research at The University of Texas at Austin, 7*(21), 1–121. www.utexas.edu/research/ctr

Kim, D., Naaman, A. E., & El-Tawil, S. (2008). Comparative flexural behavior of four fiber reinforced cementitious composites. *Cement and Concrete Composites, 30*(10), 917–928. doi:10.1016/j.cemconcomp.2008.08.002

Kim, S. B., Yi, N. H., Kim, H. Y., Kim, J. H. J., & Song, Y. C. (2010). Material and structural performance evaluation of recycled PET fiber reinforced concrete. *Cement and Concrete Composites, 32*(3), 232–240. doi:10.1016/j.cemconcomp.2009.11.002

Kim, Y. Y., Fischer, G., Lim, Y. M., & Li, V. C. (2004). Mechanical performance of sprayed engineered cementitious composite using wet-mix shotcreting process for repair applications. *Materials Journal, 101*(1), 42–49.

Kim, Y. Y., Fischer, G., & Li, V. C. (2004). Performance of bridge deck link slabs designed with ductile engineered cementitious composite. *ACI Structural Journal, 101*(6), 792–801. doi:10.14359/13454

Kong, H. J., Bike, S. G., & Li, V. C. (2003). Development of a self-consolidating engineered cementitious composite employing electrosteric dispersion/stabilization. *Cement and Concrete Composites, 25*(3), 301–309. doi:10.1016/S0958-9465(02)00057-4

Krishnaraja, A. R., & Kandasamy, S. (2018). Flexural performance of hybrid engineered cementitious composite layered reinforced concrete beams. *Periodica Polytechnica. Civil Engineering, 62*(4), 921–929. doi:10.3311/PPci.11748

Krystek, M., Pakulski, D., Patroniak, V., Górski, M., Szojda, L., Ciesielski, A., & Samorì, P. (2019). High-Performance Graphene-based Cementitious Composites. *Advancement of Science*, *6*(9), 1801195. 1002/advs.201801195 PMID:31065510

Kumar, D., Kumar, A., Amit, K., & Banerjee, P. (2023). Role of Nano Composite in Additive Technologies: A State of Art Review. *Additive Manufacturing in Multidisciplinary Cooperation and Production*, 145–154.

Kunieda, M., & Rokugo, K. (2006). Recent progress on HPFRCC. *Journal of Advanced Concrete Technology*, *4*(1), 19–33. doi:10.3151/jact.4.19

Lavanya, G., & Jegan, J. (2015). Evaluation of relationship between split tensile strength and compressive strength for geopolymer concrete of varying grades and molarity. *International Journal of Applied Engineering Research: IJAER*, *10*(15), 35523–35527.

Lee, B. Y., Lee, Y., Kim, J. K., & Kim, Y. Y. (2010). Micromechanics-Based fiber-bridging analysis of strain-hardening cementitious composite accounting for fiber distribution. *CMES -. Computer Modeling in Engineering & Sciences*, *61*(2), 111–132.

Lei, D. Y., Guo, L. P., Chen, B., Curosu, I., & Mechtcherine, V. (2019). The connection between microscopic and macroscopic properties of ultra-high strength and ultra-high ductility cementitious composites (UHS-UHDCC). Compos. Part B Eng., 164, 144–157. doi:10.1016/j.compositesb.2018.11.062

Lepech, M. D., & Li, V. C. (2009). Application of ECC for bridge deck link slabs. *Materials and Structures/Materiaux et Constructions, 42*(9), 1185–1195. doi:10.1617/s11527-009-9544-5

Lepech, M. D., & Li, V. C. (2010). Sustainable pavement overlays using engineered cementitious composites. *International Journal of Pavement Research and Technology, 3*(5), 241-250.

Lepech, V. L. (2008). Design of green engineered cementitious composites for improved sustsinability. *ACI Materials Journal, 105*(6), 567–575.

Lepech, M. D., & Li, V. C. (2008). Large scale processing of Engineered Cementitious Composite. *ACI Materials Journal, 105*, 358–366.

Lepech, M. D., & Li, V. C. (2009). Water permeability of engineered cementitious composites. *Cement and Concrete Composites*, *31*(10), 744–753. doi:10.1016/j.cemconcomp.2009.07.002

Lepech, M. D., & Li, V. C. (2010). Sustainable pavement overlays using Engineered Cementitious Composites. *International Journal of Pavement Research and Technology*, *3*(5), 241–250.

Lepech, M. D., Li, V. C., Robertson, R. E., & Keoleian, G. A. (2008). Design of green engineered cementitious composites for improved sustainability. *ACI Materials Journal*, *105*(6), 567–575. doi:10.14359/20198

Lepech, M., & Li, V. C. (2005). Design and field demonstration of ECC link slabs for jointless bridge decks. *Proceedings of ConMat*, *5*, 22–24.

Letelier, V., Henríquez-Jara, B. I., Manosalva, M., Parodi, C., & Ortega, J. M. (2019). Use of waste glass as a replacement for raw materials in mortars with a lower environmental impact. *Energies*, *12*(10), 1974. Advance online publication. doi:10.3390/en12101974

Li, V. (2006). Article. *Structure Magazine*, 45-48.

Li, V. C. (1997). Engineered Cementitious Composites (Ecc) – Tailored Composites through Micromechanical Modeling. Canadian Society for Civil Engineering.

Li, V. C. (1998). Engineered Cementitious Composites (ECC) – Tailored Composites through Micromechanical Modeling. In Fiber Reinforced Concrete: Present and the Future (pp. 64–97). Academic Press.

Li, V. C. (2008). Engineered cementitious composite (ecc): Material, structural, and durability performance. Concrete Construction Engineering Handbook, 1001–1048.

Li, V. C. (2008). Engineered Cementitious Composites (ECC) – Material, Structural, and Durability Performance. In Concrete Construction Engineering Handbook. CRC Press.

Li, V. C. (2008). *Engineered cementitious composites (ECC) material, structural, and durability performance*. Academic Press.

Li, V. C. (2008). Engineered cementitious composites (ECC) material, structural, and durability performance. In Concrete Construction Engineering Handbook. CRC Press.

Li, V. C. (2019). Engineered cementitious composites (ECC). *Engineered Cementitious Composites (ECC): Bendable Concrete for Sustainable and Resilient Infrastructure*, 286–290.

Li, V. C., & Maalej, M. (1996). Tensile strain-hardening behavior of PVA/ECC. In *Proceedings of the First International RILEM Symposium on Ultra High Performance Concrete and High-Performance Construction Materials* (pp. 183-192). Academic Press.

Li, V. C., Fischer, G., Kim, Y. Y., Lepech, M., Qian, S., Weimann, M., & Wang, S. (2003). Field Demonstration of Durable Link Slabs for Jointless Bridge Decks Based on Strain-Hardening Cementitious Composites. Report for Michigan Department of Transportation RC-1438.

Li, V. C., Lepech, M., & Wang, S. (2004). Development of green engineered cementitious composites for sustainable infrastructure systems. *International Workshop on Sustainable Development and Concrete Technology, 1*(September), 181–191. http://core.kmi.open.ac.uk/download/pdf/11346106.pdf#page=192

Li, H., Huang, W., & Zhang, Y. (2020, February). Durability of ultra-high performance concrete with different fiber types under carbonation and chloride attack. *Cement and Concrete Composites*, *112*, 22–31.

Li, H., Leung, C. K. Y., Xu, S., & Cao, Q. (2009). Potential use of strain hardening ECC in permanent formwork with small scale flexural beams. *Journal of Wuhan University of Technology. Materials Science Edition*, *24*(3), 482–487. doi:10.1007/s11595-009-3482-5

Li, J., Li, Z., & Li, X. (2022, January). Effect of curing temperature on the chloride resistance and microstructure of ultra-high performance concrete. *Construction & Building Materials*, *303*, 124166.

Li, J., & Yang, E. H. (2017). Macroscopic and microstructural properties of engineered cementitious composites incorporating recycled concrete fines. *Cement and Concrete Composites*, *78*, 33–42. doi:10.1016/j.cemconcomp.2016.12.013

Li, L., Ma, X., & Liu, D. (2021, December). Effect of mineral admixture on the durability of ultra-high performance concrete under freeze-thaw and carbonation. *Construction & Building Materials*, *302*, 124089.

Li, M. (2014). 7 - Engineered cementitious composites for bridge decks. In Y. J. Kim (Ed.), *Advanced Composites in Bridge Construction and Repair* (pp. 177–209). Woodhead Publishing. doi:10.1533/9780857097019.2.177

Li, M., & Li, V. C. (2009). Influence of material ductility on performance of concrete repair. *ACI Materials Journal*, *106*(5), 419–428. doi:10.14359/51663142

Li, M., & Li, V. C. (2011). High-early-strength engineered cementitious composites for fast, durable concrete repair-material properties. *ACI Materials Journal*, *108*(1), 3–12. doi:10.14359/51664210

Li, M., & Li, V. C. (2011a). Cracking and healing of engineered cementitious composites under chloride environment. *ACI Materials Journal*, *108*(3), 333–340.

Li, M., & Li, V. C. (2011b). High-Early-Strength Engineered Cementitious Composites for Fast, Durable Concrete Repair-Material Properties. *ACI Materials Journal*, *108*(1), 3–12.

Li, M., & Li, V. C. (2012). Rheology, Fiber Dispersion, and Robust Properties of Engineered Cementitious Composites. *Materials and Structures*, *46*(3), 405–420. doi:10.1617/s11527-012-9909-z

Li, M., Zhang, Y., & Li, J. (2019, December). Durability of ultra-high performance concrete reinforced with alkali-resistant glass fiber under coupling attack of freeze-thaw cycles and chloride salt. *Construction & Building Materials*, *215*, 176–186.

Lima, L., Trindade, E., Alencar, L., Alencar, M., & Silva, L. (2021). Sustainability in the construction industry: A systematic review of the literature. *Journal of Cleaner Production*, *289*, 125730. doi:10.1016/j.jclepro.2020.125730

Lim, T. H., Ellis, B. R., & Skerlos, S. J. (2023). Life cycle assessment of railway ties fabricated with ductile cementitious composites and carbonation curing. *Journal of Cleaner Production*, *413*, 137276. doi:10.1016/j.jclepro.2023.137276

Lim, Y. M., & Li, V. C. (1997). Durable repair of aged infrastructures using trapping mechanism of engineered cementitious composites. *Cement and Concrete Composites*, *19*(4), 373–385. doi:10.1016/S0958-9465(97)00026-7

Lin, Y., & Du, H. (2020). Graphene Reinforced Cement Composites: A Review. *Construction & Building Materials*, *265*, 120312. doi:10.1016/j.conbuildmat.2020.120312

Li, P., Liu, J., Suh, H., Zal Nezhad, E., & Bae, S. (2022). Understanding the Role of Graphene Oxide Nanoribbons–Functionalized Carbon Nanotubes–Graphene Oxide (GNFG) Complex in Enhancing the Fire Resistance of Cementitious Composites. *Construction & Building Materials*, *348*(July), 128637. doi:10.1016/j.conbuildmat.2022.128637

Liu, C., He, X., Deng, X., Wu, Y., Zheng, Z., Liu, J., & Hui, D. (2020). Application of Nanomaterials in Ultra-High Performance Concrete: A Review. *Nanotechnology Reviews*, *9*(1), 1427–1444. doi:10.1515/ntrev-2020-0107

Liu, H., Zhang, Q., Gu, C., Su, H., & Li, V. (2017a). Influence of microcrack self-healing behavior on the permeability of Engineered Cementitious Composites. *Cement and Concrete Composites*, *82*, 14–22. doi:10.1016/j.cemconcomp.2017.04.004

Liu, H., Zhang, Q., Gu, C., Su, H., & Li, V. (2017b). Self-healing of microcracks in Engineered Cementitious Composites under sulfate and chloride environment. *Construction & Building Materials*, *153*, 948–956. doi:10.1016/j.conbuildmat.2017.07.126

Liu, H., Zhang, Q., Gu, C., Su, H., & Li, V. C. (2016). Influence of micro-cracking on the permeability of engineered cementitious composites. *Cement and Concrete Composites*, *72*, 104–113. doi:10.1016/j.cemconcomp.2016.05.016

Liu, H., Zhang, Q., Li, V., Su, H., & Gu, C. (2017). Durability study on engineered cementitious composites (ECC) under sulfate and chloride environment. *Construction & Building Materials*, *133*, 171–181. doi:10.1016/j.conbuildmat.2016.12.074

Liu, S., Lu, S., Yin, L., Yan, C., Lu, L., & Zhou, J. (2022). Mechanical strength model of engineered cementitious composites with freeze–thaw damage based on pore structure evolution. *Cement and Concrete Composites*, *134*, 104706. doi:10.1016/j.cemconcomp.2022.104706

Liu, Y., Zhou, X., Lv, C., Yang, Y., & Liu, T. (2018). Use of silica fume and GGBS to improve frost resistance of ECC with high-volume fly ash. *Advances in Civil Engineering*, *2018*, 1–11. doi:10.1155/2018/7987589

Li, V. (2008). *Engineered cementitious composites (ECC)- Material, Structural and durability performance*. Concrete Construction Engineering Handbook.

Li, V. C. (1993). From mechanics to structural engineering - The design of cementitious composites for civil engineering applications. *Structural Engineering / Earthquake Engineering*, *10*, 37s–48s.

Li, V. C. (1993). From micromechanics to structural engineering the design of cementitious composites for civil engineering applications. *Journal of Structural Engineering and Earthquake Engineering*, *471*, 1–24.

Li, V. C. (1994). From micromechanics to structural engineering -the design of cementitious composites for civil engineering applications. *Structural Engineering / Earthquake Engineering*, *10*(2), 1–34.

Li, V. C. (2003). On engineered cementitious composites (ECC) a review of the material and its applications. *Journal of Advanced Concrete Technology*, *1*(3), 215–230. doi:10.3151/jact.1.215

Li, V. C. (2012). Tailoring ECC for special attributes: A review. *International Journal of Concrete Structures and Materials*, *6*(3), 135–144. doi:10.1007/s40069-012-0018-8

Li, V. C. M. L., S. W. (2004). Development of green engineered cementitious composites for sustainable infrastructure systems. Beijing-China development of green ECC for sustainable iinfrastructure systems. *Proc. International workshop on sustainable development and concrete technology.*

Li, V. C., Bos, F. P., Yu, K., McGee, W., Ng, T. Y., Figueiredo, S. C., Nefs, K., Mechtcherine, V., Nerella, V. N., Pan, J., van Zijl, G. P. A. G., & Kruger, P. J. (2020). On the emergence of 3D printable engineered, strain hardening cementitious composites (ECC/SHCC). *Cement and Concrete Research*, *132*, 106038. doi:10.1016/j.cemconres.2020.106038

Li, V. C., Horii, H., Kabele, P., Kanda, T., & Lim, Y. M. (2000). Repair and retrofit with engineered cementitious composites. *Engineering Fracture Mechanics*, *65*(2-3), 317–334. doi:10.1016/S0013-7944(99)00117-4

Li, V. C., Li, M., & Wu, C. (2001). Tensile strain-hardening behavior of PVA-ECC. *Journal of Materials in Civil Engineering*, *13*(6), 419–428.

Li, V. C., Mishra, D. K., & Wu, H.-C. (1995). Matrix design for pseudo-strain-hardening fibre reinforced cementitious composites. *Materials and Structures*, *28*(10), 586–595. doi:10.1007/BF02473191

Li, V. C., Stang, H., & Krenchel, H. (1993). Micromechanics of crack bridging in fiber reinforced concrete. *Materials and Structures*, *26*(8), 486–494. doi:10.1007/BF02472808

Li, V. C., & Wang, S. (2002). Failure mode and structural ductility of GFRP reinforced engineered cementitious composite beams. *ACI Materials Journal*, *99*(1), 11–21.

Li, V. C., Wang, S., & Wu, C. (2001). Tensile strain-hardening behavior of polyvinyl alcohol engineered cementitious composite (PVA-ECC). *ACI Materials Journal*, *98*(6), 483–492.

Li, V. C., Wang, S., & Wu, C. (2001). Tensile strain-hardening behavior or polyvinyl alcohol engineered cementitious composite (PVA-ECC). *ACI Materials Journal*, *98*(6), 483–492. doi:10.14359/10851 PMID:34682615

Li, V. C., Wang, Y., & Backer, S. (1991). A micromechanical model of tension-softening and bridging toughening of short random fiber reinforced brittle matrix composites. *Journal of the Mechanics and Physics of Solids*, *39*(5), 607–625. doi:10.1016/0022-5096(91)90043-N

Li, V. C., Wu, C., Wang, S., Ogawa, A., & Saito, T. (2002). Interface tailoring for strain-hardening polyvinyl alcohol-engineered cementitious composite (PVA-ECC). *ACI Materials Journal*, *99*(5), 463–472. doi:10.14359/12325 PMID:28338189

Li, V. C., Wu, H. C., & Wang, S. (2007). Tensile strain-hardening behavior of polyvinyl alcohol-engineered cementitious composites (PVA-ECC). *ACI Materials Journal*, *104*(3), 245–252.

Li, V. C., & Yang, E. H. (2005). Building blocks for self-healing structural composites. *ACI Materials Journal*, *102*(6), 405–412.

Li, W., Zhang, Y., Li, J., & Chen, D. (2016, September). Effect of different curing conditions on the chloride resistance of ultra-high-performance concrete. *Cement and Concrete Composites*, *72*, 64–71.

Li, X., Zhang, Y., Li, J., & Chen, D. (2019, August). Durability of ultra-high performance concrete reinforced with recycled steel fiber under freezing-thawing and chloride attack. *Construction & Building Materials*, *235*, 117500.

Li, Y., Li, X., & Xie, Z. (2022, March). Effect of metakaolin and slag powder on the durability of ultra-high performance concrete under sulfate and freeze-thaw cycles. *Construction & Building Materials*, *312*, 125189.

Li, Y., Wu, Z., & Guo, C. (2019, June). Effect of different mineral admixtures on the pore structure and permeability of ultra-high performance concrete. *Construction & Building Materials*, *199*, 696–704.

Li, Z., Li, J., & Wang, J. (2021, June). Durability of ultra-high performance concrete with supplementary cementitious materials subjected to freezing and thawing cycles. *Construction & Building Materials*, *288*, 123065.

Li, Z., Lu, B., Feng, J., Zhao, H., & Qian, S. (2023). Development of Engineered Cementitious Composites/Strain-hardening Cementitious Composites (ECC/SHCC) with waste granite fine powders. *Construction & Building Materials*, *409*, 133883. doi:10.1016/j.conbuildmat.2023.133883

Lu., X. S. (2014). Impact of high temperature on the compressive strength of ECC. *Advances in Materials Science and Engineering*.

Lu, D., & Zhong, J. (2022). Carbon-based nanomaterials engineered cement composites: A review. *Journal of Infrastructure Preservation and Resilience*, *3*(1), 1–20. doi:10.1186/s43065-021-00045-y

Luo, Z., Zhi, T., Liu, X., Yin, K., Pan, H., Feng, H., Song, Y., & Su, Y. (2023). Effects of Different Nanomaterials on the Early Performance of Ultra-High Performance of Ultra-High Performance Concrete (UHPC): C-S-H Seeds and Nano-Silica. *Cement and Concrete Composites*, *142*, 105211. doi:10.1016/j.cemconcomp.2023.105211

Lu, Z., Xing, X., & Li, J. (2019, July). Effect of Alkali-resistant glass fiber on the chloride resistance and microstructure of ultra-high performance concrete. *Construction & Building Materials*, *233*, 1–9.

Maalej, M., Ahmed, S. F., & Paramasivam, P. (2003). Corrosion durability and structural response of functionally-graded concrete beams. *Journal of Advanced Concrete Technology*, *1*(3), 307–316. doi:10.3151/jact.1.307

Maalej, M., Chhoa, C. Y., & Quek, S. T. (2010). Effect of cracking, corrosion and repair on the frequency response of RC beams. *Construction & Building Materials*, *24*(5), 719–731. doi:10.1016/j.conbuildmat.2009.10.036

Maalej, M., & Leong, K. S. (2005). Engineered cementitious composites for effective FRP-strengthening of RC beams. *Composites Science and Technology*, *65*(7-8), 1120–1128. doi:10.1016/j.compscitech.2004.11.007

Maalej, M., & Li, V. C. (1995). Introduction of strain-hardening engineered cementitious composites in design of reinforced concrete flexural members for improved durability. *Structural Journal*, *92*(2), 167–176.

Maalej, M., Quek, S. T., Ahmed, S. F. U., Zhang, J., Lin, V. W. J., & Leong, K. S. (2012). Review of potential structural applications of hybrid fiber Engineered Cementitious Composites. *Construction & Building Materials*, *36*, 216–227. doi:10.1016/j.conbuildmat.2012.04.010

Ma, H., Qian, S., & Li, V. C. (2019). Tailoring Engineered Cementitious Composite with Emulsified Asphalt for High Damping. *Construction & Building Materials*, *201*, 631–640. doi:10.1016/j.conbuildmat.2018.12.222

Ma, H., Qian, S., & Zhang, Z. (2014). Effect of self-healing on water permeability and mechanical property of medium-early-strength engineered cementitious composites. *Construction & Building Materials*, *68*, 92–101. doi:10.1016/j.conbuildmat.2014.05.065

Ma, H., Qian, S., Zhang, Z., Lin, Z., & Li, V. C. (2015). Tailoring Engineered Cementitious Composites with local ingredients. *Construction & Building Materials*, *101*(Part 1), 584–595. doi:10.1016/j.conbuildmat.2015.10.146

Ma, H., Yi, C., & Wu, C. (2021). Review and outlook on durability of engineered cementitious composite (ECC). *Construction & Building Materials*, *287*, 122719. doi:10.1016/j.conbuildmat.2021.122719

Ma, H., & Zhang, Z. (2020). Paving an engineered cementitious composite (ECC) overlay on concrete airfield pavement for reflective cracking resistance. *Construction & Building Materials*, *252*, 119048. doi:10.1016/j.conbuildmat.2020.119048

Maheswari, B. U., Imambi, S. S., Hasan, D., Meenakshi, S., Pratheep, V., & Boopathi, S. (2023). Internet of things and machine learning-integrated smart robotics. In Global Perspectives on Robotics and Autonomous Systems: Development and Applications (pp. 240–258). IGI Global. doi:10.4018/978-1-6684-7791-5.ch010

Malathi, J., Kusha, K., Isaac, S., Ramesh, A., Rajendiran, M., & Boopathi, S. (2024). IoT-Enabled Remote Patient Monitoring for Chronic Disease Management and Cost Savings: Transforming Healthcare. In Advances in Explainable AI Applications for Smart Cities (pp. 371–388). IGI Global.

Manirathnam, A., Manikandan, M. D., Prakash, R. H., Kumar, B. K., & Amarnath, M. D. (2021). Experimental analysis on solar water heater integrated with Nano composite phase change material (SCi and CuO). *Materials Today: Proceedings*, *37*, 232–240. doi:10.1016/j.matpr.2020.05.093

Manish, A., & Kewalramani, O. A. (2016). Engineered cementitious composites for modern civil engineering structures in hot arid coastal climatic conditions. *International high-performance built environment conference- A sustainable built environment conference.*

Mansoori, A., & Behfarnia, K. (2020). Evaluation of the effects of nanomaterials on durability of engineered cementitious composites exposed to the aggressive environment. *Journal of Composite Materials, 54*(14), 1807–1817. doi:10.1177/0021998319887208

Mateos, A. J. H., Paniagua, J. C., & Paniagua, F. (2015). Development of Improved Guidelines and Designs for Thin Whitetopping : Literature Review. *University of California Pavement Research Center,* 1–31. www.its.ucdavis.edu

Ma, X., Xu, L., Cai, J., Han, J., Zhu, W., & Pan, J. (2023). Study on the performances of fire-resistive engineered cementitious composites (ECCs). *Journal of Building Engineering, 67*, 105949. doi:10.1016/j.jobe.2023.105949

Mazzoli, A., Corinaldesi, V., Donnini, J., Di Perna, C., Micheli, D., Vricella, A., Pastore, R., Bastianelli, L., Moglie, F., & Mariani Primiani, V. (2017, February). 2018. "Effect of Graphene Oxide and Metallic Fibers on the Electromagnetic Shielding Effect of Engineered Cementitious Composites.". *Journal of Building Engineering, 18*, 33–39. doi:10.1016/j.jobe.2018.02.019

Mehmet Serkan Kırgız. (2019). *Smart Nanoconcretes and Cement-Based Materials In Nano Size Particle Packing for Nanoconcretes and Cement Based Materials: Mathematical Models.* Theory, and Technology.

Meng, D., Huang, T., Zhang, Y. X., & Lee, C. K. (2017). Mechanical behaviour of a polyvinyl alcohol fibre reinforced engineered cementitious composite (PVA-ECC) using local ingredients. *Construction & Building Materials, 141*, 259–270. doi:10.1016/j.conbuildmat.2017.02.158

Merli, R., Preziosi, M., Acampora, A., Lucchetti, M. C., & Petrucci, E. (2020). Recycled fibers in reinforced concrete: A systematic literature review. *Journal of Cleaner Production, 248*, 119207. doi:10.1016/j.jclepro.2019.119207

Metaxa, Z. S., Tolkou, A. K., Efstathiou, S., Rahdar, A., Favvas, E. P., Mitropoulos, A. C., & Kyzas, G. Z. (2021). Nanomaterials in Cementitious Composites: An Update. *Molecules (Basel, Switzerland), 26*(5), 1430. doi:10.3390/molecules26051430 PMID:33800797

Michael, D. V. (2008). Large-scale processing of engineered cementitious composites. *ACI Materials Journals, 105.*

Minard, A. (n.d.). Bendable concrete heals itself-just ass water. *National Geographic News.*

Minard, A/ (2009). Bendable Concrete Heals Itself -- Just Add Water. *National Geographic News.*

Mindess, S. (2019). Sustainability of concrete. In *Developments in the Formulation and Reinforcement of Concrete* (pp. 3–17). Woodhead Publishing. doi:10.1016/B978-0-08-102616-8.00001-0

Mitra, A., Mandal, S. K., & Rai, R. N. (2023). Synthesis and characterisation of magnesium-based nano-composite–A review. *Proceedings of the Institution of Mechanical Engineers, Part E: Journal of Process Mechanical Engineering*. 10.1177/09544089231167758

Mohammed, B. S., Achara, B. E., & Liew, M. S. (2018). The influence of high temperature on microstructural damage and residual properties of nano-silica-modified (NS-modified) self-consolidating engineering cementitious composites (SC-ECC) using response surface methodology (RSM). *Construction & Building Materials*, *192*, 450–466. doi:10.1016/j.conbuildmat.2018.10.114

Mohammed, B. S., Khed, V. C., & Liew, M. S. (2018). Optimization of hybrid fibres in engineered cementitious composites. *Construction & Building Materials*, *190*, 24–37. doi:10.1016/j.conbuildmat.2018.08.188

Mohanty, A., Venkateswaran, N., Ranjit, P., Tripathi, M. A., & Boopathi, S. (2023). Innovative Strategy for Profitable Automobile Industries: Working Capital Management. In Handbook of Research on Designing Sustainable Supply Chains to Achieve a Circular Economy (pp. 412–428). IGI Global.

Mohanty, A., Jothi, B., Jeyasudha, J., Ranjit, P., Isaac, J. S., & Boopathi, S. (2023). Additive Manufacturing Using Robotic Programming. In *AI-Enabled Social Robotics in Human Care Services* (pp. 259–282). IGI Global. doi:10.4018/978-1-6684-8171-4.ch010

Mueller, S., Ranjbarian, M., & Mechtcherine, V. (2019). Fatigue behavior of strain-hardening cement-based composites—From the single fiber level to real-scale application. *Structural Concrete*, *20*(4), 1231–1242. doi:10.1002/suco.201900032

Müller, S., & Mechtcherine, V. (2018). *Use of Strain-Hardening Cement-Based Composites (SHCC) in Real Scale Applications BT*. In V. Mechtcherine, V. Slowik, & P. Kabele (Eds.), *Strain-Hardening Cement-Based Composites* (pp. 690–700). Springer Netherlands.

Murali, B., Padhi, S., Patil, C. K., Kumar, P. S., Santhanakrishnan, M., & Boopathi, S. (2023). Investigation on hardness and tensile strength of friction stir processing of Al6061/TiN surface composite. *Materials Today: Proceedings*.

Naaman, A. E., & Reinhardt, H. W. (2006). Proposed classification of HPFRC composites based on their tensile response. *Materials and Structures/Materiaux et Constructions, 39*(5), 547–555. doi:10.1617/s11527-006-9103-2

Nagaraj, S., & Shanmugam, P. (2022). Mechanical properties of engineered cementitious composites (ECC) incorporating different mineral admixtures and fiber: A review. *Journal of Building Pathology and Rehabilitation.*, *7*(1), 40. doi:10.1007/s41024-022-00182-1

Nanda, A. K., Sharma, A., Augustine, P. J., Cyril, B. R., Kiran, V., & Sampath, B. (2024). Securing Cloud Infrastructure in IaaS and PaaS Environments. In Improving Security, Privacy, and Trust in Cloud Computing (pp. 1–33). IGI Global. doi:10.4018/979-8-3693-1431-9.ch001

Naveeenkumar, N., Rallapalli, S., Sasikala, K., Priya, P. V., Husain, J., & Boopathi, S. (2024). Enhancing Consumer Behavior and Experience Through AI-Driven Insights Optimization. In *AI Impacts in Digital Consumer Behavior* (pp. 1–35). IGI Global. doi:10.4018/979-8-3693-1918-5.ch001

Nehdi, M. L., & Ali, M. (2019). Experimental and numerical study of engineered cementitious composites with strain recovery under impact loading. *Applied Sciences (Basel, Switzerland)*, *9*(5), 994. doi:10.3390/app9050994

Noorvand, H., Arce, G. A., & Hassan, M. M. (2021). Evaluation of the effects of engineered cementitious composites (ECC) plasticity on concrete pavement performance. *The International Journal of Pavement Engineering*, *0*(0), 1–13. doi:10.1080/10298436.2021.1954180

Noorvand, H., Arce, G., Hassan, M., Rupnow, T., & Mohammad, L. N. (2019). Investigation of the Mechanical Properties of Engineered Cementitious Composites with Low Fiber Content and with Crumb Rubber and High Fly Ash Content. *Transportation Research Record: Journal of the Transportation Research Board*, *2673*(5), 418–428. doi:10.1177/0361198119837510

Oluokun, F. A., Burdette, K. J., & Deatherage, J. H. (1991). Splitting tensile strength and compressive strength at early ages. *ACI Materials Journal*, *88*(2), 115–121.

Owusu-Antwi, E. B., Khazanovich, L., & Titus-Glover, L. (1998). Mechanistic-based model for predicting reflective cracking in asphalt concrete-overlaid pavements. *Transportation Research Record: Journal of the Transportation Research Board*, *1629*(1), 234–241. doi:10.3141/1629-26

Ozbay, E. K. (2013). Dual effectiveness of freezing-thawing and sulfate attack on high volume slag incorporated ECC. *Composites B Engineering*, *45*(1), 1384-1390.

Özbay, E., Sahmaran, M., Lachemi, M., & Yücel, H. E. (2013). Effect of microcracking on frost durability of high-volume-fly-ash-and slag-incorporated engineered cementitious composites. *ACI Materials Journal*, *110*(3), 259–267.

Ozturk, M., Karaaslan, M., Akgol, O., & Korkut, U. (2020). Mechanical and Electromagnetic Performance of Cement Based Composites Containing Different Replacement Levels of Ground Granulated Blast Furnace Slag, Fly Ash, Silica Fume and Rice Husk Ash. *Cement and Concrete Research*, *136*(January), 106177. doi:10.1016/j.cemconres.2020.106177

Ozturk, O., & Ozyurt, N. (2022). Sustainability and cost-effectiveness of steel and polypropylene fiber reinforced concrete pavement mixtures. *Journal of Cleaner Production*, *363*, 132582. doi:10.1016/j.jclepro.2022.132582

Öztürk, O., Yıldırım, G., Keskin, Ü. S., Siad, H., & Şahmaran, M. (2020). Nano-tailored multi-functional cementitious composites. *Composites. Part B, Engineering*, *182*, 107670. doi:10.1016/j.compositesb.2019.107670

Pachiappan, K., Anitha, K., Pitchai, R., Sangeetha, S., Satyanarayana, T., & Boopathi, S. (2024). Intelligent Machines, IoT, and AI in Revolutionizing Agriculture for Water Processing. In *Handbook of Research on AI and ML for Intelligent Machines and Systems* (pp. 374–399). IGI Global.

Pakravan, H. R., Jamshidi, M., & Latifi, M. (2018). The effect of hydrophilic (polyvinyl alcohol) fiber content on the flexural behavior of engineered cementitious composites (ECC). *Journal of the Textile Institute*, *109*(1), 79–84. doi:10.1080/00405000.2017.1329132

Palaniappan, M., Tirlangi, S., Mohamed, M. J. S., Moorthy, R. S., Valeti, S. V., & Boopathi, S. (2023). Fused Deposition Modelling of Polylactic Acid (PLA)-Based Polymer Composites: A Case Study. In Development, Properties, and Industrial Applications of 3D Printed Polymer Composites (pp. 66–85). IGI Global.

Panda, S., Sarkar, P., & Davis, R. (2020). Effect of Water-Cement Ratio on Mix Design and Mechanical Strength of Copper Slag Aggregate Concrete. *IOP Conference Series: Materials Science and Engineering,* 936. 10.1088/1757-899X/936/1/012019

Park, P., Jones, R., Castillo, L., Vallangca, M., & Cantu, F. (2020). *Engineered Cementitious Composites (ECC) for Applications in Texas.* https://library.ctr.utexas.edu/hostedpdfs/utrgv/0-7030-1.pdf

Paul, A., Thilagham, K., KG, J., Reddy, P. R., Sathyamurthy, R., & Boopathi, S. (2024). Multi-criteria Optimization on Friction Stir Welding of Aluminum Composite (AA5052-H32/B4C) using Titanium Nitride Coated Tool. Engineering Research Express.

Paul, J., Kadirgama, K., Samykano, M., Pandey, A., & Tyagi, V. (2022). A comprehensive review on thermophysical properties and solar thermal applications of organic nano composite phase change materials. *Journal of Energy Storage*, *45*, 103415. doi:10.1016/j.est.2021.103415

Pavlu, T., Kocí, V., & Hájek, P. (2019). Environmental assessment of two use cycles of recycled aggregate concrete. *Sustainability (Basel)*, *11*(21), 6185. Advance online publication. doi:10.3390/su11216185

Peng, K. D., Zeng, J.-J., Huang, B.-T., Huang, J.-Q., Zhuge, Y., & Dai, J.-G. (2022). Bond Performance of FRP Bars in Plain and Fiber-Reinforced Geopolymer under Pull-out Loading. *Journal of Building Engineering*, *57*(June), 104893. doi:10.1016/j.jobe.2022.104893

Perumal, R. (2014). Correlation of compressive strength and other engineering properties of high-performance steel fiber–reinforced concrete. *Journal of Materials in Civil Engineering*, *27*(1), 4014114. doi:10.1061/(ASCE)MT.1943-5533.0001050

Pramila, P., Amudha, S., Saravanan, T., Sankar, S. R., Poongothai, E., & Boopathi, S. (2023). Design and Development of Robots for Medical Assistance: An Architectural Approach. In Contemporary Applications of Data Fusion for Advanced Healthcare Informatics (pp. 260–282). IGI Global.

Pranav, S., Aggarwal, S., Yang, E. H., Sarkar, A. K., Singh, A. P., & Lahoti, M. (2020). Alternative materials for wearing course of concrete pavements: A critical review. *Construction & Building Materials*, *236*, 117609. doi:10.1016/j.conbuildmat.2019.117609

Pranav, S., Lahoti, M., Shan, X., Yang, E. H., & Muthukumar, G. (2022). Economic input-output LCA of precast corundum-blended ECC overlay pavement. *Resources, Conservation and Recycling*, *184*(September), 106385. doi:10.1016/j.resconrec.2022.106385

Praveenkumar, K. (2023). Retaining the Particle Packing Approach and Its Application in Developing the Cement Composites towards Sustainability. *Journal of Building Pathology and Rehabilitation*, *8*(1), 26. Advance online publication. doi:10.1007/s41024-023-00271-9

Pul, S. (2008). Experimental investigation of tensile behaviour of high strength concrete. *Indian Journal of Engineering and Materials Sciences*, *15*, 467–472.

Qaidi, S. M., Dinkha, Y. Z., Haido, J. H., Ali, M. H., & Tayeh, B. A. (2021). Engineering properties of sustainable green concrete incorporating eco-friendly aggregate of crumb rubber: A review. *Journal of Cleaner Production*, *324*, 129251. doi:10.1016/j.jclepro.2021.129251

Qasem, K. M., Khan, S., Chinnam, S., Saleh, H. A., Mantasha, I., Zeeshan, M., Manea, Y. K., & Shahid, M. (2022). Sustainable fabrication of Co-MOF@ CNT nano-composite for efficient adsorption and removal of organic dyes and selective sensing of Cr (VI) in aqueous phase. *Materials Chemistry and Physics*, *291*, 126748. doi:10.1016/j.matchemphys.2022.126748

Qian, S. Z., Zhou, J., de Rooij, M. R., Schlangen, E., Ye, G., & van Breugel, K. (2009). Self-healing behaviour of strain hardening cementitious composites incorporating local waste materials. *Cement and Concrete Composites*, *31*(9), 613–621. doi:10.1016/j.cemconcomp.2009.03.003

Qin, Zhang, Yin, Di, Xu, & Xu. (2019). Use of high strength, high ductility engineered cementitious composites. *J. Build. Eng.*, *32*.

Qin, F., Zhang, Z., Yin, Z., Di, J., Xu, L., & Xu, X. (2020). Use of high strength, high ductility engineered cementitious composites (ECC) to enhance the flexural performance of reinforced concrete beams. *Journal of Building Engineering*, *32*, 101746. doi:10.1016/j.jobe.2020.101746

Rahamathunnisa, U., Sudhakar, K., Padhi, S., Bhattacharya, S., Shashibhushan, G., & Boopathi, S. (2024). Sustainable Energy Generation From Waste Water: IoT Integrated Technologies. In Adoption and Use of Technology Tools and Services by Economically Disadvantaged Communities: Implications for Growth and Sustainability (pp. 225–256). IGI Global.

Ramadoss, P. (2014). Combined effect of silica fume and steel fiber on the splitting tensile strength of high-strength concrete. *International Journal of Civil Engineering*, *12*, 96–103.

Raphel, J. M. (1984). Tensile strength of concrete. *ACI Journal Proceedings*, *81*(2), 158–165.

Rashid, M. A., Mansur, M. A., & Paramasivam, P. (2002). Correlations between mechanical properties of high-strength concrete. *Journal of Materials in Civil Engineering*, *14*(3), 230–238. doi:10.1061/(ASCE)0899-1561(2002)14:3(230)

Ratnayake, K. A. S. D., Yu, J., Mishra, D. K., & Leung, C. K. Y. (2020). Optimization of low-cost polyvinyl alcohol (Pva) fiber reinforced engineered cementitious composites (ECC) for practical applications. *Indian Concrete Journal*, *94*(6), 39–48.

Ravisankar, A., Sampath, B., & Asif, M. M. (2023). Economic Studies on Automobile Management: Working Capital and Investment Analysis. In Multidisciplinary Approaches to Organizational Governance During Health Crises (pp. 169–198). IGI Global.

Ravisankar, A., Shanthi, A., Lavanya, S., Ramaratnam, M., Krishnamoorthy, V., & Boopathi, S. (2024). Harnessing 6G for Consumer-Centric Business Strategies Across Electronic Industries. In AI Impacts in Digital Consumer Behavior (pp. 241–270). IGI Global.

Redon, C., Li, V. C., Wu, C., Hoshiro, H., Saito, T., & Ogawa, A. (2001). Measuring and Modifying Interface Properties of PVA Fibers in ECC Matrix. *Journal of Materials in Civil Engineering*, *13*(6), 399–406. doi:10.1061/(ASCE)0899-1561(2001)13:6(399)

RighiD. P.da CostaF. B. P.GraeffÂ. G.da Silva FilhoL. C. P. (2017). Tensile behaviour and durability issues of Engineered Cementitious Composites with Rice Husk Ash. *Revista Materia*, *22*(2). https://doi.org/ doi:10.1590/s1517-707620170002.0182

Rilem, T. C. (2001). Test and design methods for steel fibre reinforced concrete: Uni-axial tension test for steel fibre reinforced concrete. In Materials and Structures/Materiaux et Constructions (Vol. 34, pp. 3–6). Academic Press.

Rokugo, K. (2018). Strain-hardening cement-based composites. In SHCC44 (Vol. 15, pp. 671–680). doi:10.1007/978-94-024-1194-2

Rokugo, K., & Kanda, T. (2013). Strain Hardening Cement Composites: Structural Design and Performance. In Rilem Tac (Vol. 6). Academic Press.

Rong, M., Yu, W., & Wu, J. (2020, July). Durability of ultra-high-performance concrete under the combined action of freeze-thaw cycles and chloride salt attack. *Construction & Building Materials*, *266*, 121010.

Sabapathy, L., Mohammed, B. S., Al-Fakih, A., Wahab, M. M. A., Liew, M. S., & Amran, Y. H. M. (2020). Acid and Sulphate Attacks on a Rubberized Engineered Cementitious Composite Containing Graphene Oxide. *Materials (Basel)*, *13*(14), 3125. doi:10.3390/ma13143125 PMID:32668788

Safuta, M., Ciesielski, A., & Samori, P. (2023). Controlling the Formation of Electroactive Graphene-based Cementitious Composites: Towards Structural Health Monitoring of Civil Structures. *Chemistry (Weinheim an der Bergstrasse, Germany)*, *29*(71), e202301816. doi:10.1002/chem.202301816 PMID:37725392

Sahmaran, M., Lachemi, M., Hossain, K. M. A., Ranade, R., & Li, V. C. (2009). Influence of aggregate type and size on ductility and mechanical properties of engineered cementitious composites. *ACI Materials Journal*, *106*(3), 308–316. doi:10.14359/56556

Sahmaran, M., Lachemi, M., Hossain, K. M., & Li, V. C. (2009a). Internal curing of engineered cementitious composites for prevention of early age autogenous shrinkage cracking. *Cement and Concrete Research*, *39*(10), 893–901. doi:10.1016/j.cemconres.2009.07.006

Sahmaran, M., Lachemi, M., & Li, V. C. (2009b). Assessing the durability of engineered cementitious composites under freezing and thawing cycles. *Journal of ASTM International*, *6*(7), 85–102. doi:10.1520/JAI102406

Sahmaran, M., Lachemi, M., & Li, V. C. (2010). Assessing Mechanical Properties and Microstructure of Fire-Damaged Engineered Cementitious Composites. *ACI Materials Journal*, *107*(3), 297–304.

Sahmaran, M., Li, M., & Li, V. C. (2007). Transport properties of engineered cementitious composites under chloride exposure. *ACI Materials Journal*, *104*(6), 604–611.

Sahmaran, M., & Li, V. C. (2007). De-icing salt scaling resistance of mechanically loaded engineered cementitious composites. *Cement and Concrete Research*, *37*(7), 1035–1046. doi:10.1016/j.cemconres.2007.04.001

Sahmaran, M., & Li, V. C. (2008). Durability of mechanically loaded engineered cementitious composites under highly alkaline environments. *Cement and Concrete Composites*, *30*(2), 72–81. doi:10.1016/j.cemconcomp.2007.09.004

Şahmaran, M., & Li, V. C. (2009). Durability properties of micro-cracked ECC containing high volumes fly ash. *Cement and Concrete Research*, *39*(11), 1033–1043. doi:10.1016/j.cemconres.2009.07.009

Sahmaran, M., Li, V. C., & Andrade, C. (2008). Corrosion resistance performance of steel-reinforced engineered cementitious composite beams. *ACI Materials Journal*, *105*(3), 243.

Sahmaran, M., Özbay, E., Yücel, H. E., Lachemi, M., & Li, V. C. (2012). Frost resistance and microstructure of Engineered Cementitious Composites: Influence of fly ash and micro poly-vinyl-alcohol fiber. *Cement and Concrete Composites*, *34*(2), 156–165. doi:10.1016/j.cemconcomp.2011.10.002

Sahmaran, M., Yücel, H. E., Yildirim, G., Al-Emam, M., & Lachemi, M. (2014). Investigation of the Bond between Concrete Substrate and ECC Overlays. *Journal of Materials in Civil Engineering*, *26*(1), 167–174. doi:10.1061/(ASCE)MT.1943-5533.0000805

Said, S. H., & Razak, H. A. (2015). The effect of synthetic polyethylene fiber on the strain hardening behavior of engineered cementitious composite (ECC). *Materials & Design*, *86*, 447–457. doi:10.1016/j.matdes.2015.07.125

Sakulich, A. R., & Li, V. C. (2011). Nanoscale Characterization of Engineered Cementitious Composites (ECC). *Cement and Concrete Research*, *41*(2), 169–175. doi:10.1016/j.cemconres.2010.11.001

Salami, B. A., Mukhtar, F., Ganiyu, S. A., Adekundle, S., & Saleh, T. A. (2023). Graphene-based Concrete: Synthesis Strategies and Reinforcement Mechanisms in Graphene-based Cementitious Composites (Part 1). *Construction & Building Materials*, *396*, 132296. doi:10.1016/j.conbuildmat.2023.132296

Samikannu, R., Koshariya, A. K., Poornima, E., Ramesh, S., Kumar, A., & Boopathi, S. (2022). Sustainable Development in Modern Aquaponics Cultivation Systems Using IoT Technologies. In *Human Agro-Energy Optimization for Business and Industry* (pp. 105–127). IGI Global.

Sampath, B., & Haribalaji, V. (2021). Influences of Welding Parameters on Friction Stir Welding of Aluminum and Magnesium: A Review. *Materials Research Proceedings*, *19*(1), 322–330.

Sampath, B., Naveenkumar, N., Sampathkumar, P., Silambarasan, P., Venkadesh, A., & Sakthivel, M. (2022). Experimental comparative study of banana fiber composite with glass fiber composite material using Taguchi method. *Materials Today: Proceedings*, *49*, 1475–1480. doi:10.1016/j.matpr.2021.07.232

Sampath, B., Pandian, M., Deepa, D., & Subbiah, R. (2022). Operating parameters prediction of liquefied petroleum gas refrigerator using simulated annealing algorithm. *AIP Conference Proceedings*, *2460*(1), 070003. doi:10.1063/5.0095601

Saraireh, D., Walls, S., Suryanto, B., Starrs, G., & McCarter, W. J. (2018). The influence of multiple micro-cracking on the electrical impedance of an engineered cementitious composite. In RILEM Bookseries (Vol. 15). doi:10.1007/978-94-024-1194-2_34

Saravanan, M., Vasanth, M., Boopathi, S., Sureshkumar, M., & Haribalaji, V. (2022). Optimization of Quench Polish Quench (QPQ) Coating Process Using Taguchi Method. *Key Engineering Materials*, *935*, 83–91. doi:10.4028/p-z569vy

Sarode, P. U., Raul, J. G., Gaikwad, A. S., & Gawande, S. (2023). Metal Matrix Nanocomposites: A Brief Overview. *Nanoscience & Nanotechnology-Asia*, *13*(2), 24–42. doi:10.2174/2210681213666230301152349

Sathish, T., Sunagar, P., Singh, V., Boopathi, S., Al-Enizi, A. M., Pandit, B., Gupta, M., & Sehgal, S. S. (2023). Characteristics estimation of natural fibre reinforced plastic composites using deep multi-layer perceptron (MLP) technique. *Chemosphere*, *337*, 139346. doi:10.1016/j.chemosphere.2023.139346 PMID:37379988

Selvakumar, S., Adithe, S., Isaac, J. S., Pradhan, R., Venkatesh, V., & Sampath, B. (2023). A Study of the Printed Circuit Board (PCB) E-Waste Recycling Process. In Sustainable Approaches and Strategies for E-Waste Management and Utilization (pp. 159–184). IGI Global.

Selvakumar, S., Shankar, R., Ranjit, P., Bhattacharya, S., Gupta, A. S. G., & Boopathi, S. (2023). E-Waste Recovery and Utilization Processes for Mobile Phone Waste. In *Handbook of Research on Safe Disposal Methods of Municipal Solid Wastes for a Sustainable Environment* (pp. 222–240). IGI Global. doi:10.4018/978-1-6684-8117-2.ch016

Sengeni, D., Padmapriya, G., Imambi, S. S., Suganthi, D., Suri, A., & Boopathi, S. (2023). Biomedical waste handling method using artificial intelligence techniques. In *Handbook of Research on Safe Disposal Methods of Municipal Solid Wastes for a Sustainable Environment* (pp. 306–323). IGI Global. doi:10.4018/978-1-6684-8117-2.ch022

Senthil, T., Puviyarasan, M., Babu, S. R., Surakasi, R., Sampath, B., & Associates. (2023). Industrial Robot-Integrated Fused Deposition Modelling for the 3D Printing Process. In Development, Properties, and Industrial Applications of 3D Printed Polymer Composites (pp. 188–210). IGI Global.

Shah, Y. I., Hu, Z., & Yao, P. (2022). Interface Bond Strength Performance of Overlap Joints within the Corrugated Pipes Used in Prefabricated Bridges. *Advances in Structural Engineering*, *25*(6), 1240–1253. doi:10.1177/13694332211072324

Shamsaei, E., & Basquiroto de Souza, F. (2018). Graphene-Based Nanosheets for Stronger and More Durable Concrete: A Review. *Construction & Building Materials*, *183*, 642–660. doi:10.1016/j.conbuildmat.2018.06.201

Shanmugasundaram & Praveenkumar. (2022). Mechanical Properties of Engineered Cementitious Composites Incorporating different mineral admixtures and fibre – A Review. *Journal of Building Pathology and Rehabilitation*, 1–15.

Shanmugasundaram, Praveenkumar, Gayathiri, & Divya. (2022). Prediction on Compressive Strength of Engineered Cementitious Composites Using Machine Learning Approach. *Construction and Building Materials, 342*(PA), 127933. . doi:10.1016/j.conbuildmat.2022.127933

Shanmugasundaram, N., & Praveenkumar, S. (2021). Influence of Supplementary Cementitious Materials, Curing Conditions and Mixing Ratios on Fresh and Mechanical Properties of Engineered Cementitious Composites – A Review. *Construction & Building Materials*, *309*(October), 125038. doi:10.1016/j.conbuildmat.2021.125038

Shanmugasundaram, N., & Praveenkumar, S. (2023). Adhesive characteristics of novel greener engineered cementitious composite with conventional concrete substrate. *Construction & Building Materials*, *407*(September), 133591. doi:10.1016/j.conbuildmat.2023.133591

Sharma, M., Sharma, M., Sharma, N., & Boopathi, S. (2024). Building Sustainable Smart Cities Through Cloud and Intelligent Parking System. In *Handbook of Research on AI and ML for Intelligent Machines and Systems* (pp. 195–222). IGI Global.

Sherir, M. (2012). *Fracture energy, fatigue & creep properties of engineered cementitious composites incorporating fly ash/slag with different aggregates* [MSc Thesis]. Civil Engineering Department, Ryerson University, Toronto, Canada.

Shi, H., Zhang, Y., & Li, J. (2018, October). Influence of nanosilica on the corrosion resistance of ultra-high performance concrete. *Construction & Building Materials*, *164*, 407–414.

Shi, H., Zhang, Y., Li, J., & Chen, D. (2020, October). Corrosion resistance of ultra-high-performance concrete with fly ash and silica fume under coupled chloride and carbonation environments. *Journal of Materials Science*, *55*(27), 12348–12366.

Shi, L., Lin, S. T. K., Lu, Y., Ye, L., & Zhang, Y. X. (2018). Artificial neural network based mechanical and electrical property prediction of engineered cementitious composites. *Construction & Building Materials*, *174*, 667–674. doi:10.1016/j.conbuildmat.2018.04.127

Shi, L., Liu, X., & Wang, W. (2019, September). Effects of silica fume and steel fiber on the compressive and tensile properties of ultra-high-performance concrete under different curing regimes. *Journal of Materials in Civil Engineering*, *31*(9).

Shimizu, K. T. K. (2006). Shear behaviour of PVA-ECC beams. *International RILEM Workshop on high performance fiber reinforced cementitious composites in structural applications.*

Shoba, K. B., & Asha, P. (2019). Relationship between compressive strength and ultrasonic pulse velocity of engineered cementitious composites inducing rice husk ash. *International Journal of Recent Technology and Engineering*, *7*(6), 682–685.

Shoji, D., He, Z., Zhang, D., & Li, V. C. (2022). The greening of engineered cementitious composites (ECC): A review. *Construction & Building Materials*, *327*, 126701. doi:10.1016/j.conbuildmat.2022.126701

Shumuye, E. D., Li, W., Fang, G., Wang, Z., Liu, J., & Zerfu, K. (2023). Review on the durability of eco-friendly engineering cementitious composite (ECC). *Case Studies in Construction Materials*, *19*, e02324. doi:10.1016/j.cscm.2023.e02324

Shumuye, E. D., Li, W., Liu, J., Wang, Z., Yu, J., & Wu, H. (2022). Self-healing recovery and micro-structural properties of slag/fly-ash based engineered cementitious composites under chloride environment and tidal exposure. *Cement and Concrete Composites*, *134*, 104789. doi:10.1016/j.cemconcomp.2022.104789

Siad, H., Lachemi, M., Sahmaran, M., & Hossain, K. M. A. (2017). Mechanical, Physical, and Self-Healing Behaviors of Engineered Cementitious Composites with Glass Powder. *Journal of Materials in Civil Engineering*, *29*(6), 1–12. doi:10.1061/(ASCE)MT.1943-5533.0001864

Siad, H., Lachemi, M., Sahmaran, M., Mesbah, H. A., & Hossain, K. A. (2018). Advanced Engineered Cementitious Composites with Combined Self-Sensing and Self-Healing Functionalities. *Construction & Building Materials*, *176*, 313–322. doi:10.1016/j.conbuildmat.2018.05.026

Siad, H., Lachemi, M., Sahmaran, M., Mesbah, H. A., & Hossain, K. M. A. (2018). Use of recycled glass powder to improve the performance properties of high volume fly ash-engineered cementitious composites. *Construction & Building Materials*, *163*, 53–62. doi:10.1016/j.conbuildmat.2017.12.067

Singh, M., Saini, B., & Chalak, H. (2019). Performance and composition analysis of engineered cementitious composite (ECC)–A review. *Journal of Building Engineering*, *26*, 100851. doi:10.1016/j.jobe.2019.100851

Singh, M., Saini, B., & Chalak, H. D. (2020). Long term evaluation of engineered cementitious composite containing stone slurry powder. *Construction & Building Materials*, *264*, 120183. doi:10.1016/j.conbuildmat.2020.120183

Singh, S., & Munjal, P. (2020). Engineered cementitious composite and its applications. *Materials Today: Proceedings*, *32*, 797–802. Advance online publication. doi:10.1016/j.matpr.2020.03.743

Smadi, M., & Migdady, E. (1991). Properties of high strength tuff light weight aggregate concrete. *Cement and Concrete Composites*, *13*(2), 129–135. doi:10.1016/0958-9465(91)90008-6

Snoeck, D., & De Belie, N. (2012). Mechanical and self-healing properties of cementitious composites reinforced with flax and cottonised flax and compared with polyvinyl alcohol fibres. *Biosystems Engineering*, *111*(4), 325–335. doi:10.1016/j.biosystemseng.2011.12.005

Soe, K. T., Zhang, Y. X., & Zhang, L. C. (2013). Material properties of a new hybrid fibre-reinforced engineered cementitious composite. *Construction & Building Materials*, *43*, 399–407. doi:10.1016/j.conbuildmat.2013.02.021

Soe, K. Z., Zhang, Y. X., & Zhang, L. C. (2013b). Impact resistance of hybrid-fiber engineered cementitious composite panels. *Composite Structures*, *104*, 320–330. doi:10.1016/j.compstruct.2013.01.029

Sridhar, R. (2022). Durability study on engineered cementitious composites with hybrid fibers under sulfate and chloride environments. *Cleaner Materials*, *5*, 100121. doi:10.1016/j.clema.2022.100121

Srikakulam, L. M., & Khed, V. C. (2020). Experimental investigation on the strength parameters of rubberized engineered cementitious composite with M sand. *Materials Today: Proceedings*, *27*, 1230–1234. doi:10.1016/j.matpr.2020.02.146

Subedi, S., Arce, G. A., Hassan, M. M., Barbato, M., & Mohammad, L. N. (2021). Effect of raw sugarcane bagasse ash as sand replacement on the fiber-bridging properties of engineered cementitious composites. *Transportation Research Record: Journal of the Transportation Research Board*, *2675*(11), 1028–1042. doi:10.1177/03611981211023762

Subha, S., Inbamalar, T., Komala, C., Suresh, L. R., Boopathi, S., & Alaskar, K. (2023). A Remote Health Care Monitoring system using internet of medical things (IoMT). *IEEE Explore*, 1–6.

Suthiwarapirak, P. M., Matsumoto, T., & Kanda, T. (2004). Multiple cracking and fiber bridging characteristics of engineered cementitious composites under fatigue flexure. *Journal of Materials in Civil Engineering*, *16*(5), 433–443. doi:10.1061/(ASCE)0899-1561(2004)16:5(433)

Syamala, M., Komala, C., Pramila, P., Dash, S., Meenakshi, S., & Boopathi, S. (2023). Machine Learning-Integrated IoT-Based Smart Home Energy Management System. In *Handbook of Research on Deep Learning Techniques for Cloud-Based Industrial IoT* (pp. 219–235). IGI Global. doi:10.4018/978-1-6684-8098-4.ch013

Taha, M. M. (2008). Mechanical, fracture and microstructural investigations of rubber concrete. *Journal of Materials in Civil Engineering*, *20*(10), 640–649. doi:10.1061/(ASCE)0899-1561(2008)20:10(640)

Tan, G., Zhu, Z., Wang, W., & He, X. (2022). A fractal-based approach for cracking characterization and whole process prediction exploration of PP fiber reinforced ECC containing sustainable ingredients. *Construction & Building Materials*, *318*, 126015. doi:10.1016/j.conbuildmat.2021.126015

Tan, G., Zhu, Z., Wang, W., Wu, C., Ou, J., Cui, G., & Zhang, D. (2021). Flexural ductility and crack-controlling capacity of polypropylene fiber reinforced ECC thin sheet with waste superfine river sand based on acoustic emission analysis. *Construction & Building Materials*, *277*, 122321. Advance online publication. doi:10.1016/j.conbuildmat.2021.122321

Tan, T. G., Szalachowski, P., & Zhou, J. (2022). Challenges of post-quantum digital signing in real-world applications: A survey. *International Journal of Information Security*, *21*(4), 937–952. doi:10.1007/s10207-022-00587-6

Thomas, J., & Ramaswamy, A. (2007). Mechanical properties of steel fiber-reinforced concrete. *Journal of Materials in Civil Engineering*, *19*(5), 385–392. doi:10.1061/(ASCE)0899-1561(2007)19:5(385)

Ting, M. Z. Y., Wong, K. S., Rahman, M. E., & Meheron, S. J. (2021). Deterioration of marine concrete exposed to wetting-drying action. *Journal of Cleaner Production*, *278*, 123383. doi:10.1016/j.jclepro.2020.123383

Turk, K., & Demirhan, S. (2013). The mechanical properties of engineered cementitious composites containing limestone powder replaced by microsilica sand. *Canadian Journal of Civil Engineering*, *40*(2), 151–157. doi:10.1139/cjce-2012-0281

Turk, K., & Nehdi, M. L. (2021). Flexural toughness of sustainable ECC with high-volume substitution of cement and silica sand. *Construction & Building Materials*, *270*, 121438. Advance online publication. doi:10.1016/j.conbuildmat.2020.121438

Ullah, S., Zheng, J., Din, N., Hussain, M. T., Ullah, F., & Yousaf, M. (2023). Elliptic Curve Cryptography; Applications, challenges, recent advances, and future trends: A comprehensive survey. *Computer Science Review*, *47*, 100530. doi:10.1016/j.cosrev.2022.100530

Veeranjaneyulu, R., Boopathi, S., Kumari, R. K., Vidyarthi, A., Isaac, J. S., & Jaiganesh, V. (2023). Air Quality Improvement and Optimisation Using Machine Learning Technique. *IEEE Explore*, 1–6.

Venkateswaran, N., Vidhya, K., Ayyannan, M., Chavan, S. M., Sekar, K., & Boopathi, S. (2023). A Study on Smart Energy Management Framework Using Cloud Computing. In 5G, Artificial Intelligence, and Next Generation Internet of Things: Digital Innovation for Green and Sustainable Economies (pp. 189–212). IGI Global. doi:10.4018/978-1-6684-8634-4.ch009

Vennila, T., Karuna, M., Srivastava, B. K., Venugopal, J., Surakasi, R., & Sampath, B. (2022). New Strategies in Treatment and Enzymatic Processes: Ethanol Production From Sugarcane Bagasse. In Human Agro-Energy Optimization for Business and Industry (pp. 219–240). IGI Global.

Verma, R., Christiana, M. B. V., Maheswari, M., Srinivasan, V., Patro, P., Dari, S. S., & Boopathi, S. (2024). Intelligent Physarum Solver for Profit Maximization in Oligopolistic Supply Chain Networks. In *AI and Machine Learning Impacts in Intelligent Supply Chain* (pp. 156–179). IGI Global. doi:10.4018/979-8-3693-1347-3.ch011

Vijayakumar, G. N. S., Domakonda, V. K., Farooq, S., Kumar, B. S., Pradeep, N., & Boopathi, S. (2024). Sustainable Developments in Nano-Fluid Synthesis for Various Industrial Applications. In Adoption and Use of Technology Tools and Services by Economically Disadvantaged Communities: Implications for Growth and Sustainability (pp. 48–81). IGI Global.

Wafa, F. F., & Ashour, S. A. (1992). Mechanical properties of high-strength fiber reinforced concrete. *Materials Journal.*, *89*, 449–455.

Wang, S., & Li, V. (2003). Light Weigth Engineered Cementitious Composites (ECC). In *Fourth int'l RILEM workshop on high performance fiber reinforced cement composites (HPFRCC 4)* (pp. 379–390).

Wang, C., Li, H., & Zhang, Y. (2019, June). Durability of ultra-high performance concrete with different fibers under combined attack of carbonation and chloride. *Construction & Building Materials*, *200*, 34–41.

Wang, G., Yang, C., Pan, Y., Zhu, F., Jin, K., Li, K., & Nanni, A. (2019). Shear behaviors of RC beams externally strengthened with engineered cementitious composite layers. *Materials (Basel)*, *12*(13), 2163. doi:10.3390/ma12132163 PMID:31284446

Wang, G., Zhang, Y., & Li, J. (2019, January). Influence of steel fiber on chloride penetration and microstructure of ultra-high performance concrete. *Construction & Building Materials*, *217*, 276–284.

Wang, H., Zhang, Y., & Li, J. (2019, June). Durability of ultra-high performance concrete under the coupling attack of freeze-thaw cycles and carbonation. *Construction & Building Materials*, *198*, 365–373.

Wang, Q., Xiong, Z., Zhang, J., Fang, Z., Lai, M., & Ho, J. (2023a). Impact of polyvinyl alcohol fiber on the full life-cycle shrinkage of cementitious composite. *Journal of Building Engineering*, *63*, 105463. doi:10.1016/j.jobe.2022.105463

Wang, Q., Zhang, J., & Ho, J. C. M. (2020). Zeolite to improve strength-shrinkage performance of high-strength engineered cementitious composite. *Construction & Building Materials*, *234*, 117335. doi:10.1016/j.conbuildmat.2019.117335

Wang, S., & Li, V. C. (2006). High-early-strength engineered cementitious composites. *ACI Materials Journal*, *103*(2), 97–105. doi:10.14359/15260

Wang, T., Zhang, D., Zhu, H., Ma, B., & Li, V. C. (2022). Durability and self-healing of engineered cementitious composites exposed to simulated sewage environments. *Cement and Concrete Composites*, *129*, 104500. doi:10.1016/j.cemconcomp.2022.104500

Wang, W., Wu, H., Ma, Z., & Wu, R. (2020). Using eco-friendly recycled powder from CDW to prepare strain hardening cementitious composites (SHCC) and properties determination. *Materials (Basel)*, *13*(5), 1143. Advance online publication. doi:10.3390/ma13051143 PMID:32143483

Wang, W., Zhang, Y., & Li, J. (2019, April). Effect of slag powder and fly ash on the chloride diffusion coefficient of ultra-high performance concrete. *Construction & Building Materials*, *226*, 789–799.

Wang, W., Zhang, Y., & Li, J. (2020, December). Effect of recycled glass powder on the permeability and durability of ultra-high performance concrete. *Construction & Building Materials*, *253*, 119067.

Wang, X., Ding, S., Qiu, L., Ashour, A., Wang, Y., Han, B., & Ou, J. (2022). Improving Bond of Fiber-Reinforced Polymer Bars with Concrete through incorporating Nanomaterials. *Composites. Part B, Engineering*, *239*, 109960. doi:10.1016/j.compositesb.2022.109960

Wang, Y., Guo, R., Fu, C., Pan, T., Lin, R., & Zhang, S. (2023b). Effect of calcined zeolite on the shrinkage and mechanical properties of lightweight engineering cementitious composites. *Journal of Building Engineering*, *77*, 107480. doi:10.1016/j.jobe.2023.107480

Wang, Y., Liu, F., Yu, J., Dong, F., & Ye, J. (2020). Effect of polyethylene fiber content on physical and mechanical properties of engineered cementitious composites. *Construction & Building Materials*, *251*, 118917. doi:10.1016/j.conbuildmat.2020.118917

Wang, Z., Sun, P., Hu, Y., & Han, S. (2023c). Crack morphology tailoring and permeability prediction of polyvinyl alcohol-steel hybrid fiber engineered cementitious composites. *Journal of Cleaner Production*, *383*, 135335. doi:10.1016/j.jclepro.2022.135335

Wei, J., Zhang, M., Wang, Y., Qiao, S., Zhang, H., & Li, X. (2021). Synergistic Optimization of Thermoelectric Performance in Cementitious Composites by Lithium Carbonate and Carbon Nanotubes. *International Journal of Energy Research*, *45*(2), 2460–2473. doi:10.1002/er.5940

Weimann, M. B., & Li, V. C. (2003). Drying Shrinkage and Crack Width of Engineered Cementitious Composites (Ecc). In *Brittle Matrix Composites 7*. Institute of Fundamental Technological Research. doi:10.1533/9780857093103.37

Wille, K., El-Tawil, S., & Naaman, A. E. (2014). Properties of strain hardening ultra high performance fiber reinforced concrete (UHP-FRC) under direct tensile loading. *Cement and Concrete Composites*, *48*, 53–66. doi:10.1016/j.cemconcomp.2013.12.015

Winkelman, T. J. (2005). Whitetopping Performance in Illinois. Illinois Department of Transportation.

Wu, C., & Li, V. C. (2017). CFRP-ECC hybrid for strengthening of the concrete structures. *Composite Structures*, *178*, 372–382. doi:10.1016/j.compstruct.2017.07.034

Wu, C., Pan, Y., & Ueda, T. (2018). Characterization of the abrasion resistance and the acoustic wave attenuation of the engineered cementitious composites for runway pavement. *Construction & Building Materials*, *174*, 537–546. doi:10.1016/j.conbuildmat.2018.04.152

Wu, H. L., Zhang, D., Du, Y. J., & Li, V. C. (2020). Durability of engineered cementitious composite exposed to acid mine drainage. *Cement and Concrete Composites*, *108*, 103550. doi:10.1016/j.cemconcomp.2020.103550

Xiao, H., Zhang, N., Li, G., Zhang, Y., Wang, Y., Wang, Y., & Zhang, Y. (2023). Graphene-Iron Ore Tailings–Based Cementitious Composites with High Early Flexural Strength. *Materials (Basel), 16*(1), 327. Advance online publication. doi:10.3390/ma16010327 PMID:36614666

Xu, Huang, & Dai. (2023). *Utilization of Artificial Geopolymer Aggregates in High-Strength Engineered Cementitious Composites (HS-ECC)*. Academic Press.

Xu, B. W., & Shi, H. S. (2009). Correlations among mechanical properties of steel fiber reinforced concrete. *Construction & Building Materials, 23*(12), 3468–3474. doi:10.1016/j.conbuildmat.2009.08.017

Xu, L. Y., Huang, B. T., Lan-Ping, Q., & Dai, J. G. (2022). Enhancing long-term tensile performance of Engineered Cementitious Composites (ECC) using sustainable artificial geopolymer aggregates. *Cement and Concrete Composites, 133*, 104676. doi:10.1016/j.cemconcomp.2022.104676

Xu, M., Song, S., Feng, L., Zhou, J., Li, H., & Li, V. C. (2021). Development of basalt fiber engineered cementitious composites and its mechanical properties. *Construction & Building Materials, 266*, 121173. doi:10.1016/j.conbuildmat.2020.121173

Xu, Z., Huang, Z., Liu, C., Deng, H., Deng, X., Hui, D., Zhang, X., & Bai, Z. (2021). Research Progress on Key Problems of Nanomaterials-modified Geopolymer Concrete. *Nanotechnology Reviews, 10*(1), 779–792. doi:10.1515/ntrev-2021-0056

Yang, Y., Lepech, M. D., & Li, V. C. (2005). Self-Healing of Engineered Cementitious Composites Under Cyclic Wetting and Drying. In *International Workshop on Durability of Reinforced Concrete Under Combined Mechanical and Climatic Loads* (*Vol. 1*, Issue 1, pp. 231–242). Academic Press.

Yang, E. H., & Li, V. C. (2010). Strain-hardening fiber cement optimization and component tailoring by means of a micromechanical model. *Construction & Building Materials, 24*(2), 130–139. doi:10.1016/j.conbuildmat.2007.05.014

Yang, E. H., Yang, Y., & Li, V. C. (2007). Use of high volumes of fly ash to improve ECC mechanical properties and material greenness. *ACI Materials Journal, 104*(6), 620–628.

Yang, E. Y. (2007). Use of high volume of fly ash to improve ECC mechanical properties and material greenness. *ACI Materials Journal, 104*(6), 620–628.

Yang, G., Yu, J., & Luo, Y. (2019). Development and Mechanical Performance of Fire-Resistive Engineered Cementitious Composites. *Journal of Materials in Civil Engineering, 31*(5), 1–10. doi:10.1061/(ASCE)MT.1943-5533.0002666

Yang, S., Zhang, Y., & Li, J. (2019, December). Influence of metakaolin on the durability of ultra-high performance concrete under coupling attack of freeze-thaw and chloride. *Construction & Building Materials, 210*, 132–140.

Yang, S., Zhang, Y., Li, J., & Chen, D. (2019, March). Durability of ultra-high-performance concrete with different sizes and contents of limestone powder. *Construction & Building Materials, 197*, 247–256.

Yan, Y., Lu, Y., Zhao, Q., & Li, S. (2023). Flexural behavior of pre-damaged and repaired reinforced concrete beams with carbon fiber reinforced polymer grid and engineered cementitious composite. *Engineering Structures*, *277*, 115390. doi:10.1016/j.engstruct.2022.115390

Yao, Li, & Lu. (2021). *Development of Engineered Cementitious Composites Using Sea Sand and Metakaolin.* Academic Press.

Yao, Q., Li, Z., Lu, C., Peng, L., Luo, Y., & Teng, X. (2021). Development of Engineered Cementitious Composites Using Sea Sand and Metakaolin. *Frontiers in Materials*, *8*(August), 1–13. doi:10.3389/fmats.2021.711872

Yao, Q., Teng, X., Lu, C., Sun, H., Mo, J., & Chen, Z. (2023). Influence of accelerated chloride corrosion on mechanical properties of sea sand ECC and damage evaluation method based on nondestructive testing. *Journal of Building Engineering*, *63*, 105520. doi:10.1016/j.jobe.2022.105520

Ye, B., Zhang, Y., Han, J., & Pan, P. (2019). Effect of water to binder ratio and sand to binder ratio on shrinkage and mechanical properties of High-strength Engineered Cementitious Composite. *Construction & Building Materials*, *226*, 899–909. doi:10.1016/j.conbuildmat.2019.07.303

Yıldırım, G. (2020). Self-Sensing Capability of Engineered Cementitious Composites: Effects of Aging and Loading Conditions. *Construction & Building Materials*, 231.

Yoo, D.-Y., Taekgeun, O., & Banthia, N. (2022). Nanomaterials in Ultra-High-Performance Concrete (UPHC) – A Review. *Cement and Concrete Composites*, *134*, 104730. doi:10.1016/j.cemconcomp.2022.104730

Yucel, H. E., Jashami, H., Sahmaran, M., Guler, M., & Yaman, I. O. (2013). Thin ECC overlay systems for rehabilitation of rigid concrete pavements. *Magazine of Concrete Research*, *65*(2), 108–120. doi:10.1680/macr.12.00022

Yu, J., Chen, Y., & Leung, C. K. Y. (2019). Mechanical performance of Strain-Hardening Cementitious Composites (SHCC) with hybrid polyvinyl alcohol and steel fibers. *Composite Structures*, *226*(July), 111198. Advance online publication. doi:10.1016/j.compstruct.2019.111198

Yu, J., & Leung, C. K. Y. (2017). Strength Improvement of Strain-Hardening Cementitious Composites with Ultrahigh-Volume Fly Ash. *Journal of Materials in Civil Engineering*, *29*(9), 05017003. Advance online publication. doi:10.1061/(ASCE)MT.1943-5533.0001987

Yu, J., Lin, J., Zhang, Z., & Li, V. C. (2015). Mechanical performance of ECC with high-volume fly ash after sub-elevated temperatures. *Construction & Building Materials*, *99*, 82–89. doi:10.1016/j.conbuildmat.2015.09.002

Yu, J., Lu, C., Leung, C. K. Y., & Li, G. (2017). Mechanical properties of green structural concrete with ultrahigh-volume fly ash. *Construction & Building Materials*, *147*, 510–518. doi:10.1016/j.conbuildmat.2017.04.188

Yu, J., Wu, H.-L., & Leung, C. K. Y. (2020). Feasibility of using ultrahigh-volume limestone-calcined clay blend to develop sustainable medium-strength Engineered Cementitious Composites (ECC). *Journal of Cleaner Production*, *262*, 121343. doi:10.1016/j.jclepro.2020.121343

Yu, K., Dai, J.-G., Lu, Z.-D., & Leung, C. K. Y. (2015). Mechanical properties of engineered cementitious composites subjected to elevated temperature. *Journal of Materials in Civil Engineering*, *27*(10), 04014268. doi:10.1061/(ASCE)MT.1943-5533.0001241

Yu, K., McGee, W., Ng, T. Y., Zhu, H., & Li, V. C. (2021). 3D-printable engineered cementitious composites (3DP-ECC): Fresh and hardened properties. *Cement and Concrete Research*, *143*, 106388. doi:10.1016/j.cemconres.2021.106388

Yu, K., Wang, Y., Yu, J., & Xu, S. (2017). A strain-hardening cementitious composites with the tensile capacity up to 8%. *Construction & Building Materials*, *137*, 410–419. doi:10.1016/j.conbuildmat.2017.01.060

Zhai, S., Pang, B., Liu, G., Zhang, Y., Xu, K., She, W., & Zhang, Y. (2021). Investigation on Preparation and Multifunctionality of Reduced Graphene Oxide Cement Mortar. *Construction & Building Materials*, *275*, 122119. doi:10.1016/j.conbuildmat.2020.122119

Zhang, C., Guan, X., Li, J., Li, Y., & Lu, R. (2023a). Coupling effect of cementitious capillary crystalline waterproof material and exposure environments on self-healing properties of engineered cementitious composites (ECC). *Journal of Building Engineering*, *63*, 105471. doi:10.1016/j.jobe.2022.105471

Zhang, C., Lu, R., Li, Y., & Guan, X. (2021). Effect of crystalline admixtures on mechanical, self-healing and transport properties of engineered cementitious composite. *Cement and Concrete Composites*, *124*, 104256. doi:10.1016/j.cemconcomp.2021.104256

Zhang, C., Sun, X., & Song, Y. (2022, May). Experimental study on the durability of ultra-high performance concrete under coupling effects of chloride ion and freeze-thaw cycle. *Construction & Building Materials*, *320*, 126370.

Zhang, D., Yu, J., Wu, H., Jaworska, B., Ellis, B. R., & Li, V. C. (2020). Discontinuous micro-fibers as intrinsic reinforcement for ductile Engineered Cementitious Composites (ECC). *Composites. Part B, Engineering*, *184*, 107741. doi:10.1016/j.compositesb.2020.107741

Zhang, J. M., Maalej, M., & Quek, S. T. (2007). Performance of hybrid-fiber ECC blast/shelter panels subjected to drop weight impact. *Journal of Materials in Civil Engineering*, *19*(10), 855–863. doi:10.1061/(ASCE)0899-1561(2007)19:10(855)

Zhang, J., & Li, V. C. (2002). Monotonic and fatigue performance in bending of fiber-reinforced engineered cementitious composite in overlay system. *Cement and Concrete Research*, *32*(3), 415–423. doi:10.1016/S0008-8846(01)00695-0

Zhang, J., Wang, Z., & Ju, X. (2013). Application of ductile fiber reinforced cementitious composite in jointless concrete pavements. *Composites. Part B, Engineering*, *50*, 224–231. doi:10.1016/j.compositesb.2013.02.007

Zhang, J., Zhang, Y., & Li, J. (2019, February). Durability of ultra-high performance concrete reinforced with basalt fibers under carbonation and chloride attack. *Construction & Building Materials*, *220*, 236–246.

Zhang, L., Bian, M., Xiao, Z., Wang, X., & Han, B. (2023). A Comprehensive Review of Cementitious Composites modified with Nano Silica: Fabrication, Microstructures, Properties and Applications. *Construction & Building Materials*, *409*, 133922. doi:10.1016/j.conbuildmat.2023.133922

Zhang, L., Chen, X., & Li, Q. (2022, January). Investigation of the durability of ultra-high performance concrete with fly ash and slag powder under chloride and freeze-thaw cycles. *Construction & Building Materials*, *307*, 125212.

Zhang, L., Li, L., Ashour, A., Ding, S., & Han, B. (2024). Monitoring Damage of Concrete Beams via Self-Sensing Cement Mortar Coating with Carbon Nanotube-Nano Carbon Black Composite Fillers. *Journal of Intelligent Material Systems and Structures*, *5*(6), 633-648. doi:10.1177/1045389X23122112

Zhang, P., Wang, K., Wang, J., Guo, J., & Ling, Y. (2021). Macroscopic and Microscopic Analyses on Mechanical Performance of Metakaolin/Fly Ash Based Geopolymer Mortar. *Journal of Cleaner Production*, *294*, 126193. doi:10.1016/j.jclepro.2021.126193

Zhang, R., Matsumoto, K., Hirata, T., Ishizeki, Y., & Niwa, J. (2015). Application of PP-ECC in beam-column joint connections of rigid-framed railway bridges to reduce transverse reinforcements. *Engineering Structures*, *86*, 146–156. doi:10.1016/j.engstruct.2015.01.005

Zhang, S., Li, V. C., & Ye, G. (2020). Micromechanics-guided development of a slag/fly ash-based strain-hardening geopolymer composite. *Cement and Concrete Composites*, *109*, 103510. doi:10.1016/j.cemconcomp.2020.103510

Zhang, S., Yao, Z., Shi, J., & Ding, X. (2019, January). Influence of water-cement ratio and mineral admixtures on the permeability and microstructure of ultra-high-performance concrete. *Construction & Building Materials*, *192*, 98–107.

Zhang, W., Zheng, C., Li, Z., Jin, H., Liu, J., Zhu, J., Liu, W., & Xing, F. (2022). Investigation on mechanical properties improvement of seawater engineered cementitious composites (ECC) using FA/LC2. *Construction & Building Materials*, *345*, 128271. doi:10.1016/j.conbuildmat.2022.128271

Zhang, X., Wang, X., & Zhang, X. (2019, May). Mechanical and durability properties of ultra-high performance concrete with recycled glass powder. *Construction & Building Materials*, *197*, 534–544.

Zhang, Z. G. (2013). Influence of crumb rubber on the mechanical behaviour of engineered cementitious composites. *Proceedings of the VIII International Conference on fracture mechanics of concrete structures*.

Zhang, Z. G., Ma, H., & Qian, S. (2015). Investigation on properties of ECC incorporating crumb rubber of different sizes. *Journal of Advanced Concrete Technology*, *31*(5), 241–251. doi:10.3151/jact.13.241

Zhang, Z., Gao, Y., Qin, F., Sun, F., & Huang, Y. (2023c). Mechanical properties of sustainable high strength ECC with substitution of cement by limestone powder. *Case Studies in Construction Materials*, *19*, e02434. doi:10.1016/j.cscm.2023.e02434

Zhang, Z., Liu, J. C., Xu, X., & Yuan, L. (2020). Effect of sub-elevated temperature on mechanical properties of ECC with different fly ash contents. *Construction & Building Materials*, *262*, 120096. doi:10.1016/j.conbuildmat.2020.120096

Zhang, Z., Liu, S., Yang, F., Weng, Y., & Qian, S. (2021). Sustainable high strength, high ductility engineered cementitious composites (ECC) with substitution of cement by rice husk ash. *Journal of Cleaner Production*, *317*, 128379. doi:10.1016/j.jclepro.2021.128379

Zhang, Z., Qian, S., Liu, H., & Li, V. C. (2017). Ductile concrete material with self-healing capacity for jointless concrete pavement use. *Transportation Research Record: Journal of the Transportation Research Board*, *2640*(1), 78–83. doi:10.3141/2640-09

Zhang, Z., Qian, S., & Ma, H. (2014). Investigating mechanical properties and self-healing behavior of micro-cracked ECC with different volume of fly ash. *Construction & Building Materials*, *52*, 17–23. doi:10.1016/j.conbuildmat.2013.11.001

Zhang, Z., Yu, J., Qin, F., Huang, Y., & Sun, F. (2023b). Mechanical and self-healing properties of calcium-sulfoaluminate-cement-based engineered cementitious composites (ECC). *Journal of Building Engineering*, *77*, 107512. doi:10.1016/j.jobe.2023.107512

Zhang, Z., Yuvaraj, A., Di, J., & Qian, S. (2019). Matrix design of light weight, high strength, high ductility ECC. *Construction & Building Materials*, *210*, 188–197. doi:10.1016/j.conbuildmat.2019.03.159

Zhang, Z., & Zhang, Q. (2017). Self-healing ability of Engineered Cementitious Composites (ECC) under different exposure environments. *Construction & Building Materials*, *156*, 142–151. doi:10.1016/j.conbuildmat.2017.08.166

Zhang, Z., & Zhang, Q. (2018). Matrix tailoring of Engineered Cementitious Composites (ECC) with non-oil-coated, low tensile strength PVA fiber. *Construction & Building Materials*, *161*, 420–431. doi:10.1016/j.conbuildmat.2017.11.072

Zhang, Z., Zhang, Q., Qian, S., & Li, V. C. (2015). Low E modulus early strength engineered cementitious composites material development for ultrathin whitetopping overlay. *Transportation Research Record: Journal of the Transportation Research Board*, *2481*(1), 41–47. doi:10.3141/2481-06

Zhao, A., Yang, J., & Yang, E. H. (2015). Self-cleaning engineered cementitious composites. *Cement and Concrete Composites*, *64*, 74–83. doi:10.1016/j.cemconcomp.2015.09.007

Zhao, J., Yan, C., Liu, S., Zhang, J., Li, S., & Yan, Y. (2020). Effect of solid waste ceramic on uniaxial tensile properties and thin plate bending properties of polyvinyl alcohol engineered cementitious composite. *Journal of Cleaner Production*, *268*, 122329. doi:10.1016/j.jclepro.2020.122329

Zhao, Y., Chen, X., Wang, Z., Zhang, Y., & Leng, Y. (2023). An experimental study on the influence of fly ash and crumb rubber on the mechanical property and permeability of Engineered Cementitious Composite (ECC). *Journal of Building Engineering*, *76*, 106998. doi:10.1016/j.jobe.2023.106998

Zhao, Y., Wu, Z., & Zhang, G. (2022, March). Effect of polyethylene terephthalate fiber on the durability of ultra-high performance concrete under sulfate and freeze-thaw cycles. *Construction & Building Materials*, *312*, 125169.

Zhenbo, W., Jun, Z., Jiahe, W., & Zhengjie, S. (2015). Tensile performance of polyvinyl alcohol-steel hybrid fiber reinforced cementitious composite with impact of water to binder ratio. *Journal of Composite Materials*, *49*(18), 2169–2186. doi:10.1177/0021998314542450

Zhou, J., Qian, S., Beltran, M. G. S., Ye, G., Van Breugel, K., & Li, V. C. (2010). Development of engineered cementitious composites with limestone powder and blast furnace slag. *Materials and Structures*, *43*(6), 803–814. doi:10.1617/s11527-009-9549-0

Zhou, S., Liu, W., & Yan, Y. (2020, January). Influence of carbon fiber on the mechanical properties and durability of ultra-high performance concrete. *Construction & Building Materials*, *258*, 119609.

Zhou, Y., Gong, G., Huang, Y., Chen, C., Huang, D., Chen, Z., & Guo, M. (2021). Feasibility of incorporating recycled fine aggregate in high performance green lightweight engineered cementitious composites. *Journal of Cleaner Production*, *280*, 124445. doi:10.1016/j.jclepro.2020.124445

Zhou, Y., Gong, G., Xi, B., Guo, M., Xing, F., & Chen, C. (2022). Sustainable lightweight engineered cementitious composites using limestone calcined clay cement (LC3). *Composites. Part B, Engineering*, *243*, 110183. doi:10.1016/j.compositesb.2022.110183

Zhou, Y., Xi, B., Yu, K., Sui, L., & Xing, F. (2018). Mechanical properties of hybrid ultra-high performance engineered cementitous composites incorporating steel and polyethylene fibers. *Materials (Basel)*, *11*(8), 1448. doi:10.3390/ma11081448 PMID:30115842

Zhou, Z., Gu, D., Zhang, Y., & Li, J. (2020, March). Effect of slag powder and nanosilica on chloride penetration resistance of ultra-high performance concrete. *Construction & Building Materials*, *261*, 120143.

Zhu, H., Wang, T., Wang, Y., Hu, W. H., & Li, V. C. (2023b). Feasibility of structural retrofit concrete pipelines using limestone calcined clay cement Engineered Cementitious Composites (LC3 ECC). *Engineering Structures*, *289*, 116305. doi:10.1016/j.engstruct.2023.116305

Zhu, H., Yu, K., & Li, V. C. (2021b). Sprayable engineered cementitious composites (ECC) using calcined clay limestone cement (LC3) and PP fiber. *Cement and Concrete Composites*, *115*, 103868. doi:10.1016/j.cemconcomp.2020.103868

Zhu, H., Zhang, D., Wang, Y., Wang, T., & Li, V. C. (2021a). Development of self-stressing engineered cementitious composites (ECC). *Cement and Concrete Composites*, *118*, 103936. doi:10.1016/j.cemconcomp.2021.103936

Zhu, J., Li, G., Feng, C., Wang, L., & Zhang, W. (2019). Effect of Delaminated MXene ($Ti_3C_2$) on the Performance of Cement Paste. *Journal of Nanomaterials*, 2019, 3074206. doi:10.1155/2019/3074206

Zhu, J.-X., Xu, L.-Y., Huang, B.-T., Weng, K.-F., & Dai, J.-G. (2022). Recent developments in Engineered/Strain-Hardening Cementitious Composites (ECC/SHCC) with high and ultra-high strength. *Construction & Building Materials*, *342*, 127956. doi:10.1016/j.conbuildmat.2022.127956

Zhu, M., Chen, B., Wu, M., & Han, J. (2023a). Preparation and mechanical characterization of cost-effective low-carbon engineered cementitious composites with seawater and sea-sand. *Cement and Concrete Composites*, *136*, 104883. doi:10.1016/j.cemconcomp.2022.104883

Zhu, Y., Yang, Y., & Yao, Y. (2012). Autogenous self-healing of engineered cementitious composites under freeze–thaw cycles. *Construction & Building Materials*, *34*, 522–530. doi:10.1016/j.conbuildmat.2012.03.001

Zhu, Z., Tan, G., Zhang, W., & Wu, C. (2020). Preliminary analysis of the ductility and crack-control ability of engineered cementitious composite with superfine sand and polypropylene fiber (SSPP-ECC). *Materials (Basel)*, *13*(11), 1–15. doi:10.3390/ma13112609 PMID:32521673

Ziada, M., & Erdem, S. (2023). Strain hardening green cementitious composites reinforced with nanoparticles: Mechanical and microstructural properties and high temperature effect. *Case Studies in Construction Materials*, *18*, e02033. doi:10.1016/j.cscm.2023.e02033

Zollo, R. (1997). Fiber-reinforced concrete: An overview after 30 years of development. *Cement and Concrete Composites*, *19*(2), 107–122. doi:10.1016/S0958-9465(96)00046-7

# About the Contributors

**S. Praveenkumar** obtained his BE Civil Engineering with Best outgoing student award from Sri Krishna college of Technology, ME Computer Methods and application in Structural Engineering from Coimbatore Institute of Technology, Coimbatore. He completed PhD in civil engineering at Anna University, Chennai. He is recognized as supervisor for guiding PhD scholars of Anna University under faculty of Civil Engineering in the areas of Concrete Technology, Materials Characterization and Structural Engineering. He is the life time member of professional bodies such as Institution of Engineers, Indian Society for Technical education, Indian Concrete Institute, International Association of Engineers, Indian Society for Construction Materials and Structures, International Association for Automation and Robotics in construction, Indian Association for Computational Mechanics, Indian Association for Structural Engineering. He has published more than 35 papers in Scopus indexed journals as well as Journals listed in annexure-1 of Anna University with high impact factor. He has presented 27 papers in national and international conference organised by reputed institutions like IITs, NITs, Research labs etc. He has attended various FDP, Short term course, workshop and seminar organised by IITs, NITs. He received a grant of 2.75 lakhs from UGC, New Delhi under minor research project scheme for the period of 2017-2019 towards the project title "Synergic effect of agricultural wasters on mechanical properties of high-performance concrete". He also has received grants from various government agencies such as CSIR, UGC, MoEs, etc. for organizing workshop, seminar. FDP and training programme to worth of 5 lakhs. He is institutional in setting up a laboratory named "Advanced Concrete Research Laboratory (ACRL)" at PSG College of Technology, Coimbatore with advanced equipment's to study the behavior of construction materials with respect to field applications. He is currently guiding 3 Full time and 2 Part time PhD scholar under Anna University, Chennai. He has guided 20 PG projects and 15 UG projects. He had various responsibilities in department level such as Anna university coordinator, placement coordinator and so on. He is student chapter coordinator for Indian Concrete Institute (ICI), PSGCT. He is ISO auditor for answer script auditing at PSGCT. He has involved in preparation of SAR for NBA accreditation towards both UG and PG programmes. He has delivered more than 20 keynote lectures in FDP, Training programmes, workshop, seminar organised by various institutions in and around TamilNadu. He is reviewer in reputed journals of Elsevier, Sage and Springer publications. He is guest editor and editorial board member in reputed journals in field of civil engineering. Currently working in the following research areas: Engineered Cementitious composite, nano-silica based mortar and concrete, graphene based mortar and concrete, various optimization techniques, microstructure of construction materials, Fibre reinforced hollow concrete blocks.

* * *

**Sampath Boopathi** is an accomplished individual with a strong academic background and extensive research experience. He completed his undergraduate studies in Mechanical Engineering and pursued his postgraduate studies in the field of Computer-Aided Design. Dr. Boopathi obtained his Ph.D. from Anna University, focusing his research on Manufacturing and optimization. Throughout his career, Dr. Boopathi has made significant contributions to the field of engineering. He has authored and published over 180 research articles in internationally peer-reviewed journals, highlighting his expertise and dedication to advancing knowledge in his area of specialization. His research output demonstrates his commitment to conducting rigorous and impactful research. In addition to his research publications, Dr. Boopathi has also been granted one patent and has three published patents to his name. This indicates his innovative thinking and ability to develop practical solutions to real-world engineering challenges. With 17 years of academic and research experience, Dr. Boopathi has enriched the engineering community through his teaching and mentorship roles. He has served in various engineering colleges in Tamilnadu, India, where he has imparted knowledge, guided students, and contributed to the overall academic development of the institutions. Dr. Sampath Boopathi's diverse background, ranging from mechanical engineering to computer-aided design, along with his specialization in manufacturing and optimization, positions him as a valuable asset in the field of engineering. His research contributions, patents, and extensive teaching experience exemplify his expertise and dedication to advancing engineering knowledge and fostering innovation.

**Haytham F. A. Isleem** is an associate professor in structural engineering, the Department of Construction Management, Qujing Normal University, Yunnan, China. He received his master's and Ph.D. degrees from the Harbin Institute of Technology, in 2015 and 2019 respectively. He did his Postdoctoral research experience at the Department of Civil Engineering at Tsinghua University from 2019 to 2021. Dr. Isleem is actively engaged in research on materials and concrete structures, and course development related to materials in civil engineering and the application of artificial intelligence in civil engineering projects. Dr. Isleem is the author of 40 SCI articles.

**Amirtavarshini K. S.** is currently a Post Graduate Scholar pursuing M.E. Structural Engineering from the Coimbatore Institute of Technology, Coimbatore. She previously completed her B.Tech Civil Engineering from Amrita Vishwa Vidyapeetham, Coimbatore. Her previously done research topics include concrete, concrete structures, transportation infrastructure and traffic management. Areas of interest are concrete and steel structures.

**S. Logeswaran** is Assistant Professor Department of Civil Engineering, KPR Institute of Engineering and Technology Avinashi road, Arasur, Coimbatore - 641407. Coimbatore Tamilnadu.

**Balpreet Madan** is Assistant Professor, Department of Art & Design, School of Design, Architecture & Planning Sharda University, Knowledge park -III, Plot no. 32 Greater Noida, Uttar Pradesh, 201310.

**Sureshkumar Myilsamy** completed his undergraduate in Mechanical Engineering and postgraduate in the field of Engineering Design. He completed his Ph.D. from Anna University, Chennai, Tamil Nādu, India.

**Jagadesh Palanisamay** is working as Assistant Professor, Department of Civil Engineering, Coimbatore Institute of Technology, Coimbatore. He has published more than 30 international journals in the area related to Civil Engineering. His area of interest are supplementary cementitious materials, cement, mortar, concrete properties, material characterization, ECC, etc.

**Pushpendra Rai** is Associate Professor, Department of Science & Humanities, K J Somaiya College of Engineering Somaiya Vidyavihar University, Vidyavihar, Mumbai.

**Praveen Rathod** is Assistant Professor, Department of Mechanical Engineering Vishwakarma Institute of Information Technology.

**Sanjay Kumar Singh** is Professor, Chemistry Department of Applied Sciences and Humanities, Institute of Engineering and Technology.

**Ozlem Ustundag** is currently studying as a researcher in the Department of Civil Engineering at Istanbul University-Cerrahpasa. Dr. Ustundag received his BS, MS degrees in Civil Engineering from Istanbul Technical University and PhD degree in Civil Engineering from Istanbul University-Cerrahpasa. During the graduate education, she served as a Research Assistant in the Department of Civil Engineering. Between 2005 to 2012. Her study subjects are construction materials, composite materials and industrial wastes.

**Rafael Vargas-Bernal** received a bachelor's degree in Communications and Electronics Engineering from the University of Guanajuato in 1995 Mexico, and the degrees of Master of Science and Doctorate in Science with a Specialty in Electronics from the National Institute of Astrophysics, Optics and Electronics (INAOE) in 1997 and 2000 from Mexico, respectively. Since September 2000, he has been a professor-researcher at the University of Guanajuato and the Higher Technological Institute of Irapuato (ITESI), Mexico, where he has established himself as a researcher in electronics and materials engineering. He has authored 19 articles in journals, 50 chapters in books, and more than 100 conference articles. He has edited three books for IntechOpen. He is a member of the National System of Researchers (SNI-Mexico). He regularly serves as a reviewer of scientific articles in RSC, Elsevier, MDPI, Wiley, Taylor & Francis, and ACS, as well as Standards in Semiconductor Equipment and Materials International (SEMI). His research interests include two-dimensional materials, nanomaterials, space materials, composite materials, gas sensors, and biosensors.

**Sampada Viraj Dravid** is Assistant Professor, Department of Mechanical Engineering Vishwakarma Institute of Information Technology.

# Index

# F



# G



# M



# N



# O



# P



# R



# S

# T



# U

www.ingramcontent.com/pod-product-compliance
Lightning Source LLC
LaVergne TN
LVHW081345300926
849161LV00008B/2066
* 9 7 8 1 6 6 8 4 8 1 8 2 0 *